RECREATING NEWTON:
NEWTONIAN BIOGRAPHY AND THE MAKING OF NINETEENTH-CENTURY HISTORY OF SCIENCE

SCIENCE AND CULTURE IN THE NINETEENTH CENTURY

Series Editor: Bernard Lightman

TITLES IN THIS SERIES

Styles of Reasoning in the British Life Sciences:
Shared Assumptions, 1820–1858
James Elwick

FORTHCOMING TITLES

The Transit of Venus Enterprise in Victorian Britain
Jessica Ratcliff

Medicine and Modernism: A Biography of Sir Henry Head
L. S. Jacyna

Science and Eccentricity: Collecting, Writing and Performing Science for Early
Nineteenth-Century Audiences
Victoria Carroll

www.pickeringchatto.com/scienceculture

RECREATING NEWTON:
NEWTONIAN BIOGRAPHY AND THE MAKING OF NINETEENTH-CENTURY HISTORY OF SCIENCE

BY

Rebekah Higgitt

LONDON
PICKERING & CHATTO
2007

Published by Pickering & Chatto (Publishers) Limited
21 Bloomsbury Way, London WC1A 2TH

2252 Ridge Road, Brookfield, Vermont 05036-9704, USA

www.pickeringchatto.com

All rights reserved.
No part of this publication may be reproduced,
stored in a retrieval system, or transmitted in any form or by any means,
electronic, mechanical, photocopying, recording, or otherwise
without prior permission of the publisher.

© Pickering & Chatto (Publishers) Limited 2007
© Rebekah Higgitt 2007

BRITISH LIBRARY CATALOGUING IN PUBLICATION DATA
Higgitt, Rebekah
Recreating Newton: Newtonian biography and the making of nineteenth-century history of science. – (Science and culture in the nineteenth century)
1. Newton, Isaac, Sir, 1642–1727 2. Brewster, David, Sir, 1781–1868 3. Scientists – Biography – History and criticism 4. Science – Great Britain – History – 17th century – Historiography 5. Biography as a literary form 6. Great Britain – Intellectual life – 19th century
I. Title
530'.092

ISBN-13: 9781851969067

This publication is printed on acid-free paper that conforms to the American National Standard for the Permanence of Paper for Printed Library Materials.

Typeset by Pickering & Chatto (Publishers) Limited
Printed in the United Kingdom at Athenaeum Press Ltd, Gateshead

CONTENTS

Acknowledgments	vii
List of Illustrations and Tables	ix
Introduction	1
Background	2
Science and Genius	6
Sources for Newtonian Biography	8
Outline of Contents	12
Conclusion	16
1 Jean-Baptiste Biot's 'Newton' and its Translation (1822–1829)	19
Biot's 'Newton' and the Laplacian Programme	20
Biot's 'Newton': Light, Priority, Madness and Religion	23
Newton for the Workers? The SDUK and Biography	30
Translating Biot's 'Newton'	35
Conclusion	42
2 David Brewster's *Life of Sir Isaac Newton* (1831): Defending the Hero	43
Brewster's *Life of Newton*	44
Contradictions: Brewster on Genius and Baconianism	47
The *Life of Newton* and the Reform of Science	50
Responses to Brewster's *Life of Newton*	59
Conclusion	67
3 Francis Baily's *Account of the Revd. John Flamsteed* (1835)	69
The Flamsteed/Newton Controversy Revisited	70
A Select Audience	80
Published Responses	88
Baily's Reply	94
Conclusion	97
4 Newtonian Studies and the History of Science 1835–1855	99
Stephen Rigaud's Historical Writings	101
Antiquarians, Archivists, Librarians and Historians of Science	106
Joseph Edleston's *Correspondence of Newton and Cotes* (1850)	110
Augustus De Morgan's Historical Writings	116

	Morality and 'Impartial' History	124
	Conclusion	127
5	David Brewster's *Memoirs of Sir Isaac Newton* (1855): The 'regretful witness'	129
	The Gestation of Brewster's *Memoirs*	130
	The *Memoirs* and the History of Science	134
	Controversies: The Second Volume of the *Memoirs*	138
	Newton's Personality in the *Memoirs* and its Reviews	149
	Conclusion	157
6	The 'Mythical' and the 'Historical' Newton	159
	Placing Newton on his Pedestal: The Grantham Statue (1858)	160
	Newton: His Friend: And His Niece (1853–1870): Misreadings and Reassessment	163
	'*Newton dépossédé!*': The Affair of the Pascal Forgeries (1867–1870)	171
	The British Response to the Pascal Forgeries	175
	Conclusion	184
Conclusion		187
Notes		195
Appendix: Translations of Quotations from Biot's 'Newton' in Chapter 1		247
Works Cited		253
Index		275

ACKNOWLEDGMENTS

I would like to thank Rob Iliffe for his invaluable assistance in the development and writing of the dissertation on which this book is based, and Andrew Warwick for reading and commenting on my work at a critical stage. My thanks go also to those whose ideas and suggestions have been of benefit, including Will Ashworth, Janet Browne, Geoffrey Cantor, Serafina Cuomo, David Edgerton, Patricia Fara, Bernard Lightman, Andrew Mendelsohn, Simon Schaffer, Jim Secord, Jon Topham, Richard Yeo, the participants of the 2002 Poetics of Scientific Biography Workshop and the anonymous referees. Ken Alder, Matthias Dörries and Steven Shapin receive my gratitude for allowing me to see copies of their unpublished work. I am also particularly indebted to Charles Withers for his comments and support during the period of revision.

The assistance of archivists at a number of repositories has been much appreciated, especially that of Peter Hingley at the Royal Astronomical Society, Gill Furlong at University College London, Adam Perkins at Cambridge University Library and Colin Harris at the Bodleian Library, Oxford University.

Those who have provided practical help in the writing of this book deserve particular gratitude, especially Caroline Higgitt for translations from the French and John Higgitt for translations from the Latin. The general support provided by these individuals (who happen to be my parents) and by Dominic Sutton has been essential to the completion of this project.

Lastly, I acknowledge the assistance, companionship and support of my contemporaries while at the London Centre for the History of Science, Technology and Medicine: Terence Banks, Leigh Bregman, Sabine Clarke, Raquel Delgado-Moreira, Karl Galle, John Heard, Louise Jarvis, Jenny Marie, Guy Ortolano, Georgia Petrou and Jessica Reinisch.

For John Higgitt, 1947–2006

LIST OF ILLUSTRATIONS AND TABLES

Figure 1.	Brewster, *Life of Newton*, title page and frontispiece	46
Figure 2.	'Discordance between Theory and Practice'	73
Figure 3.	Francis Baily, after Thomas Phillips	77
Figure 4.	Edleston, *Correspondence of Newton and Cotes*, 'Synoptical View of Newton's Life'	112
Figure 5.	Edleston, *Correspondence of Newton and Cotes*, 'Notes'	113
Figure 6.	'Coat of Arms of the Royal Society'	122
Figure 7.	The British Association'	137
Figure 8.	'Sir Isaac Newton's Courtship'	152
Figure 9.	Isaac Newton by Godfrey Kneller, *c.* 1689	154
Figure 10.	Edleston, *Correspondence of Newton and Cotes*, frontispiece	155
Figure 11.	Isaac Newton by Godfrey Kneller, 1702	156
Figure 12.	'Inauguration of the Statue of Sir Isaac Newton'	161
Figure 13.	Forged and genuine examples of Pascal's handwriting	173

Table 1. List of recipients of the *Account of Flamsteed*   82

INTRODUCTION

> The history of astronomy has numerous points of contact with the general history of mankind; and it concerns questions which interest a wider class than professed astronomers.
> Sir George Cornewall Lewis[1]

This book examines how Isaac Newton's reputation was utilized, and altered, by British men of science in biographies and historical studies published between 1820 and 1870.[2] A detailed analysis of these works and the contexts in which they were produced demonstrates the contemporary significance of these portraits for the scientific community. It is, therefore, among a number of recent 'Reputational studies' which argue that representations of historical figures reflect the circumstances in which they are created and that the reputations of such figures can be used to legitimate current interests.[3] Because of the fundamental importance of Newton as a scientific icon, uses of his posthumous reputation, whether in science, religion, biography, poetry, art or more popular genres, have long been subjected to analysis. However, this book focuses on the increase of knowledge about Newton's life and character within a fifty-year period and thus offers a far more detailed examination of the motivations and influences of writers on Newton than any of these previous works. The period under consideration is significant for three reasons. First, it saw a sudden expansion in the amount of material relating to Newton that was available to researchers and readers; second, it saw a series of debates in which Newton's personal and scientific character was either central or used as a resource; and third, it was a period that saw important changes for science and its practitioners. These texts appeared against the background of the increasing professionalization, specialization and secularization of science and it is not coincidental that a period that saw the creation of modern science also featured an identifiable debate about the life and character of the most famous of British natural philosophers.

Background

Some writers have identified a 'second scientific revolution' as occurring around the turn of the nineteenth century, ushering in a recognizably 'modern' form of science. The period covered by this book was one of growing specialization for practitioners of an increasingly mathematicized and objectified science. It saw the creation of new scientific disciplines and radical transformations in the existing sciences. By the 1820s, the 'analytical revolution', which brought Continental mathematical techniques to Britain, was almost complete. Also transmitted were the techniques and vision of mathematical physics that produced the wave theory of light, which gained ascendancy in Britain during the 1830s. The use of new mathematical techniques in astronomical theory led to notable triumphs for both Newton's theory and its subsequent enlargement, including the successful prediction of the orbit of Neptune in 1846. Astronomy also saw the development of a new rigour in observation and standardized international co-operation. Such techniques meant that increasing numbers of individuals with diverse skills were included among the scientific community. This transformation in the role of the practitioner of science was symbolized by the coining of the word 'scientist' in the 1830s but begged questions about what qualities were most appropriate for this new figure, who represented an increasingly fragmented field.[4]

The large number of specialist scientific societies that appeared in the early nineteenth century is another indicator of these developments, as is the rise of specialist journals and disciplinary divisions within more popular works such as encyclopaedias. Such factors have been read as indicating the professionalization of science during the nineteenth century, although applications of this term have been criticized in recent decades. Studies of institutions that have been seen as signposts on the path of professionalization, such as the British Association for the Advancement of Science (1831) and the Geological Survey (1835), have shown the very different motivations of those most directly involved in their foundation and have underlined the continuing dominance of an amateur and gentlemanly ethos.[5] A more recent and nuanced study, by Ruth Barton, investigates how men of science chose to define themselves and their community, and convincingly demonstrates the complexity of the issue. There was, however, a clear sense of the existence of a scientific community and corresponding notions of inclusion and exclusion.[6] One means by which both this wider group and the disciplinary and other communities of which it consisted were consolidated was through the invention of a scientific tradition, and disciplinary histories 'proliferated as part of the process of staking out boundaries and establishing legitimacy'.[7] Also required were heroic, emulative forbears and the notion of a national scientific heritage able to rival that of the Continent.

The nineteenth century has long been discussed in terms of the relationship between the scientific enterprise and religious belief and has been characterized as a time when the 'investigation of nature was changed from a "godly" to a secular activity'.[8] Within the British context, particular attention has been given to the tradition of natural theology and its decline in the second half of the century. Early in the century, however, the tradition received a new impetus with the Evangelical Revival and an intensification of religious feeling and practice in the wake of the French Revolution. Newton's science was a key element of eighteenth- and nineteenth-century natural theology. Equally, Newton himself – his religious faith and positive personal characteristics – was a resource. As Susan Cannon has said, 'Sheltered under Newton's great name, science and religion had developed a firm alliance in England, symbolized by that very British person, the scientific parson of the Anglican Church'.[9] Historians of science have in addition demonstrated the extent to which natural theology existed to support the political status quo and the establishment of the Anglican Church rather than to legitimate science.[10] The adherence of important scientific figures to orthodox religious values was a key element in this defence.

It was against this background that the publications examined in this book appeared. However, the period has been dictated by the boundaries of an identifiable debate about the life and character of the most famous of British natural philosophers that was, in turn, largely shaped by the publication of hitherto little-known or unknown materials. This book therefore considers the reciprocal relationship between Newtonian studies and the development of a new expertise in the history of science that drew on developments in contemporary historiography, especially in the critical use of manuscript sources. The increase of knowledge about Newton did not occur in isolation but echoed wider developments in historical and biographical writing. The nineteenth century's fascination for and utilization of history has frequently been acknowledged, as the past began to be 'cherished as a heritage that validated and exalted the present'.[11] This interest in the past was linked to a new belief in progress and unprecedented recent change. John Stuart Mill's 'The Spirit of the Age' (1831) argued that the idea of comparing the past and present could only have become popular at a time when people had become conscious of living in a changing world and looked to the past as a guide to future development.[12] With science viewed as the most clearly progressive of human activities, its history became a topic for study in the hope that lessons could be learned and further successes ensured.

While some historians hoped that, like a science, the study of history might reveal general laws, there was an opposing trend that also claimed authority from comparison with the sciences. Rather than searching for patterns and laws, history was to be a collective enterprise, based on the gathering of historical 'facts' and the study of the particular. In the 1860s, historians, beginning to enter the academic world, pointed to the German school of history, and especially Leopold von Ranke, as their guide for having taught the importance of the critical reading of primary sources.[13] While Ranke's interest in the availability, use and care of source materials was not as innovative as was sometimes claimed, he did come to represent a new historical style.[14] Although the position of the former as a 'founding father' of academic history was largely created in retrospect, from the 1830s Ranke and Barthold Niebuhr were frequently referred to in Britain with esteem. However, an interest in historical texts came before widespread knowledge of German historical writing, as demonstrated both by a burgeoning market for autograph manuscripts and by initiatives to make the nation's archives available to the public. Although not uncontested, the presentation of increasing amounts of archival evidence was, from the beginning of the century, seen as the most valuable means of understanding past events and lives.[15]

Biography became the dominant genre in history of science, and its flowering from the late eighteenth century has received particular attention from historians.[16] However, commentators have frequently been impatient of nineteenth-century biography, seeing it as lacking either historical credibility or artistic merit, abandoning the good example of earlier works like Boswell's *Johnson* in favour of uninspired *Lives and Letters* or hagiography. The former of these trends, which saw the inclusion of large amounts of manuscript material within biographies, was celebrated in the *Edinburgh Encyclopaedia* as a means by which 'the narrative of the historian is supported, and elucidated'.[17] The latter trend, the presentation of the subject as a moral exemplar, has been described by historians as universal within nineteenth-century biography. The tension between these two factors, especially when the contents of the manuscripts undermined the story's moral, has been noted, as has the acceptability of a resolution involving the suppression of difficult evidence. Recently, biographies of scientific figures have received particular attention, and academics who have produced biographies of scientists have meditated on the benefits and dangers of their approach.[18] Others have studied biography in order to highlight its importance in the creation of a collective identity, the justification of the scientific enterprise and the changing and competing identities of scientific heroes. This approach has demonstrated that biographies of men of science and histories of science can be invaluable tools for revealing the author's views about the scientific enterprise, but it can blind the historian to reading such works as contributions to a nascent field of the history of science.

In general the history of science produced before the subject was professionalized in the twentieth century has received inadequate consideration.[19] While the potential of examining early writings has been recognized there has tended to be a focus on ambitious conceptions of the progress of science. Historians have therefore given prominence to the ideas of writers such as Auguste Comte and William Whewell, treating their work in isolation from other approaches. This book therefore aims to highlight an understudied style of history of science, which focused on manuscript sources, bibliography and narrow topics rather than narrative. Not only were more individuals engaged in this kind of enterprise but it was relied upon by writers such as Whewell, who carried out little original research. However, analyses of Whewell's historical work have produced useful discussions regarding, for example, the relationship between history and biography, showing that, while biographies could explore the individual's scientific character 'as a means of showing its conformity with existing models of virtuous behaviour and for explicating its distinctive features', histories frequently emphasized the role of scientific method and progress. However, Whewell's history contained 'biographical' concepts, such as 'the relation between intellectual and moral character', and the works considered in this book also muddy the distinction.[20] Those that are furthest from straightforward life narratives, for example published collections of correspondence, might still demonstrate an overriding interest in personal character.

Ideas about biography and histories of science have been included within studies that explore how Newton's reputation was forged. Of greatest significance is Richard Yeo's valuable essay on images of Newton between 1760 and 1860, which identifies the main strands in the debates about Newton, neatly summed up in a title that links perception of genius to ideas about scientific method and personal morality.[21] Patricia Fara's recent *Newton: The Making of Genius* gives the 'afterlife' of Newton more sustained examination in a popular format. As the title suggests, she also explores the intermeshed history of ideas regarding scientific genius. Both works are immensely useful in understanding the background to the debates under consideration here but, because they cover a broad period and topic, they do not give detailed consideration to the reasons why particular individuals expended time on researching and writing about Newton's life.[22] Their work suggests that, if more space is devoted to the examination of these motivations, an enormous amount can be revealed regarding the individual's position within the scientific community, their understanding of the manner in which science advances and their beliefs about the place of science within contemporary culture. In both accounts, however, the emphasis is on the changing perception of genius that developed with the later eighteenth-century interest in the individual and originality. The narratives, therefore, hinge at the turn of the eighteenth and nineteenth centuries. While this development is undeniably important to understanding the writings considered in this book, it is not the crux of the narrative.

The British debate about Newton, commencing in the late 1820s, helped construct 'a new image of scientific genius, with Newton as its central example'.[23]

It was, however, a by-product of the already-awakened interest in understanding the life and discoveries of a generally acknowledged genius. By shifting the focus of the narrative to this later period, my story is dictated instead by the processes of historical research. This book is, therefore, about the development of expertise in writing about Newton and is much less concerned with popular or artistic portrayals than other 'reputational' studies. Although a number of the works under consideration were aimed at a popular audience, even these were written with a sophisticated knowledge of previous accounts and current developments in the field of Newtonian scholarship. Because of the release of an unprecedented amount of information from manuscript collections, the gap between popular and 'historical' understandings of Newton's life widened dramatically in this fifty-year period. This information was mediated by individuals who had detailed knowledge of the period in which Newton lived and worked and who were in communication with each other regarding the available sources. From this point of view, other studies about the reputations of deceased men of science pay too little attention to the practice of writing history and biography and frequently treat biographical writings in isolation from related developments within the field of history of science. In this book I show that interest in Newton led the way in writing about the history of science in Britain, for he was the first figure to be discussed in such depth and in relation to such a wide range of sources. *Recreating Newton* reveals why the contributions to the debates over Newton's reputation were, in these fifty years, conducted in this manner and why the status that Newton was commonly accorded at the beginning of the century was defended by some and undermined by others. Individuals from both groups were, for differing motives, to become the first community of experts in Newtonian scholarship.

Science and Genius

This book highlights the themes of the use of Newton's reputation in support of various interests within the scientific community, the increasing use of his archives and the role of political and religious commitments in defining attitudes to the revelation of foibles in the illustrious dead. In addition, the writings on Newton examined in the following chapters elucidate another significant theme that relates to the nature of science and how it advances. Consideration of a figure such as Newton begs the question: are scientific discoveries the result of a moment of inspiration or the product of the application of a scientific method? Related questions are: is scientific theory or practical observation and experimentation more important to scientific progress? Is science a solitary or a communal enterprise? Is individual character and morality or the adherence to a set of communal norms more admirable in the man of science? More widely, we might ask if the answers to such questions are altered by the branch of science under consideration, or if different fields or different tasks require different types of ability. During the early and mid-nineteenth century these questions were widely debated and were made all the more contentious by the recent evolution in the understanding of the word 'genius'.

The importance to Newton's posthumous reputation of the eighteenth-century evolution of the understanding of creativity and 'genius' has been highlighted by Yeo and Fara. Conversely, they note the extent to which Newton's image affected the developing concept of genius. By the latter half of the eighteenth century, the term had come to imply an innate quality of mind: it '*grows*, it is not *made*'.[24] This innate quality was thus likely to become apparent in childhood and it was, indeed, frequently connected with the vigour of youth rather than the experience of age. While some writers emphasized 'poetic' over 'philosophical' genius, the moral and natural philosopher Alexander Gerard discussed both, claiming 'A GENIUS for science is formed by *penetration*, a genius for the arts, by *brightness*'. To Gerard, 'Diligence and acquired abilities may assist or improve genius: but a fine imagination alone can produce it'.[25] This individual imagination was the key element of the new conception of genius, and the suggestion that this was true of philosophic genius had important implications for scientific methodology. If imagination is accorded a role in the process of discovery, the individual scientist is given greater status but the concept of a universally applicable methodology is undermined. Likewise, if discovery is attributed to inspiration or an imaginative leap, the pedagogical utility of a genius's biography is decreased. However, in Gerard's understanding, although a methodology existed to enable the collection of facts, it was the imagination, controlled by judgment, that made connections and drew analogies from those facts. He used the story of Newton and the apple as an example of the philosophic genius at work, allowing him to make the leap from ordinary circumstance to universal concept.[26]

Others, however, rejected this attribution of scientific progress to the individual and his imagination. Joseph Priestley believed that genius had little role to play in discovery, and promoted science as an egalitarian enterprise, comprehensible to all.[27] He claimed that Newton deliberately obscured his path to discovery, making it seem mysterious and inaccessible:

> Were it possible to trace the succession of ideas in the mind of Sir Isaac Newton, during the time he made his greatest discoveries, I make no doubt but our amazement at the extent of his genius would a little subside. But if, when a man publishes discoveries, he, either through design, or through habit, omit the intermediary steps by which he himself arrived at them; it is no wonder that his speculations confound others, and that the generality of mankind stand amazed at his reach of thought.[28]

Priestley therefore considered Newton's texts elitist and useless for teaching science. A related fear surrounding the concept of genius was the possibility that it would discourage ordinary men from striving to better themselves while convincing the gifted that they need not work to achieve their potential.[29]

Discussions about genius and methodology had clear moral implications. On the one hand, if success was due to the painstaking application of a particular method, this dedication was to be admired and imitated. On the other, an individual who made a discovery in a moment of inspiration might be assumed to have a connection with the Creator. If a moral example existed here, it must be assumed that the genius lived an exemplary life that made him worthy of such an honour, and Newton was portrayed within the British natural theological tradition as a paragon of all virtues with a god-like understanding of nature. However, by the beginning of the nineteenth century, the image of the genius was increasingly problematic. Although the Romantic movement might involve a rejection of science, the image of the Romantic, poetic genius was also applied to the scientific genius.[30] Older ideas of the great philosopher's other-worldliness, melancholy, absent-mindedness or eccentricity were reinterpreted within newer frameworks, where genius might involve dissoluteness, drunkenness and even madness.[31] These were commonly seen to be an accompaniment to, and sometimes even a cause of, inspiration, and might be linked to the notion that creation demanded personal sacrifice. By the 1830s such phenomena were discussed as medical symptoms of either an overdevelopment of the mental at the expense of the physical, or an inherent weakness of born geniuses. J. M. Gully, later Charles Darwin's doctor, lectured on this theme in 1830 and displayed the ambiguities surrounding this concept. How far, he asked, are we 'called upon to admire and esteem the brilliancy of genius and talent' and how far are we 'authorized to despise and condemn its infirmities'?[32] The nineteenth-century revision of Newton's character began with revelations that he had suffered such infirmities, thus raising the spectre of this dark side of genius.

Sources for Newtonian Biography

No full-length biography of Newton was produced in the eighteenth century, but those that appeared in biographical dictionaries and encyclopaedias were based largely on the 'Éloge' produced by Bernard le Bovier de Fontenelle in his capacity as secretary to the Académie des Sciences.[33] The main source for this account was a memoir by John Conduitt, the husband of Newton's niece Catherine (née Barton). The 1728 English translation of Fontenelle's 'Éloge' went through five printings and this, together with the debt owed to it by later accounts, made it the best-known account of Newton's life until the 1830s.[34] Rupert Hall has published several eighteenth-century biographies of Newton that demonstrate this lack of originality and adherence to standard biographical formulae. Some did include additional material – Thomas Birch's 1738 article for the *General Dictionary* contained a significant amount of correspondence from the collection of the Earl of Macclesfield, the Royal Society and elsewhere – but this was not analysed or used to modify the account.[35]

These articles repeated a basic narrative of Newton's life, heavily influenced by standard ideas about the lives of thinkers inherited from classical and Renaissance models. Newton, the posthumous child born on Christmas Day 1642, was described as having shown 'early tokens of an uncommon genius' that made him unsuited to the work of managing the family estate at Woolsthorpe. He was presented as an autodidact, even after his arrival in Cambridge:

> A desire to know whether there was anything in judicial astrology first put him upon studying mathematics; he discovered the emptiness of that study, as soon as he erected a figure, for which purpose he made use of two or three problems in Euclid, which he turned to by means of an index, and did not then read the rest, looking upon it as a book containing only plain and obvious things. He went at once to Descartes Geometry and made himself master of it, by dint of genius and application, without going through the usual steps, or having the assistance of any other person.

The major discoveries of the heterogeneity of white light, the method of fluxions and universal gravitation were placed around 1665/6 and he 'had laid the foundation of all his discoveries before he was twenty-four years old'. The famous apple anecdote was reported by Catherine Conduitt: 'in the year 1665 when he retired to his own estate, on account of the plague, he first thought of his system of gravity, which he hit upon by observing the fall of an apple from a tree'.[36] Conduitt did not mention the delay in Newton's announcement of his discoveries, but his dislike of publication and preference for a quiet life were mentioned by Fontenelle and later writers.[37]

Because early sources for Newton's biography – the Conduitts, William Stukeley, Henry Pemberton – knew Newton in later life, there was a greater focus on Newton as Master of the Mint and President of the Royal Society, who, in London, 'always lived in a very handsome generous manner, tho' without ostentation or vanity; always hospitable, & upon proper occasions, gave Splendid entertainments'. He was, however, also said to be 'generous and charitable without bounds' with a 'contempt of his own money' but a 'scrupulous frugality of that wch belonged to the publick, or to any society he was entrusted for'.[38] This portrait of the public man is tempered by a brief portrait of Newton the scholar. We are told that even in London 'he was hardly ever alone without a pen in his hand & a book before him – & in all the studies he undertook he had a perseverance & patience equal to his sagacity & invention'. Setting the pattern for the early biographers of Newton, Conduitt made little of the other studies that Newton undertook, noting merely that at Cambridge Newton had:

> spent the greatest part of his time in his closet & when he was tired with his severer studies of Philosophy his only releif [*sic*] & amusement was going to some other study, as History Chronology Divinity & Chymistry[,] all wch he examined & searched thoroughly as appears by the many papers he has left on those subjects.[39]

The final section of Conduitt's memoir follows the pattern of classical eulogy by including a peroration, which traditionally summarized the emulative qualities that might be associated with the subject, whether or not in strict accordance with the truth, or indeed the preceding pages. In this case Conduitt provided an extravagant description of the commendation of Newton's work by the Princess of Wales, a note of Newton's great humility, his 'meekness and sweetness', 'innate modesty and simplicity' and the conclusion that 'his whole life was one continued series of labour, patience, charity, generosity, temperance, piety, goodness, & all other virtues, without a mixture of any vice whatsoever'.[40] Before a description of his final illness, naturally endured with patience and fortitude, we are told that both physically and mentally Newton had remained in remarkable health and 'to the last had all his senses & faculties strong & vigorous & lively & continued writing & studying many hours a day'.[41]

Conduitt's original memoir, together with a few other papers from the collection of Newtonian manuscripts held by the Earl of Portsmouth, was first published in 1806 by Edmund Turnor, an antiquary and MP whose family had bought Woolsthorpe Manor in 1732.[42] John Conduitt had been dissatisfied with Fontenelle's 'Éloge', calling it 'a very imperfect attempt', adding 'I fear he had neither abilities nor inclination to do justice to that great man, who had eclipsed the glory of [the French] hero Descartes'.[43] His response was to collect material to furnish a more suitable biography, but it was never completed. Turnor's publication included Stukeley's response to Conduitt's request for information, extracts from the Royal Society's Journal Books, and the record of 'A remarkable and curious conversation' with Newton.[44] Turnor's book therefore recorded at least some key anecdotes about Newton's early life, including reports of the mechanical devices Newton made as a child, of him as a 'sober, silent, thinking lad' and of his preference for reading to rural labour.[45] Turnor also added in a footnote Conduitt's record of some of Newton's words which were oft repeated in the nineteenth century as displaying the true Christian philosopher:

> I do not know what I may appear to the world; but to myself I seem to have been only like a little boy, playing on the sea-shore, and diverting myself, in now and then finding a smoother pebble or a prettier shell than ordinary, whilst the great ocean of truth lay all undiscovered before me.[46]

Likewise Turnor quoted from the first of Newton's letter's to Richard Bentley, in which he not only claimed his successes were 'due to nothing but industry and patient thought' but also that he 'had an eye upon such principles as might work with considering men for the belief of a Deity'. However, one dark note was perhaps sounded by the strangeness of the views that Conduitt reported in the 1725 'curious conversation'. Although Conduitt began by stating that on that day Newton's 'head [was] clearer, and memory stronger than I had known them for some time', this seemed to undermine the claim in his 'Memoir' that Newton suffered no weakening of his faculties and to back rumours that he had ceased to understand his own book.[47]

As scholarship, Turnor's book provided a reasonably accurate transcription of what were considered important documents, included some useful footnotes and was informative regarding sources. It was not part of his brief to analyse the contents of the manuscripts that he printed; they told their own story, and this did not, as in Birch's article, conflict with a formulaic narrative or contain obvious factual discrepancies. It was to be the task of writers in the following decades to attempt to include such material within a revised narrative of Newton's life. Turnor's publication, which demonstrated a reverence for manuscripts, places and objects connected to Newton and an interest in his formative years, is illustrative of contemporary attitudes towards the memory of great men. The Portsmouth Papers were deemed to be of 'public importance', of inherent interest and requiring neither analysis nor narrative.[48] Turnor was clearly also desirous of advertising Newton's connection with Woolsthorpe and Grantham and, by extension, with himself.[49] This reverence for great men and their remains must be understood within the context of the newly developed emphasis on individuality and originality. It lent a new interest to personal recollections of that increasingly mysterious creature, the gifted individual. Stukeley's anecdotes of Newton's youth were well received at a time when promise of childhood and the effects of early experience began to form an important part of biography, while Conduitt's report of a conversation with Newton carried the impression of actual contact with the elderly sage.

This interest in the manuscript record of Newton gathered pace through the nineteenth century. Biographers of Newton were to add to these existing accounts through the discovery or rediscovery of a range of sources, a process which largely forms the narrative of this book. The main collections of correspondence, scientific papers and notebooks were to be found at the Portsmouth Estate, Hurtsbourne Park, and at Trinity College, Cambridge, and began to be examined much more fully and systematically from the 1830s. To these were added items relating to or reporting on Newton among the papers of his contemporaries. These included the manuscripts of John Flamsteed, the first Astronomer Royal, at the Royal Observatory in Greenwich and the collections in the hands of Lord King (correspondence of John Locke), the Earl of Macclesfield (including correspondence and mathematical papers collected by William Jones) and Lord Braybrooke (correspondence of Samuel Pepys). These, together with a variety of smaller collections and individual items, provided new information about Newton's life or, very often, acted to confirm previously existing rumours. The publishing of such material to investigate aspects of Newton's heritage that had been, at least among certain circles, long suspected as problematic was the key feature of the period under consideration. These resources were given a new status that challenged that of existing narratives. Their importance was ultimately confirmed by their incorporation within the collections of large institutions, a process that began in the later nineteenth century, most significantly with the arrival of the scientific portion of the Portsmouth Papers at Cambridge University Library and the cataloguing of the whole collection.

Outline of Contents

The following six chapters are arranged chronologically. Four take a single publication as their focus, examining the origins of each work, the novelties they introduced, the authorial aims and concerns, and their reception. Chapters 4 and 6 both deal with a number of biographical and historical writings, interpreted as either part of a broader historical movement or as contributions to a particular debate. The authors under examination form an important part of the subject matter of this book. They were active and often well-known members of the scientific community whose influential opinions frequently reflected issues of immediate concern. As Fara has suggested, the 'story of Newton's shifting reputations is inseparable from the rise of science itself'.[50] However, the authors' responses to the unfolding Newtonian archive could also be personal and emotional. In reviewing two modern biographies of Newton, B. J. T. Dobbs wrote, 'Newton has become something of a Rorschach inkblot test or a thematic apperception test for historians. What we already have in our psyches and intellects we tend to find in Newton.'[51] The following chapters bear out both of these statements and, in much greater depth than previous studies, demonstrate that the scientific, personal, religious and political concerns of writers on Newton are reflected in their publications.

The story begins with Jean-Baptiste Biot's article on Newton in the *Biographie universelle* (1822) and its English translation, published by the Society for the Diffusion of Useful Knowledge (SDUK; 1829).[52] This was the first significant retelling of Newton's life and the first to contain evidence regarding Newton's putative breakdown in 1692–3. The problematic and recently developed notion of scientific genius, and its presentation to different audiences, is central to this chapter. This work promoted a Romanticized image of Newton that, once translated, proved to be controversial and potentially awkward as a production of the utilitarian SDUK in London. Biot (1774–1862) was a key figure within the Parisian scientific establishment and a member of the circle surrounding Pierre-Simon Laplace, a group that had achieved conspicuous successes in extending Newton's work but that was undergoing an eclipse in the 1820s. Biot's biography therefore illustrates the use of Newton's reputation to support a particular scientific approach. In addition he presented Newton as a consistent advocate of the corpuscular theory of light at a time when he felt this was under increasing attack from the supporters of the alternative wave theory of light. Because of this aspect of his work it was welcomed in Britain by advocates of the Laplacians and the corpuscular theory. These included Henry (later Lord) Brougham (1778–1868), the Whig politician who was founder and Chairman of the SDUK.

The controversial nature of Biot's biography led David Brewster (1781–1868), a close friend of Brougham, to respond with *The Life of Sir Isaac Newton* (1831).[53] Brewster, an Edinburgh-based researcher in the field of optics, wished to defend Newton from Biot's 'attack' because of personal reverence but also because of the religious and moral implications of Newton's 'madness'. As well as countering Biot's interpretation of the evidence, Brewster used his account of Britain's premier scientific hero to contribute to the contemporary campaign against the 'Decline of Science'. As a result the biography contains some contradictions regarding Brewster's image of Newton and his ideas on the progress and support of science. He believed that the process of scientific discovery involved the inspiration of unique minds but was at pains to point out that successful discoverers should be useful members of society rather than cloistered scholars. Likewise, he played an important part in the formation of the British Association for the Advancement of Science (BAAS) but rejected the 'Baconianism' that it came to represent. Brewster's views on all these points were informed by his own career disappointments, which fed his view that he, Newton and British science in general had been neglected by the authorities. The emphasis on these areas was recognized by his reviewers, who also chided him for his uncritical hero-worship of Newton. In countering this, the reviewers, who can be linked to the reformist and non-denoninational SDUK, advocated an 'impartial' and source-based approach to history.

Chapter 3 centres on the 1835 *Account of the Revd. John Flamsteed*, by Francis Baily (1774–1844), the President of the Royal Astronomical Society (RAS). This publication consisted largely of the manuscript correspondence and papers of Flamsteed, which depicted Newton in a radically different manner from earlier biographies and was to be a major impetus to subsequent research.[54] Baily's interest in Flamsteed, and his acceptance of Flamsteed's criticisms of Newton, were provoked by his appreciation of Flamsteed's careful, book-keeping approach to astronomy. Flamsteed's virtues in this area were echoed by Baily's own approach to astronomy – and to historical research, for he appropriated the objective techniques of scientific data-recording to the presentation of a controversial historical subject. However, the seventeenth-century argument between Flamsteed and Newton had a wider contemporary relevance that revealed divisions between the scientific constituency represented by the RAS, at which Baily aimed his book, and that which centred on Oxbridge and the unreformed Royal Society. The different abilities of Newton and Flamsteed and the values attached to these – individual genius or laborious collective enterprise – were key elements of the debates. However, the letters sent to Baily regarding the *Account of Flamsteed* suggest that responses were also dictated by political and religious commitments. Those who approved of Baily's publication, together with those who criticized Brewster's *Life of Newton*, indicate a reformist/radical critique of the idolization of Newton. The chief tactic at their disposal was the dissemination of documents that undermined that idealized image.

A number of publications that were fundamental to the increase of knowledge about Newton are discussed in Chapter 4, which places all the writings examined in this book within the context of the developing expertise in the history of science. There is a discrepancy in the existing literature, which has devoted significantly more attention to broad, narrative histories than to the primary-source based works of writers such as Baily, Stephen Rigaud (1774–1839), Joseph Edleston (*c.* 1816–95) and the contributors to the short-lived Historical Society of Science (founded in 1840). Their publications brought new evidence to readers but refrained from developing grand schemes regarding scientific development and frequently avoided all interpretation and theorizing. Their focus on original sources was in tune with contemporary developments in general historiography, but can also be viewed as a particularly 'scientific' technique. In his critical studies, Augustus De Morgan (1806–71) likewise insisted on the need for citing original authorities and an 'impartial' approach. His stance, like that of Brewster's reviewers and Baily's supporters, reflected his religious Nonconformity and political reformism. While rejecting the overt moralizing of hagiographical biography, these approaches embodied their own moral values, whether in the 'inductivist' approach demonstrated by Baily, Rigaud and Edleston or the 'impartial' judgments of De Morgan.

Despite the links between Baily, Rigaud, Edleston and De Morgan as experts on the life of Newton and the sources for the history of science, they had very different political and religious opinions, which are revealed in their attitudes to Newton. Baily, a former stock-broker, was the only one of these four who was not university educated. It was only after making his fortune in the business world that he was able to retire and concentrate on his scientific work and the welfare of the RAS. His approach to astronomy, as to history, was similar to that of Rigaud, but they existed in very different milieus. Baily lived among metropolitan circles of middle-class reformers and Nonconformists while Rigaud was Savilian Professor of Astronomy at Oxford and a conservative in both politics and religion. Like William Whewell (1794–1866), the Master of Trinity College, Cambridge, Rigaud felt an urgent need to protect Newton from the revelations that were published in the 1830s. Edleston, a Fellow of Trinity College who later became a vicar in County Durham, likewise aimed to protect the reputation of his alma mater's most illustrious inhabitant. De Morgan was also educated at Trinity but moved to London, where he became a Fellow of the RAS and a close friend of Baily. He was Professor of Mathematics at the University of London, the non-denominational, or 'godless', response to the monopoly on learning of Anglican Oxford and Cambridge. In his teaching and writing, De Morgan was devoted to the idea of the separation of science and scholarship from religious interests.

The writings of Baily, De Morgan and others encouraged Brewster to mount his defences once more. Chapter 5 therefore considers his extended life of Newton, which appeared as the two-volume *Memoirs of the Life, Writings and Discoveries of Sir Isaac Newton* (1855).[55] Brewster's correspondence with other 'experts' on the life of Newton provides valuable information about his original research and his desire to find material with which he could defend Newton against these perceived attacks. His attempts to tackle problematic areas in Newton's biography have been commended by historians, but these points had all been raised first by others. Brewster examined the Portsmouth Papers in the hope of refuting these writings but found instead that much of the contents of this archive required him to make painful admissions. This standard biographical work must therefore be understood as the product of an individual's struggle between fidelity to an idealized image and to the historical sources, that is, between suppression and revelation. However, while Brewster occasionally resorted to suppression, the new evidence contained in the *Memoirs* – particularly from the Portsmouth Papers – ensured that the depiction of Newton within the reviews of this book was significantly different from that of 1831.

By the second half of the nineteenth century, the gap between the 'historical' Newton, created by the expert community, and the 'mythical' Newton, celebrated by men of science and the public at large, was clear. De Morgan criticized the latter in print and attempted to propagate a more nuanced picture, opposing the hero-worship that led to the erection of Newton's statue in Grantham. His campaign against Brewster and all those who adopted Newton as a hero and moral exemplar was continued in his *Newton: His Friend: And His Niece* (written c. 1856–70), which has been misread by most historians, who have erroneously presented it as a defence of Newton's morals. The battle was once more revived in a response to the uninformed reaction to an apparent attack on Newton's scientific status in 1867, when a series of forged documents was revealed that claimed Pascal had discovered universal gravitation. This occasion provoked the last words of both De Morgan and Brewster on Newton and saw the final flaring of their sporadic argument. As well as highlighting the ability of 'experts' to deal with anomalous manuscripts, this episode serves to demonstrate that more popular views of Newton were not significantly affected by either the biographies produced over the preceding decades or by new historical standards. The rhetorical importance of the 'mythical' Newton proved – and still proves – to be more appealing and enduring than carefully constructed historical accounts.

Conclusion

Recreating Newton provides examples of individuals whose biographical and historical works linked Newton's authority to their own positions within contemporary scientific debates. A particularly strong case can be made regarding the writings of Biot, and subsequently Brewster, which sought to maintain the prominence of a corpuscular theory of light at a time when this was being successfully challenged, first in France and then in Britain. It likewise demonstrates in greater detail than other studies that debates regarding scientific methodology and the role of individual genius were played out in these and the other works under consideration. However, it shows that the question of Newton's personal morality was at the root of the controversy and that the primary choice facing his biographers was whether to continue the convention of Newtonian eulogy or to provide a critique of this tendency. The move towards Newton's archives was not a necessary product of the debates about his character and the nature of his genius and this book shows how and why this became the chief resource of the disputants. The growth of Newtonian studies, which contributed significantly to subsequent developments in the history of science, was stimulated initially by those who objected to eulogy and thereafter maintained by both defenders and critics of an idealized image of Newton.[56]

The attitude of the writer to Newton's position within the natural theological tradition was a key element in dictating this decision. He remained an icon for Oxbridge, Anglican science but, as Nonconformists began to achieve official toleration – through the 1828 repeal of the Test and Corporation Acts and the 1829 Catholic Emancipation Act – and developed educational niches such as University College London (UCL), this could not remain uncontested. Attempts to divorce Newton from establishment interests led to a rejection of the assumption that he had led an incomparably virtuous life. This peculiar interest in Newton's character lessened later in the century, when the question of the moral and religious authority of individuals was of less immediate import. This was the result of three key factors. First, the major revelations from the

Newtonian archives had ceased in the 1850s.[57] Second, as scientific practitioners became more numerous and were more frequently trained and paid for their work, trust was increasingly placed in scientific methodologies and techniques rather than the individual's personal integrity. Third, the breakdown of natural theology both lessened Newton's importance as a resource in the support of Church and State and the need for opposition to this strategy.

This book is unusual in being able to point to a number of works that criticized the use of the biographical subject as a moral exemplar. It therefore challenges simplistic accounts of nineteenth-century biography and suggests that alternative moral strategies could replace the provision of an emulative hero. Such accounts should also be modified by a greater awareness of the links between biography and history, especially when considering biographies of a historical figure. This book highlights the use of new styles of historical writing, especially the printing and criticism of manuscript sources, within publications that can be broadly described as biographies.[58] The conscious transferral of such methods to biography gave works authority but might also be used to place the individual within a historical setting, to undermine an existing idealized image, or to argue against critics. Ultimately, although faith was placed in the possibility of reaching the 'truth' about the past through the archives, it is clear that manuscripts could be used to support widely differing positions.

Therefore, as well as attempting to understand these publications on Newton as products of individuals with particular scientific or personal concerns, they are considered as examples of differing trends within the history of science. In many cases the loyalties of the author might be to the development of that field rather than to the promotion of a particular view of contemporary science and its practitioners. Although there was antagonism between those using Newton's authority to support contradictory positions, these men were in frequent communication, seeing themselves as engaged in the same enterprise. *Recreating Newton* exposes the existence of networks of 'experts' in the history of science and highlights a neglected contribution to nineteenth-century historical-scientific writing by drawing attention to those who focused on the publication of manuscript sources, redefining the Newtonian collections of colleges and private individuals as objects of national importance. These men, like the first 'professional' historians identified by Phillippa Levine, included librarians and archivists, and, just as Levine's group differed from much-studied narrative historians such as Thomas Babington Macaulay and Thomas Carlyle, their backgrounds and concerns were different to those of better-known contributors to the field.[59]

Although Newton remained an iconic figure in the later nineteenth century, available to scientific and non-scientific communities alike, research by biographers and historians of science had created an alternative understanding of his character and achievement. This more complex depiction, which included the darker side of Newton's character and a more sophisticated grasp of the intellectual context in which his ideas developed, served to undermine assumptions regarding the relationship between genius and morality and between the successful pursuit of science and divine favour. This secularization of his legacy was welcomed by those who wished to break the link between Newton and Anglicanism but left an open question about how Newton's achievement and the history of science should be interpreted in the future. As in general history, specialists chose to pay greater attention to details and specifics and, once Newton's scientific papers were made available to scholars in Cambridge, research efforts were directed into uncovering the details of his discoveries. Although biographical forms remained important in the history of science, efforts were increasingly devoted to understanding Newton's ideas within their intellectual or social context. His moral character, or his personality, has remained an object of fascination but it is no longer matter for impassioned debate.

1 JEAN-BAPTISTE BIOT'S 'NEWTON' AND ITS ENGLISH TRANSLATION (1822–1829)

> 'Great men,' they say, 'have slender wits,'
> At least, they're subject to strange fits
>   Of absentness of mind:
> And while they give to planets laws,
> In their behaviour wond'rous flaws
>   In breeding we shall find.
>   'The Philosopher's Faux-Pas' (1824)[1]

The published versions of Newton's life story in the eighteenth century, being largely based on Fontenelle's 'Éloge' (1727), were strikingly similar. While new material had been published, for example by Birch and Turnor, this was yet to be incorporated into a biographical narrative. It was for the following century to interpret manuscript evidence, the crucial factor behind the disputes over Newton's character and biography discussed in the following chapters. Unlike Turnor's *Collections*, Jean-Baptiste Biot's article on Newton, published in the *Biographie universelle* (1822), incorporated new material into a significant reinterpretation of Newton's life and work. It has, therefore, been called 'the first modern critical study of Newton's life and career'.[2] It was the first biography to point to the possibility that Newton suffered a breakdown around the years 1692–3, and therefore to suggest that his genius might have been attended by problems. In doing this Biot both responded and contributed to debates over the meaning and manifestations of genius that have been described in the Introduction. Consideration of the contents of Biot's biography must therefore focus on the controversial topics of Newton's breakdown and his role in the dispute with Gottfried Wilhelm Leibniz over who had priority in the invention of the calculus. However, to understand Biot's approach fully it is also necessary to consider his essay in the context of the waning influence of the previously dominant 'Laplacian Programme', to which he was connected. Biot saw the research of the Laplacians as falling within a Newtonian tradition and accused the rising generation of turning their backs on the heritage that he celebrated in this text.

The English translation of Biot's article was published by the SDUK in 1829. Despite the hostility of some British writers, this Society considered an almost literal translation suitable for the education of the working classes at which their tracts were aimed. This was despite the SDUK's cautious attitude towards biography and the difficulties connected with presenting genius, especially flawed genius, for the edification of their readers. Although lacking detailed information about the intentions of the translator or the SDUK with regard to this publication, the differences between the original article and the translation throw some light on Society's attitude to the text. However, it is significant that Henry Brougham, who was Chairman of the SDUK and took a close interest in the Society's publications, was, like Biot, a supporter of a corpuscular theory of light. It is possible that in relation to the religious implications of Biot's text the opinions of an opposing group within the SDUK proved dominant, but it is clear too that an unflinching consideration of well-attested facts was deemed appropriate. However, it is evident that the translation, which proved to be something of a watershed, was far more contentious than the Society anticipated.

Biot's 'Newton' and the Laplacian Programme

Biot, one of a group that had achieved conspicuous successes in extending Newton's work, was well known to the international scientific community. From the first decade of the century his main area of research was optics, a field then undergoing a radical transformation that has been identified by Kuhn as a scientific revolution.[3] Biot played a significant role in the first stage of its development, in which optics became part of mathematical physics. As an advocate of a materialistic, particulate theory of light he was, however, on the losing side when the field was transformed by Augustin Jean Fresnel's wave theory, increasingly accepted by French physicists in the early 1820s. As well as transforming optical science, this theory 'posed the first serious difficulty in the action-at-a-distance world view that had dominated European physical science from the time of Newton'.[4] Biot was one of the most loyal followers of Laplace, whose patronage ensured he was first a member of the Arcueil Circle and then of the Parisian scientific establishment.[5] Robert Fox has identified the work of Laplace's circle as a coherent and controlled research programme, which dominated French physical science between 1805 and 1815.[6] However, over the following decade, the group lost its dominance in teaching, leadership of the Académie des Sciences and its control of scientific periodicals. According to Fox, 'by the mid-1820s, the intricate structure of Laplacian physical science had collapsed, leaving just a few increasingly isolated diehards to pursue the chimera that the program and its attendant beliefs were then generally recognised to be'.[7] Biot is widely identified as one of those diehards, and the appearance of his biography of Newton therefore coincided with a moment of personal and professional significance.

Biot was an appropriate choice as author of an article on Newton, having a profound admiration for Newton's achievement and working in what he considered to be the Newtonian tradition. The Laplacians aimed, broadly, to confirm and enlarge the world-view developed in Laplace's *Mécanique céleste*, descriptively dubbed 'the short-range force paradigm'.[8] At its core was the attempt to explain physical phenomena in terms of short-range intermolecular forces. This explicitly Newtonian programme worked in analogy with the force of gravitation and with Newton's hints about the ether, light and heat. Laplace claimed:

> the phenomena of expansion, heat, and vibrational motion in gases are explained in terms of attractive and repulsive forces which act only over insensible distances ... All terrestrial phenomena depend on forces of this kind, just as celestial phenomena depend on universal gravitation. It seems to me that the study of these forces should now be the chief goal of mathematical philosophy.[9]

In emphasizing the use of mathematics to reduce experimental data to simple laws, they also consciously followed a Newtonian methodology. Despite their devotion to the concept of particles and forces, the group claimed to adhere to Newton's '*hypotheses non fingo*', and Laplace declared that the 'true object of the physical sciences is not the search for primary causes but the search for laws according to which phenomena are produced'.[10]

Biot presented Newton's method as a model to the readers of his *Traité de physique mathématique et expérimentale* (1816), using the example of the theory of fits of easy transmission and reflection, which attempted to explain the appearance of bright and dark fringes of light round the edges of thin plates of glass. Most commentators found Newton's treatment of this phenomenon unsatisfactory and preferred to gloss over the topic or ignore it entirely.[11] Biot, however, devoted more space to this theory than Newton had, describing how it was 'discovered by his measurements, and fixed the simple laws which the alternations of reflection and transmission follow at perpendicular incidence' and how Newton then undertook to 'determine them experimentally at oblique incidences, in order to have a complete idea of the phenomena'. This was not a simple task, but Biot asked 'all persons of good faith who have thought about this admirable part of the optics' whether it would have been possible to reach such accuracy of results in any other way. Biot took this approach as paradigmatic, and he likewise aimed to find mathematical 'laws which represent the phenomena accurately' or physical properties, such as Newton's fits, 'which reproduce them faithfully'.[12]

Geoffrey Cantor has warned against over-simplifying Newton's legacy and interpreting the history of optics as a battle between 'Newtonians' and 'anti-

Newtonians'.[13] Newtonianism meant different things to different men: a variety of theories found support in Newton's works and claims not to feign hypotheses could hide a wide range of theoretical commitments. Admitting this, Biot's loyalty to his interpretation of Newtonianism appears to have been reinforced by the challenge of the wave theory of light. Biot seems to have believed that he was working with a model for the propagation of light that had been utilized by Newton and which was also, despite his apparently 'proto-positivistic' approach, the best approximation of reality.[14] This is suggested by a paper of 1806, written by Biot and François Arago, in which it was claimed that Newton had '*proved* that [the] change of direction [in refraction] was owing to an attraction which bodies exercise upon the elements of light'.[15] Likewise, Biot not only considered Newton's theory of fits an exemplary piece of work, but also believed it to be analogous to his own theory of oscillating light molecules, or 'mobile polarization'. While Newton did not commit himself to a theoretical explanation, Biot did not hesitate to suggest that 'all the phenomena which depend on the fits of easy reflection and transmission could be represented with the most perfect fidelity by attributing to light molecules two poles, one attractive, the other repulsive'.[16] The phenomena of the polarization of light and double refraction were seen as an area of research likely to support the corpuscular theory and, between 1812 and 1818, Biot developed materialist theories to account for them.[17] However, by 1823 Fresnel, with Arago, whose change of allegiance was fundamental to the success of the new theory, had succeeded in explaining these phenomena by means of a wave theory.[18]

Fresnel's challenge had begun in 1815, when he sent his first memoir on the diffraction of light to the Académie. He achieved success first with Arago's complimentary report on this paper and then by winning a prize for another paper on diffraction in 1819. The judges of this competition included Laplace and Biot, which suggests the quality of Fresnel's work, but Biot acknowledged the successes of the new theory only grudgingly. While admitting that the undulatory theory was 'up to now the only one with which one can explain the particularities of diffraction', he added 'one feels that it offers rather a representation of the phenomena than a rigorous mechanical theory'. He went on: 'That is why it would be a beautiful and important discovery to match this phenomenon with ideas about the materiality of light which give such clear notions and such precise measures of so many other motions of light rays'.[19] By the time this statement was published, Biot was living largely in self-imposed exile after becoming increasingly isolated personally and professionally.[20] Biot and Arago had, after their collaboration, engaged in a long, though sporadic, feud. This had erupted once again in 1821, adding bitterness to the divide that was developing between the supporters of Laplace and the new faction, which had been gaining institutional footholds and the editorships of important journals. The moment of truth came for Biot in 1822 when Fresnel replaced him as an editor of the *Bulletin des sciences* and he was unexpectedly beaten by Joseph Fourier in the election for the permanent secretaryship of the Académie. From 1823 Biot stopped attending meetings at the Académie and retired from an active scientific life for nearly seven years.[21] It is clear that, at the time that he was writing his biography of Newton, Biot felt under attack and was beginning to feel that his position was untenable. However, the battle was not yet lost. Laplace was still developing the most elaborate version of his caloric theory of gases and Biot's continuing devotion to his vision of Newtonian physics was evident in his biography of Newton.[22]

Biot's 'Newton': Light, Priority, Madness and Religion

Method and Optics in Biot's 'Newton'

As would be expected, Biot's article contained both a statement of his admiration for Newton's work and a claim that Laplace and his circle were the true heirs of his legacy. Biot underlined the extent of Newton's achievement as 'le créateur de la philosophie naturelle, l'un des plus grands promoteurs de l'analyse mathématique, et le premier des physiciens qui ont jamais existé', while the crowning achievement of the *Principia* was described with the words of Laplace as having 'la prééminence sur les autres productions de l'esprit humain'.[23] Biot highlighted the advantages of Newton's methodology in his discussion of optics by comparing it with that of Robert Hooke, who had suggested that light was transmitted as vibrations in an ether. Biot acknowledged Hooke as 'un homme qui, pour le génie d'invention et l'étendue des lumières, le cédait à peine à Newton même', but claimed that he also had 'une excessive ambition de renommée' and a lack of mathematical knowledge. This was contrasted with the knowledge of pure mathematics, which was 'le grand avantage que possédait Newton, et qui assurait à ses recherches une précision et une certitude jusqu'alors inconnues dans les sciences' (p. 139). Hooke's chief error was that he, unlike Newton, could not distinguish a hypothesis from an established law of nature. His report to the Royal Society on Newton's discoveries examined the new facts 'seulement dans leurs rapports avec une hypothèse qu'il avait autrefois imaginée' (p. 140). Biot, taking on his new role as a *Rieniste* – a 'nothingist' who adopted neither theoretical position – admitted that this conception might be true, since the true nature of light was still unknown, but insisted that 'pour pouvoir être actuellement admis comme vrai et certain, il faudrait d'abord qu'il fût exactement défini dans ses détails; ensuite, qu'il fût susceptible d'être rigoureusement éprouvé par le calcul' (p. 140).

Biot was particularly concerned to assert that Newton had been consistent in his conception of the propagation of light. He admitted that there was currently doubt about the materiality of light, but claimed that it was a point that Newton 'n'a jamais mis en doute' (p. 144). Again he emphasized the theory of fits which 'ne [puisse] s'appliquer qu'à des particules matérielles' (p. 143). However, in response to the successes of the wave theory, Biot felt the need to defend Newton's experimental results along positivistic lines, claiming that the characteristics of the fits,

> sont si rigidement définis, et moulés sur les lois expérimentales avec tant d'exactitude, qu'ils subsisteraient encore sans aucun changement si l'on venait à découvrir que la lumière fût constituée d'une autre manière, par exemple, qu'elle consistât dans des ondulations propagées ... (p. 144)[24]

Despite this, Biot insisted that, although the *Opticks* did not privilege one hypothesis, Newton had retained a belief in the materiality of light. The theory that appeared in Newton's 1675 'Hypothesis of Light' was taken as Newton's lasting opinion.[25] In this 'hypothèse physique très hardie' (p. 144), light was described as a stream of heterogeneous particles which might interact with and cause vibrations in an elastic, ethereal medium. Biot stated that he alluded to this hypothesis,

> non pas dans l'intention de la défendre ou de la combattre, mais pour que l'on voie bien précisément en quoi consistaient dès cette époque les idées de Newton, et comment sans qu'elles aient en rien changé avec le temps, l'expression a pu seulement, selon les circonstances, en devenir plus ou moins explicite. (p. 144)

Again, although ending positivistically, Biot's discussion of the link between Newton's theory of fits and his own research on polarization would have been recognized as the statement of a corpuscularian. He declared that it was only with these recent discoveries that the *Opticks* had been fully appreciated (pp. 170–1). It is clear that Biot wished to point to aspects of Newton's work that could be identified with that of Laplace and himself. Knowing that at this time Biot felt that both he and his vision of Newtonian physics were under attack, it is tempting to see an analogy between Newton's conflict with Hooke and Biot's own clash with Arago and Fresnel. He could, of course, hardly charge the undulationists with the lack of mathematical sophistication he placed at Hooke's door, but he could, and did, charge them with adhering to an unproven hypothesis, suggested by speculation rather than experiment. Hooke, and others who had criticized Newton's optical paper for 'unphilosophical' reasons, were to blame for Newton's feeling of 'persecution' and unwillingness to publish his work.

Biot's Treatment of the Calculus Dispute

Despite his condemnation of Hooke, Biot was more sympathetic to both him and Newton's best-known adversary, Leibniz, than English biographers of Newton had been hitherto. He noted Hooke's positive qualities and useful ideas on gravitation and showed sympathy for Leibniz on the question of the invention of the calculus. The 'official' history, created by Newton and the Royal Society, had accused Leibniz of stealing the idea of fluxions from Newton and, after altering the notation, publishing it as his own calculus. This was a significant factor behind the British loyalty to Newton's fluxional method and ignorance of the development of the more flexible Leibnizian calculus by Continental mathematicians.[26] On the Continent, however, it was claimed as a case of parallel discovery. Newton had laid the foundations of his method by 1665, but 'onze ans plus tard, Leibnitz inventa denouveau [sic], et présenta sous une autre forme, qui est celle du calcul différentiel employé aujourd'hui' (p. 133). Biot quoted Laplace's statement that the clumsiness of Newton's fluxions was the chief defect of the *Principia*, limiting its scope and potential (p. 164). The extension of Newton's work, ascribed to French men of science, had only been made possible by the analytical methods developed from Leibniz's calculus. As with Fontenelle's 'Éloge', which had found space to eulogize Descartes in addition to its proper subject, Biot's article demonstrated how different traditions on the Continent could lead to modifications in the story of Newton's life.

Biot viewed the dispute between Newton and Leibniz with sadness but stated that it was initiated by neither. Newton was blamed for having, as so often, 'gardé long-temps et obstinément le secret de ces découvertes' and then asserting his priority, while maintaining the secret, in an obscure anagram sent to Leibniz (p. 173). This was contrasted with the 'noble loyauté de Leibnitz', whose reply openly stated his method (p. 174). However, true blame lay with Fatio de Duillier who had, in 1699, claimed that Leibniz had 'borrowed' from Newton. These words were 'le signal de l'attaque de la part des écrivains anglais' and caused the righteous indignation of Leibniz and his followers. The Royal Society, which arbitrated on the matter, was also culpable for having selected arbiters 'qui ne furent point connus, et sur le choix desquels Leibnitz ne fut nullement consulté' (pp. 175–6). In addition, through the publication of the Society's report, *Commercium Epistolicum*, the 'victors' had been allowed to dictate the history of the dispute in Britain. Biot concluded, here and in the *Biographie universelle* article on Leibniz, that, while the two great men were not the most culpable, both must be held accountable. Newton allowed rancour to get the better of him and 'il faut dire que, de son coté, Leibnitz n'avait été, ni moins passionné ni moins injuste' (p. 177). Newton was wrong to allow the publication of the *Commercium Epistolicum*, with its 'imputations méprisables', and to retract the acknowledgment he had made to Leibniz in the first edition of the *Principia*. Leibniz was blamed for entering the fray, and for encouraging supporters to do likewise, but was chiefly criticized for subsequently attacking, 'par les arguments les plus futiles et les hypothèses les plus invraisemblables, la grande et saine philosophie que Newton avait introduite dans l'étude des phénomènes de la nature'.²⁷

La douleur de Newton

In considering the reception and subsequent significance of Biot's article, far more important than either his scientific commitments or his stance towards Leibniz was his introduction of evidence that suggested Newton had suffered a mental breakdown. A letter of Christiaan Huygens had been discovered in Leyden by the physicist Jan Hendrik van Swinden:

> Le 29 mai *1694*, M. Colin, Écossais, m'a raconté que l'illustre géomètre Isaac Newton est tombé, il y a dix-huit mois, en démence, soit par suite d'un trop grand excès de travail, soit par la douleur qu'il a eue d'avoir vu consumer par un incendie son laboratoire de chimie et plusieurs manuscrits importants. M. Colin a ajouté qu'à la suite de cet accident, s'étant présenté chez l'archevêque de Cambridge, et ayant tenu des discours qui montraient l'aliénation de son esprit, ses amis se sont emparés de lui, ont entrepris sa cure, et, l'ayant tenu renfermé dans son appartement, lui ont administré, bon gré malgré, des remèdes, au moyen desquels il a recouvré la santé, de sorte qu'à présent il recommence à comprendre son livre des PRINCIPES. (p. 168)²⁸

Biot linked this report to the legend that a fire had been caused by Newton's apocryphal dog, Diamond. 'On raconte que, dans le premier saisissement d'une si grande perte, il se contenta de dire: "Oh! Diamant, Diamant, tu ne sais pas le tort que m'as fait"' (p. 168). This story had conventionally been used to demonstrate Newton's great equanimity under severe provocation.²⁹ In light of the new evidence, Biot had a different view: 'la douleur qu'il en ressentit, et que la réflexion dut rendre plus vive encore, altéra sa santé, et, à ce qu'il paraît même, si on ose le dire, troubla sa raison pendant quelque temps' (p. 168).

Biot ascribed Newton's mental collapse to the grief of losing his papers, the excessively hard work he had put into their creation, or a combination of both. As indicated above, the image of the genius had begun to include both madness and the mental, or even physical, exhaustion that followed inspiration and creation. Was Biot's Newton, then, a Romantic genius, divinely inspired and struggling to convey his ideas to mankind? This might be answered in the affirmative from his description of 'cette tête qui, pendant tant d'années s'était appliquée continument [*sic*] à des contemplations si profondes qu'elles étaient comme la dernière limite de la raison humaine' (p. 168). Biot's inclusion of the apple story is also indicative of his belief that discovery proceeds from inspiration.³⁰ This view tends to undermine the emphasis on mathematical rigour and experimental precision that accompanied the Laplacian Programme, but Biot was clearly attracted to a romanticized interpretation. He did point to Newton's own claims that his achievement was due to industry and perseverance, but these were contradicted by an image of the rapt genius: 'Quelle vive et naïve peinture du génie, attendant le moment d'inspiration!' (p. 156).

Most controversially, Biot claimed that Newton never fully recovered from the breakdown, and that this 'dérangement d'esprit' explained why Newton did not undertake any significant new scientific work after the publication of the *Principia* (p. 169). Biot's account therefore indicated that the breakdown was a serious mental affliction, paving the way for the reinterpretation of the tropes of seventeenth-century melancholy as evidence of disordered genius.[31] He implied that such traits could no longer be admired as demonstrating a lack of worldly concern, but linked them to the termination rather than the production of scientific research. However, he also suggested that Newton was a 'dysfunctional' character before the breakdown, pointing to his excessive reluctance to publish and the evidence of his odd 1669 letter to Francis Aston: 'il paraîtrait qu'il devait être fort étranger au commerce du monde' (p. 193). Biot described Newton's strange behaviour in 1713, when required to speak in Parliament in favour of the Longitude Bill, as 'presque puérile' and confirming his weakened intellect.[32] However, he also considered the possibility that it was 'l'effet d'une timidité poussée à l'excès par l'habitude d'une vie retirée et méditative' which was manifest both before and after 1692 (p. 193). The suggestion was that there was something in Newton's personality that was congenitally reclusive, if not actually disordered.

It would appear that Biot saw a connection between Newton's approach to work, and thus his genius, and both his breakdown and his long-term behaviour. He suggested that a peculiar mental strain could be created in those who had the ability to think with true profundity. He seems to have been taken with the pathos of the fact that greatness and littleness could exist side by side and felt 'on est tenté de prendre en pitié la pauvre raison humaine, et de demander à quoi sert le génie' (p. 179). Biot's clearest statement about the nature of scientific genius appeared in his 1832 review of Brewster's *Life of Newton*, when, perhaps emboldened by the further evidence for Newton's breakdown discovered since 1822, he was prepared to link Newton's breakdown to Pascal's madness:

> Telle est l'effrayante condition de l'homme: le génie et la folie peuvent exister dans son esprit à côté l'un de l'autre, et en même temps. Pascal, frappé une fois d'une grande terreur physique, croit dès lors voir toujours un abîme ouvert à ses côtés. Sa raison égarée, effrayée, lui présente des visions ascétiques, dont il fixe par écrit les incohérens détails. Il cache ces pieux dessins dans ses habits, les porte, les conserve jusqu'à son dernier jour; et, dans cet état mental, il écrit sur Dieu, sur le monde et sur l'homme les pensées les plus profondes, montrant même une observation, une appréciation infiniment judicieuse et fine des sociétés humaines, ainsi que des conditions artificielles qui les tiennent unies. Et, ce qui achève de confondre, l'expression de ces pensées est admirable par la puissance du style, par sa grandeur, sa concision.[33]

However, although Biot implied that the seeds of madness existed in genius, his picture of Newton portrayed, more precisely, a dichotomy between his inspired youth and his mundane, rather than mad, old age. This too is a commonplace of Romantic narrative, in which inspiration is a privilege of youth.

Religion in Biot's 'Newton'

After 1692 Biot's Newton had all but given up scientific work, concentrating instead on his roles at the Mint and the Royal Society and the study of theology and chronology. Biot did not attribute Newton's interest in theology to his derangement, acknowledging that it predated 1692 and suggesting that it was typical of this era (p. 189). He only noted that, by the time Newton was in his seventies, 'les lectures religieuses étaient devenues l'une de ses occupations les plus habituelles; et après qu'il s'était acquitté des devoirs de sa place, elles formaient, avec la conversation de ses amis, son unique délassement' (p. 190). For Biot it was the fact that Newton had given up scientific work that demanded explanation, not why he turned to religion. There was an implication that theology could be studied while in a weakened state of mind, but Biot's statement could be interpreted as an echo of Newton's own claim that he had turned to such studies as a means of relaxation:

> Sa tête, fatiguée par de si longs et de si profonds efforts, avait sans doute besoin d'un calme absolu et d'un entier repos. Du moins ne voit-on pas qu'il ait alors occupé le loisir de son esprit par des études sérieuses, ou cherché des distractions, soit dans les lettres, soit dans les affaires. (p. 192)

Biot's discussion of Newton's *Observations upon the ... Prophecies of Daniel and the Apocalypse of St. John* (1733) does, however, demonstrate his opinion that if Newton still had full use of his faculties he was wasting them. Despite this he considered the work in some detail, realizing that more people referred to this text than had actually read it (pp. 181–2). He judged that in most of the *Observations* 'il n'y a réellement de neuf que l'exposition précise et en quelque sorte systématique de la méthode d'interprétation', and that where Newton had gone further he had allowed himself to be carried away by his system (p. 187). Clearly, Biot's image of the discoverer of the law of universal gravitation did not sit comfortably with the contents of this tract. Worst of all, Newton seemed to have forgotten his methodology:

> On demandera sans doute comment un esprit de cette force et de cette nature, un esprit si habitué à la sévérité des considérations mathématiques, si exercé aux observations des phénomènes réels, enfin si méthodique et si sage dans ses spéculations physiques, même les plus hardies, et par conséquent si instruit des conditions auxquelles la vérité se découvre, comment, dis-je, un esprit de cet ordre a pu combiner des conjectures aussi multipliées, aussi incertaines, sans même faire attention à l'in-

vraisemblance extrême que jette dans ses interprétations la multitude infinie des concessions arbitraires dont il fait usage et sur lesquelles il les établit. (p. 188)

This otherwise incomprehensible alliance of science and religion was again explained with reference to the peculiarities of the period and the similar interests of Newton's close acquaintances.

Biot also paid attention to Newton's writing on chronology, giving it a great deal more space than any previous biographer of Newton by including, in a long footnote, an article written by Pierre-Claude-François Daunou on Newton's 'Abstract of Cronology' (1716).[34] Daunou's note, printed in small type, takes up the larger portion of six-and-a-half pages in the *Biographie universelle* and was thus a significant contribution to the biography as a whole. While he concluded that 'ce système est un très-grand fait dans l'histoire de la science chronologique', Danou did not think that the doubts Newton cast on conventional chronology were sufficient to justify the adoption of a new system. This was especially the case because Newton's chief innovation, the use of astronomical records to provide a basis for calculation, was not rigorously maintained (p. 186). Newton's tract was again presented merely as a curiosity and Biot indicated that Newton's non-scientific writings, especially those post-dating the breakdown, were of no great merit. It is possible that less space would have been devoted to them if they could not be conveniently relegated to Newton's unproductive years.

Biot's willingness to accept the possibility of a breakdown and a permanent change in Newton's creative output has been interpreted by Maurice Crosland as his attempt 'to reconcile Newton the mathematical physicist with Newton the theologian'.[35] His conclusion that these two aspects could be temporally separated highlights the Laplacians' view of the place of religion in science, and an important difference between British and French accounts of Newton. In an anecdotal exchange between Laplace and Napoleon, the former was supposed to have claimed he did not require the 'hypothesis' of God in his *Mécanique céleste*. Although apocryphal, this story nonetheless contains an element of veracity. Laplace was privately concerned about religious issues, but he maintained publicly that discussion of God and final causes played no role in scientific discourse.[36] Thus for Laplace the most awkward parts of Newton's works were the General Scholium and the Queries appended to *Opticks*, and he apparently took pleasure in being able to point out that they only appeared in later editions, and so post-dated Newton's illness.[37] It was Laplace's, rather than Biot's, understanding of the incident that proved controversial, for he asserted that Newton only turned to religion after his breakdown. In hope of confirming this hypothesis, Laplace had even asked a Swiss professor to make enquiries at Cambridge on Newton's breakdown, in order to discover 'a quelle epoche Newton a commencé a s'occuper d'objets theologiques'.[38]

While Biot's discomfort over Newton's writing on prophecy can link him with Laplace's more extreme position, his discussion of the religious passages in Newton's scientific works suggests that they did not fully agree. Quoting at some length from Query 28 of the *Opticks*, Biot concluded, 'certes, soit que l'on veuille ou non contester la conception qu'il donne de son existence, il est impossible de ne pas reconnaître dans cet admirable passage, le sentiment profond d'une ame [sic] religieuse et intimement convaincue' (p. 179).[39] Although Biot was wary of the arguments that Newton used, he thought it necessary to point repeatedly to the genuineness of his piety. Thus his analysis of Newton's theology should not be regarded as hostility to religion. Indeed, Biot was a Catholic, although he only officially rejoined the Church in 1846.[40] His beliefs meant that he was pained by the anti-Catholic 'esprit de prévention' (p. 187) that appeared in the *Observations*, but he believed that his work was not, 'comme chez d'autres écrivains protestants, un résultat dicté par l'esprit de ressentiment ou de haine; il l'expose avec tout le calme d'une conviction profonde, avec toute la simplicité d'une démonstration évidente' (p. 188). Newton's sincere belief was seen to be a mitigating factor even when most misguided, as Biot's belief in Newton apparently blinded him to his heartfelt anti-Catholicism.

I have argued that Biot's article must be read in the context of his loyalty to Laplace and, especially, the events of the years 1821–2. However, Biot had a long-term commitment to understanding the life and work of Newton, and the history of seventeenth-century science. His 'Newton' was only one of a number of entries he wrote for the *Biographie universelle*. Additionally, the debate that his work triggered ensured his continued interest as Biot felt called upon to defend his interpretation. He remained convinced that there was less to admire in the Newton who lived in London than in the Cambridge scholar. Biot reviewed the next important contributions to Newtonian biography, by David Brewster and Francis Baily, and reasserted the growing evidence for the breakdown, although he modified his claim that Newton's mind had been permanently impaired. Finally, he co-edited a new edition of the *Commercium Epistolicum* (1856), which, by comparing two early editions, showed that he had been correct to intimate that the judgment of the Royal Society committee on the calculus dispute had been less than impartial.[41]

Newton for the Workers? The SDUK and Biography

Biot's article has, rightly, been compared with English biographies to emphasize the differing scientific, political and religious attitudes of France and Britain.[42] However, in 1828, the London-based SDUK decided to publish a translation by Howard Elphinstone (1804–93).[43] Although it has been used by historians, little consideration has been given to the reasons behind its publication. This is the more surprising given its hostile reception by some. The availability of a cheap, English version of the article was a different matter to its remaining hidden in a French, multi-volume dictionary. There was, apparently, 'Little general interest' in the question of Newton's breakdown until the translation's 'wide circulation ... at once gave a notoriety to the report'. It was only when the topic was no longer 'nearly confined to the scientific world' that 'national and religious feelings were at once brought into action'.[44]

The SDUK was founded in 1826 by Henry Brougham and a group who were, like him, Whigs and Utilitarians, and frequently, unlike him, religious Dissenters.[45] The SDUK aimed to publish original, 'useful' and cheap works which avoided contentious discussions on politics and religion, for working men and the libraries of the Mechanics' Institutes.[46] It was hoped that cheap literature that was both 'useful' and 'safe' might replace 'penny dreadfuls' and radical publications.[47] This was a period of popular unrest, and there was a justified fear of revolution among the middle and upper classes. Brougham was a political and legal reformer and, as Lord Chancellor, was famous as the engineer of the Great Reform Act of 1832. These reforms, like his advocacy of working-class education, were motivated by a zeal for improvement, but also by the desire to neutralize the threat of revolution.[48] Brougham hoped to teach the worker that his route to success lay in rising within the established order rather than overthrowing it, and he apparently believed that the SDUK had been 'eminently conducive to allaying the reckless spirit which, in 1830, was leading multitudes to destroy property and break up machines'. The counter-revolutionary nature of 'improving' literature was denounced by radicals such as Cobbett, who saw SDUK tracts as educational sops devised to depoliticize the working classes, 'diverting their attention from the cause of their poverty and misery'.[49] Equally, the Society was criticized by conservatives who believed that educating the lower orders would upset the social hierarchy, especially if attempted without the stabilizing influence of religious instruction.[50]

Brougham believed strongly in the importance of scientific education, and the first treatise he wrote for the SDUK was a *Discourse on the Objects, Advantages, and Pleasures of Science* (1827). This claimed that the study of science 'elevates the faculties above low pursuits, purifies and refines the passions, and helps our reason to assuage their violence'.[51] It did this in part by holding out to the mechanic-student the possibility that he might benefit mankind through a discovery in his field, but more immediately by giving a training in reasonable thought and 'an understanding of the infinite wisdom and goodness which the Creator has displayed in his works'. Brougham claimed that the example of successful men of science was an important means of transmitting the message:

> It is surely no mean reward of our labour to become acquainted with the prodigious genius of those who have almost exalted the nature of man above its destined sphere; and, admitted to a fellowship with those loftier minds, to know how it comes to pass that by universal consent they hold a station apart, rising over all the Great Teachers of mankind, and spoken of reverently, as if NEWTON and LAPLACE were not names of mortal men.[52]

This quotation demonstrates Brougham's deep admiration for Newton, and thus we might question why his Society published a biography that caused significant problems for those who hoped to offer scientific heroes as moral exemplars and for those with a commitment to natural theology. These were, after all, Brougham's principal reasons for advocating scientific education for the working classes.[53] The appearance of the name Laplace in the quotation above offers a clue. Brougham may have consented to this publication because his sympathy with the Laplacian Programme outweighed his desire to protect Newton's memory. However, in order to understand the new context in which Biot's biography appeared, it is necessary to consider the SDUK's attitude to biography in general, and to highlight the specific benefits and dangers connected to scientific biography. Alterations were made to Biot's text to render it more suitable for its intended readers but, in general, its potentially damaging contents were allowed to stand.

The Society had always intended to publish biographies as part of their Library of Useful Knowledge, in which 'Newton' appeared, and they understood its value as attractive reading matter that might contain 'useful' information within a moral framework.[54] Their publications thus included a number of individual and collected biographies of statesmen, military men, artists and writers, as well as men of science, which aimed to illustrate tracts on the relevant period, art or science.[55] It was, however, only in late 1827 that the Society resolved that, if ready before its associated history, a biography could be published separately.[56] It was, rather, in its Library of Entertaining Knowledge that the SDUK fully embraced the popularity of the genre, with a *Gallery of Portraits* and within its successful *Penny Cyclopaedia*.[57] The Society's rules made publication a collective process. Every treatise was read and approved by at least two individuals whose recommendations were reported to the Publication Committee. Revisions, deletions and additions might be advised and would have to be implemented before the work could be published. These might involve factual errors or stylistic improvements, but were frequently made with an eye to the intended audience. Treatises had to be sufficiently elementary and clearly written to be made use of by working-class autodidacts, but it was also necessary to ensure that no immoral ideas could be drawn from their contents. The Society did publish lives of individuals who were less than exemplary, but it was essential that immoral behaviour be identified and condemned within the text.[58]

The didacticism of the SDUK's biographies is easily demonstrated by a glance at the works that appeared with Biot's 'Newton' when it was reissued in *Lives of Eminent Persons* (1833). Sarah Austin's introduction to her life of the explorer Carsten Niebuhr is worth quoting at length.[59] She suggests that the only reason he was chosen as a subject was because of the type of example he provided:

> If ever there lived a man who might safely and reasonably be held up to the people as an object of imitation, it was Carsten Niebuhr. – Not only was he a poor man, – an orphan, – born in a remote part of a remote province, far from all those facilities for acquiring knowledge, which in this age and country are poured out before the feet of the people; – he was not even gifted in any extraordinary way by nature. He was in no sense of the word a *genius*. He had, as his eminent biographer [Barthold Niebuhr] remarks, no imagination; – his power of acquiring knowledge does not seem to have been extraordinarily rapid, nor his memory singularly retentive. In all cases where the force of that will, at once steady and ardent, which enabled him to master his favourite studies, was not brought to bear, his progress was slow and inconsiderable. It is not, therefore, in any supposed intellectual advantages that we must look for the causes of his rise to eminence. They are to be found rather in the moral qualities which distinguished him, qualities attainable in a greater or less degree by men of the humblest rank, of the most homely intellect, the least favoured by situation or connexion.

This passage suggests that using a man of acknowledged genius as an exemplar was problematic. Niebuhr's plodding approach could be emulated and proved methodologically sound: 'The bent of his mind was entirely for the observation and investigation of sensible objects: abstraction and speculation were foreign to his genius, which could lay hold of nothing but the concrete'.[60] The most suitable type for emulation was that later associated with biographies by Samuel Smiles, who succeeded by their own, replicable efforts.[61]

However, John Roebuck's 'Life of Mahomet' suggested that a biography might present a different kind of example. The author dissociated himself and his readers from the prejudiced stand of earlier writers, claiming,

> We have now almost universally ceased to regard our own faith as at all concerned in the estimation that may be formed of the character, opinion, conduct, or religion of Mahomet. As our interests have become less concerned, our judgements have become more impartial. We have learned moreover that the employment of calumny and falsehood in support of any system, however admirable, is neither just nor prudent.

His portrait was not particularly sympathetic, but he did maintain that Mahomet was not, in personal dealings, cruel and that he could not be reprimanded for departing from a moral code of which he was ignorant. Roebuck's clearest moral message was in his censure of historians who were 'apt to confound matters of inference with matters of fact, [and] what they related upon testimony, with what they infer as a consequence from that testimony'.[62] Writing accurate, unprejudiced history could teach as important a lesson as could the biography of a blameless, hardworking and successful man. This implies more subtlety in the SDUK's output than commentators have been inclined to acknowledge.[63]

Lives of Eminent Persons also contains the biographies of Galileo Galilei and Johann Kepler by John Elliot Drinkwater (later Bethune).[64] Drinkwater's concerns about morality, and the moral implications of methodological choices, are clear from an 1834 letter to the Society's secretary regarding a biography of Hooke, submitted by Baden Powell, Savilian Professor of Geometry at Oxford. Drinkwater accepted that Hooke deserved treatment in an SDUK treatise as 'a very advantageous opportunity for detailing the progress of physical knowledge between Galileo & Newton, by taking advantage of his multifarious pursuits to shew the state of general contemporary knowledge ...'. He claimed that he was 'very partial to that mode of writing the history of science', but, as a referee, he objected to Powell's manuscript on moral grounds:

> I wish that Hooke's vanity, jealousy and overweening pretension had been more severely castigated by the author. Subscribing fully to the author's opinion of his genius & inventive powers, I look upon him as a most mischievous example to be held up for any but the most qualified admiration.

Drinkwater asserted 'that a more dangerous lesson cannot be taught than that great praise is to be given in reward for superficial & incomplete investigations', and suggested a remedy:

> I could wish that a strong & repeated protest against these doctrines was interwoven with this account, & that the reader should be perpetually reminded rather of what Hooke failed to do, by reason of his impatience or caprice, than of the brilliant invention shewn in his desultory ideas. His genius secured his reputation in spite of his faults, but it might be forgotten, unless the author is at pains to put it forward ...[65]

The Society approved this work only with revisions, although it was in fact never published. The relevant manuscripts unfortunately do not survive among the SDUK Papers, but the revisions undoubtedly owed much to Drinkwater's comments.

Drinkwater had practised as he later preached in his biographies of Galileo and Kepler. These detailed the contemporary state of astronomy and scientific method, and repeatedly compared the productive and sound methodology of Galileo with the hypothetical guesswork of Kepler. In his biography of Galileo he had:

> endeavoured to inculcate the safety and fruitfulness of the method followed by that great reformer in his search after the physical truth. As his success furnishes the best instance of the value of the inductive process, so the failures and blunders of his adversaries supply equally good examples of the dangers and the barrenness of the opposite course.

For Kepler it was only luck, together with his love of truth and willingness to jettison erroneous theories, that allowed him the successes he did achieve.[66] Drinkwater believed that Powell had failed to stress the similar contrast existing between the methods and morals of Hooke and Newton. However, it seems to have been the policy of the SDUK not to avoid unpleasant or problematic details in the lives of those they had decided to publish. This had been held up as a virtue by Roebuck, and Drinkwater did not suggest that Hooke should be ignored. The correct procedure was to acknowledge and admit the fault. Thus when Galileo participated in a 'virulent and unnecessary controversy', with remarks 'written in a spirit of flippant violence, such as might not be extraordinary in a common juvenile critic', he was strongly reprimanded by Drinkwater.[67]

Translating Biot's 'Newton'

The translation of Biot's 'Newton' has regularly, but mistakenly, been attributed to Brougham himself.[68] However, Brougham was closely identified with the Society, which kept a close eye on all publications from commission to composition, printing to publication. His name does not appear in the minutes of those meetings at which Elphinstone's translation was discussed, but it seems unlikely, given his reverence for Newton and his interest in optics, that he would not have read and approved the original. In addition, he wrote a review article in 1829 that addressed the evidence for Newton's breakdown.[69] While Biot's biography did not challenge the high estimation of Newton shared by Drinkwater, Brougham and, no doubt, the majority of the SDUK committee members, it did significantly alter the traditional depiction. It contained assertions that were so unacceptable to Brougham's close friend Brewster that he believed a reply necessary to 'the memory of a great man, to the feelings of his countrymen, and to the interests of Christianity itself'.[70] Yet the SDUK's members were sufficiently interested in this work to diverge from their usual policy of not publishing translations.[71]

The translator, Elphinstone, was not an SDUK member when he offered his manuscript free of charge, but, in lieu of normal payment, he was granted honorary membership in 1829. He became a full member in 1831, thereafter frequently attending committee meetings.[72] He was clearly the 'right sort' and was already known to important members of the SDUK, having been a pupil of Drinkwater's at Trinity College, Cambridge.[73] Elphinstone matriculated at Trinity in 1821, after schooling, like Brougham, at the High School in Edinburgh. He received his BA in 1826 and MA in 1829, the same year that the translation was published. In subsequent years he entered politics and the law. After standing unsuccessfully in 1832, he became MP for Hastings in 1835. In 1837 he failed, by a small margin, to gain the Liverpool seat, after which he trained to become a barrister. He was MP for Lewes between 1841 and 1847, and succeeded to his father's baronetcy in 1846.[74] Parliamentary reports in *The Times* record that Elphinstone spoke against the Corn Laws and for legal reform. His father's obituary referred to the new baronet, who was, 'it need scarcely be added, an active member of the Reform party'.[75]

Elphinstone's connection to Biot was through his father, Howard Elphinstone, a colonel in the Royal Engineers. When Biot visited Britain in 1817, he stayed in Edinburgh and set up a station in Leith from which he took pendulum measurements in co-operation with the British Ordinance Survey. He was aided by the local military engineers and their commander, Elphinstone senior, who became a close friend. Biot reported that British scientific men were very much behind the French in mathematics, but Elphinstone junior's education must have combined to make him remarkably aware of Continental techniques.[76] Above all there was the personal acquaintance with Biot, but Edinburgh, the military and Cambridge have all been identified as early locations for the spread of Continental analysis in Britain.[77] Brougham, an admirer of the system that led him through the High School and University in Edinburgh, had been, although never a sophisticated mathematician, an early convert to Continental analysis.[78] At Trinity College, Elphinstone was undoubtedly aware of the drive by the former members of the Analytical Society to introduce the new methods into the Tripos examination.[79] Although he did not become a Wrangler – those receiving the highest marks in the Mathematical Tripos – his connection to Drinkwater and, especially, George Biddell Airy, placed him in contact with those of the first generation to benefit from teaching by 'Analyticals' like George Peacock.

By the end of the 1820s, Continental mathematics was widely accepted, being championed by those who warned of Britain's decline if they failed to catch up with the rest of Europe.[80] Even working men were to be introduced to the new techniques: Brougham had solicited an account of Laplace's *Mécanique céleste* for the SDUK from Mary Somerville. Writing to her husband William in 1827, of 'that divine work', Brougham asked that she explain to 'the unlearned', 'the plan, the vast merit, the wonderful truths unfolded or methodised'.[81] Biot's name, and his connection to the Laplacians, was considered testimony in favour of the biography of Newton, one of the very few SDUK publications that acknowledged the author. Biot's name appears twice; in an introductory note and in a paragraph added to the end of the text. This directed those looking for further information to the works of Samuel Horsley, Birch, Pemberton and Colin MacLaurin, but concluded by pointing to the Continent:

> It is ... in the writings of the modern continental mathematicians, that we find the more complete development of those brilliant discoveries which have shed so much lustre on the name of Newton. It is with the works of LAPLACE, Lagrange, Biôt, Lacroix, Monge, Garnier, Poisson, DELAMBRE, Boucharlat, Carnot, Bailly, Bernouilli [*sic*], Euler, Bossut, Montucla, De Zach, Lalande, Francœur, Legendre ... Gauss, Hauy, &c. &c. that the student must become acquainted, before he can hope to attain to a thorough knowledge of the system of the universe.[82]

This author was thus shown to be among the group of Continental men of science whose extension of Newton's legacy Brougham and many others believed to epitomize correct method in science.

The subject and author of the article, therefore, naturally appealed to the SDUK, especially at a time when the Society was beginning to accept the principle of publishing stand-alone biographies. Newton was the greatest British natural philosopher and was promoted by the SDUK as the author of the greatest scientific text ever written and the founder of modern scientific methodology. However, there clearly were difficulties with Biot's article, for the translation included alterations that 'might render the treatise more adapted for the objects which the Society has in view'.[83] After the report on the translation had been discussed, the Publication Committee noted that that 'there were MSS. at Lord King's and Lord Portsmouth's; and others in the possession of Mr Whewell of Cambridge which contained interesting information concerning Sir Isaac Newton', and it was decided that these should be further investigated.[84] This suggests that the commentators felt that the life should not be published without further evidence, and it appears that the controversial issue was Newton's breakdown, for the only additional material published was a diary entry, written by Abraham de la Pryme of Trinity College, dated 3 February 1692.[85] This mentioned that Newton lost some papers in a fire, and claimed that when he 'came from chapel and had seen what was done, every one thought he would have run mad, he was so troubled thereat, that he was not himself for a month after'.[86] The SDUK Papers do not reveal whether any member saw the even more damning evidence in the possession of Lord King, about to be published in his *Life of John Locke* (1829). This included a letter from Newton to Locke, dated 16 September 1693, which apologized for having accused Locke of endeavouring to 'embroil me with women' and 'sell me an office', and for having declared "twere better if you were dead'. A second letter explained that 'when I wrote to you, I had not slept an hour a night for a fortnight together, and for five nights together not a wink'.[87]

With the apparent confirmation of Huygens's story, publication may have seemed the most advisable course, especially if the details in King's forthcoming book were known. In addition, there may have been a desire to demonstrate an impartial fidelity to sources, in line with Roebuck's stance in his 'Life of Mohamed'. Roebuck claimed that his approach was part of a recent change in historical writing and, in his *Lives of Eminent Men of Letters* (1845–6), Brougham also wrote of a necessary 'reformation in the historical character and practice'.[88] This new history was to be written with a 'philosophical spirit', to be 'sober and rational' and backed with 'constant references to authorities'. If these rules were adhered to, readers would be taught to 'entertain the proper sentiments, whether of respect or of interest, or of aversion or of indifference, for the various subjects of the narration'.[89] For Brougham it was a point of faith that an unprejudiced history would demonstrate the progress of society and the reward of virtue.[90] If this is something to which members of the SDUK generally subscribed, there would have been no difficulty in publishing something apparently as well-attested as Newton's breakdown.

The translation did make a minor alteration to the section describing Newton's breakdown, which perhaps rendered it more acceptable. The original text described Newton's mind as 'appliqué continûment à des contemplations si profondes qu'elles étaient comme la dernière limite de la raison humaine' (p. 168), but the translation omitted this description, and therefore the implication that his thoughts were of a potentially debilitating nature.[91] Overwork could cause strain, but the SDUK wished to avoid suggesting that the pursuit of science itself caused madness. This alteration again indicates the difficulties attached to writing about genius, rather than figures like Carsten Niebuhr. Another problematic section of Biot's article related to the calculus dispute, and it may have seemed fortunate that Newton's role could be confined to a period during which he was no longer at the height of his powers. In addition, the way in which Biot told this story was in conformity with the aims of the SDUK, for his condemnation of the behaviour of both Newton and Leibniz was clear. Only one section relating to their later dispute was removed from the translation. This referred to Leibniz's comparison of gravity to an occult force, and his suggestion that England had been more philosophically sound in the days of Robert Boyle, when such dubious ideas would not have been acceptable (p. 179). The SDUK promoted the Newtonian system as the antithesis of hypothetical speculation and they seem not to have wished to draw attention to this argument, despite its condemnation by Biot. Indeed, another very minor alteration in the translation is the printing in italics of the words '*without any admixture of hypothesis*'.[92]

In the context of a work published for the education of working men it is interesting to note that the longest omission from Biot's text relates to Newton's early studies. Biot repeated Newton's claim that when young he considered Euclid simple and not worth studying and declared, 'Si la chose était vraie, elle serait en effet exactement un prodige'. He then discussed the possibility of Newton having been able to deduce Euclid's propositions himself – which he believed he might have been able to do after reading the beginning of the work, but not from observation alone (pp. 129–30). This passage, like the others mentioned, may simply have been deleted because this section was thought dispensable, for the SDUK treatises were always constrained by space and expense.[93] But it may also have been considered a problematic example for men being urged to educate themselves and to whom they were trying to sell treatises on mathematics. This was hardly the kind of steady and dedicated learning the Society advocated. However, it is perhaps just as likely that the Society was dubious as to the veracity of Newton's claim which was, in any case, one that undermined the emphasis on the merits of hard work and humility.

Another reason why the men of the SDUK may have been able to accept the possibility that Newton did not recover from his breakdown was that they shared Biot's desire to explain why Newton apparently ceased to be productive. Since the Society aimed to avoid discussions on religion, and many of its members were connected to the non-sectarian University of London, there were those who probably also agreed with Biot's apparent separation of Newton's scientific and religious writing.[94] Like the University, the SDUK was suspected of being 'godless', or was, at least, correctly identified as representing a high proportion of non-Anglican members. A reviewer in the High-Church *British Critic* claimed that the current political unrest was caused by the Mechanics' Institutes, the SDUK and Unitarian philosophers, who suppress 'all reverence for sacred things'.[95] Brewster later noted that, while Biot had indicated that there was nothing in Newton's writings to justify the common suspicion that he was Antitrinitarian, 'This passage is strangely omitted in the English translation'.[96] This suggests that Brewster suspected that those responsible for the publication – ultimately the members of the Publication Committee – were themselves Unitarians.

Brougham wrote a *Discourse on Natural Theology* (1835), but his stress on natural theology was not acceptable to all SDUK members. The Society did initially publish works that emphasized the argument from design, but such topics were increasingly avoided. A cause for this is suggested by Adrian Desmond, who has demonstrated that the Society's committee was split between 'moderate Broughamites' and the 'Unitarian Left'.[97] A confrontation between the groups seems to have occurred over a possible new edition of Paley's *Natural Theology* proposed by Brougham.[98] Brougham later published the work himself and explained 'some of our colleagues justly apprehended that the adoption of it might open the door to the introduction of religious controversy among us, against our fundamental principles'.[99] This demonstrates the strength of feeling among some of the members on this topic and, revealingly, that Brougham could not dictate to the SDUK. However, Brougham himself accepted Biot's stance. Although he wished to correct the view that 'scientific men were apt to regard the study of Natural Religion as little connected with philosophical pursuits', he probably agreed that Newton's work on theology and chronology was of little merit.[100] The fact that Newton was able to gain such an insight into how the universe operated was a better demonstration that the mind was more than gross matter than was an interpretation of prophecy and, while Newton and Boyle were celebrated as 'the most zealous advocates of Natural Religion', Brougham did not mention their theological writings.[101] He stressed natural over revealed religion, wishing to demonstrate that 'Natural Theology is as much an inductive science as Physics or Natural Philosophy'.[102]

It is clear from Brougham's review of King's *Life of Locke* that he knew and approved the contents of the SDUK's translation. In this he took Huygens as the 'authority upon which this passage in the history of Newton rests', and the letters to Locke, which he reprinted, as a 'melancholy ... confirmation'. Comparing the evidence of de la Pryme with the tale of Diamond, he observed 'how wide of the truth the common version of the anecdote seems to be, which gives it as a striking instance of Newton's extreme composure and patience'.[103] Brougham suggested that the illness began in the winter of 1692–3, but quoted other letters that indicated 'a degree of irritability and suspicion' earlier in 1692. Possible causes were given as 'some bodily ailment, or some original morbid predisposition, or from too vast a burden being imposed upon' his mind. Brougham indicated that the ailment was temporary but did not hesitate to suggest, like Biot, that this evidence might 'serve to explain the otherwise remarkable fact of this illustrious person having completed all his discoveries before he attained the age of forty-five, and done nothing after that, although he lived in perfect enjoyment of health forty years longer'.[104] Brougham printed some letters which demonstrated that Newton's interest in Scripture pre-dated 1692, but made no comment about the suggestion that Newton turned increasingly to theology after this date. In general, however, this review served to further increase the circulation of Biot's thesis.

Brougham and the Theory of Light

In the final, added, paragraph of the translation the eminent French men of science listed were Laplacians and advocates of the corpuscular theory of light. There is no appearance of Fresnel, Gay-Lussac or even Arago. This is significant since Brougham is chiefly known to historians of science for his articles in the *Edinburgh Review* (1803–4) criticizing Thomas Young's explanation of diffraction by means of a wave theory of light. This episode was written into the history of science by the eventually victorious undulationists who accused Brougham of blind adherence to the Newtonian hypothesis, lacking the ability to understand the importance of Young's work, and delaying its appreciation.[105] Brougham, like Biot, believed that Newton had 'inductively discovered that matter acts on light corpuscles by forces of attraction and repulsion'.[106] Alan Shapiro suggests that the young Brougham, whose first paper in the *Philosophical Transactions* (1796) was modelled structurally on the *Opticks*, 'imagined himself as a restorer and promoter of Newtonian optics'.[107] Later in the century, when the history of Young's scientific contribution was being written, it became easy to caricature Brougham's unmathematical empiricism. He remained committed to the corpuscular theory long after Fresnel's theory was generally accepted, longer even than Biot, and was to join Brewster in a campaign against the victorious undulationists. It seems probable that Brougham had originally welcomed Biot's article as a publication that supported a corpuscular theory of light and identified it as Newton's view.

Biot's optical work was well received by Brewster and Brougham, and the former was particularly influenced by his work on polarization and double refraction.[108] However, there are problems with suggesting that Brougham approved Biot's work on the basis that it helped refute the wave theory of light. One of these is that Biot ascribed to Newton a hypothesis of light that involved an ether, a concept that Brougham found unpalatable. His dislike of premature hypothesizing was also apparent in his rejection of Newton's theory of fits, in which he saw 'the smoke of unintelligible theory'.[109] Brougham considered that such parts of Newton's work did not demonstrate his usual rigour. Like Laplace, he thought Newton's Queries the least satisfactory part of his *Opticks* – not because of their religious claims, but because they contained unfounded speculations about an ether that did not conform to the inductive process. In 1803 he wrote 'it must be remembered that the queries of Newton were given to the world at the close of the most brilliant career of solid discovery, that any mortal was ever permitted to run'. He considered them the 'sports in which such a veteran might well be allowed to relax his mind', but warned that this kind of philosophizing was 'mere idleness in the raw soldier who has never fleshed his sword' and a poor example to follow.[110] It is possible that what Brougham once saw as Newton's 'relaxation' he was happy to be able to discredit further as the result of Newton's unproductive later years.

Like Biot, Brougham continued his interest in the science, life and image of Newton. In 1839 he published what was later reissued as the first part of his *Analytical View of the Principia* (1855). His interest in Newtonian biography was revived particularly after his retirement from politics in the 1850s and he took a central role in the 1858 inauguration of a statue of Newton in Grantham. In both the *Analytical View* and his 1858 speech, 'all controversial matters [were] purposely avoided'.[111] However, his correspondence with two other biographers of Newton, Brewster and De Morgan, demonstrates an abiding interest in exactly these points. In 1847 Brewster replied to a letter from Brougham that contained 'observations on the state of Sir Isaac Newton's mind, in 1692'.[112] In the 1850s Brougham corresponded with both men on issues such as Newton's religious beliefs and the calculus dispute.[113] He believed the nation should honour Newton, and was inclined, literally and metaphorically, to put him on a pedestal, but he did not shrink from discussion of controversial points.

Conclusion

Both Biot's original article and its translation are revealing of the immediate scientific contexts in which they were produced. Biot, feeling increasingly isolated as a Laplacian and an advocate of a corpuscular theory of light, used his account of Newton's science to underline the Newtonian heritage of his methodology and his work in optics. In Britain, there was an enormous respect for Laplace and his followers, especially among reformers. This gave Biot's writing an immediate credibility with the SDUK. By the time the translation of Biot's article was published, the wave theory of light was increasingly accepted in Britain but Brougham was, like Biot, committed to a corpuscular theory that he viewed as Newton's preferred explanation. It is likely that he appreciated the biography for its statement that Newton always believed light to be material.

Biot's Newton can be seen as a version of the Romantic scientific genius that – both in personal morals and scientific methodology – was, in general, unacceptable to the SDUK as an exemplar for working men. Sadly, too little is known of Elphinstone and the details of the refereeing process of his translation to draw definitive conclusions about SDUK attitudes to Biot's article and the minor changes made in the translation. It appears certain that Brougham and the Society did not expect the biography to be as controversial as it proved. The question of Newton's illness probably seemed too well supported to be ignored, and the claim that Newton did little scientific work after his move to London was hardly new. Importantly, Biot had avoided repeating Laplace's belief that Newton's interest in theology only post-dated 1692–3. Subtle changes to his text combined to suggest that over-work rather than inspiration was the cause of Newton's illness, playing down the image of Newton as Romantic genius (see also Appendix). The SDUK did not shy from publishing biographies of people who did not live exemplary lives. Brougham believed that knowledge was in itself both useful and moral; there was no dangerous knowledge, only incorrect interpretations. Learning of a verified event in Newton's life could not be harmful, and where Newton was held culpable in the context of the calculus affair his behaviour, and that of Leibniz, was sufficiently condemned. The only thing lost was the idealized image of Newton as consistently good, mild and steady, which appears to have been something that the SDUK committee members were prepared to leave behind. As the following chapter goes on to show, the attitude of an individual to the alliance of Newtonian natural theology with the Establishment order and Church of England could be fundamental to his acceptance of a more human portrayal of Newton. The moral exemplar could be found instead in the 'impartial' acceptance of unpleasant evidence.

2 DAVID BREWSTER'S *LIFE OF SIR ISAAC NEWTON* (1831): DEFENDING THE HERO

> ... neglect is the only touchstone by which true genius is proved ...
> John Clare[1]

In her biography of her father, Sir David Brewster, Maria Gordon described his emotional response to the memory of Isaac Newton and his youthful admiration of Colin MacLaurin's tombstone, at which he gazed and 'pondered over the words, to be envied by every aspirant to scientific fame, "NEWTONE SUADENTE"'. The biography repeatedly reinforces the idea of a connection between Brewster and Newton, for example with an image of the 'biographer and loving disciple' standing before the moonlit statue of Newton at Cambridge.[2] Maria even asserted that there 'was much similarity between the genius, the characteristic individuality, and the career of both', making the unlikely claim that had they 'been contemporaries doubtless there could have been mutual warm personal sympathies'. More plausibly, she believed that 'there was something approaching to the known and personal in the affectionate admiration which Brewster ever cherished for Newton'.[3] Indeed, Brewster seemed to feel a personal injury from the perceived attack on Newton in Biot's 1822 article.

Although Brougham and the SDUK had accepted Biot's interpretation of Newton's life, Brewster viewed it as deeply threatening. He chose to respond with his own biography, *The Life of Sir Isaac Newton*, in which he analysed Biot's evidence for Newton's breakdown and introduced new material to refute the Frenchman's conclusion. This material helped to counter claims that Newton's faculties had been permanently impaired and that, after 1692–3, he studied only theology. However, it also added weight to the supposition that an illness, which at least temporarily affected Newton's mental health, had occurred. Brewster also used the *Life* to publicize a number of agendas that he considered fundamentally important to the progress of science and which reflect his writings on the 'decline of science' debate. Brewster's views on the support of science by private and government patronage were informed by both his experience of trying to forge a scientific career and his understanding of the nature of scientific genius. His failed attempts to gain employment convinced him of the failure of the existing system to recognize and nurture talent, and led him to see and identify with the neglect of genius in Newton's story. Brewster's views on how genius might better be encouraged are reflected in his plans for the BAAS, of which he was an instigator, and also permeated his presentation of Newton. He did not believe that scientific discovery occurred in the manner described by Francis Bacon or, indeed, by the 'Baconian' method as usually understood in the early nineteenth century.[4] Brewster emphasized instead the role of imagination and the innate abilities of the discoverer. However, both his image of Newton and his views on support for science involved contradictions, some of which appear irresolvable.

Brewster's *Life* received significant interest from the press, and reviews by Biot, B. H. Malkin and Thomas Galloway highlight the problems regarding its reception. Biot naturally wished to defend his interpretation of the evidence for Newton's breakdown and his belief that its consequences were permanent. He also pointed to the differences in their historiography, which he suggested were indicative of national styles. Malkin and Galloway also objected to Brewster's style, but their complaint was almost directly opposed to Biot's. While Biot felt Brewster was too pedantic in his investigation of events to allow an imaginative interpretation to bring him nearer the 'truth', Malkin and Galloway thought he was not sufficiently faithful to his evidence. Both men were members of the SDUK, which had published the English translation of Biot's 'Newton', and it is possible that they felt obliged to defend its publication. However, I suggest that their emphasis on documentary evidence and 'impartiality' in historical writing was a stance typical of many connected with the Society and, more broadly, the Nonconformist, reformist and utilitarian constituency it represented.

Brewster's *Life of Newton*

Brewster's 1831 biography was described in the advertising blurb as, 'the only extended life of the greatest of English Philosophers'. It emphasized that Brewster had 'not only sought out from resources, hitherto unknown, and inaccessible to previous writers, every fresh and novel particular regarding his life, but has given the most lucid explanation of Newton's great discoveries – and has endeavoured to render these intelligible to all classes of readers'.[5] Because of these claims, and because of Brewster's considerable scientific reputation, it was reviewed as 'a work likely to continue for some time the standard Life of Newton'.[6] This was to prove the case: in terms of length and scope it was only superseded by Brewster's second biography of Newton in 1855, and regarding popular appeal and accessibility it seems to have remained unchallenged. There were at least seven further English editions, sixteen American, two German and one French. Later commentators referred to its 'wide circulation and popularity'.[7] Like Elphinstone's translation of Biot's 'Newton', it was published as one of a series of low-priced volumes, in this case John Murray's *Family Library*. Its production standards were, however, considerably higher than the SDUK's and this 366-page duodecimo is ornamented with engravings of Newton's portrait and Woolsthorpe as frontispiece and on the title page (Figure 1). This series aimed at a breadth of audience and a commercial success that the SDUK did not, and hoped to capture the market from reformist circles.[8] Brewster wrote for both concerns, and his Whig credentials were exemplary, but this book contradicted Biot's interpretation of Newton's life as unacceptable for its moral and religious implications. Brewster was no doubt considered a good choice of author. He was 'one of the greatest and most meritorious contributors to experimental physical science' – receiving the Copley, Rumford and several Royal Medals and, in 1831, a knighthood for his research in optics – who had also written many popular expositions of science.[9] Murray staked the success of the series on the production of original works by respected authors rather than hack writers, paying them more than he could, ultimately, afford.[10]

Of the nineteen chapters that made up the *Life*, only six were directly 'biographical' with a large proportion of the remainder being devoted to the history of science.[11] However, Brewster's preface made it clear that he was responding to the SDUK's 1829 translation of Biot's 'Newton'. Biot's and Laplace's claims received treatment in two chapters, one considering the nature and timing of the illness and the second the theological writings that Biot thought post-dated the breakdown. Brewster claimed that his new evidence 'throws much light' on the event and had 'enabled me to delineate in its true character that temporary indisposition, which, from the view that has been taken of it by foreign philosophers, has been the occasion of such deep distress to the friends of science and religion'.[12] The fact that Brewster had begun 'a Life of Sir Isaac Newton for general Readers' by September 1828 suggests that he was prompted to write after learning that the SDUK had been offered the translation that would be made available to a wider and ostensibly lower-class audience.[13] Brewster took up his pen, despite recently informing Thomas Carlyle that he would 'end for ever my connexion with the Bookselling race' for 'there is no happiness in a literary life'.[14] Brewster declared:

> I feel it to be a sacred duty to the memory of that great man, to the feelings of his countrymen, and to the interests of Christianity itself, to inquire into the nature and history of that indisposition which seems to have been so much misrepresented and misapplied. (p. 227)

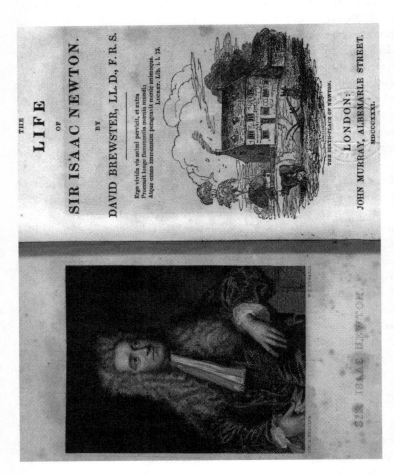

Figure 1. D. Brewster, *The Life of Sir Isaac Newton*, The Family Library, vol. 24 (London: John Murray, 1831), title page and frontispiece. Permission Special Collections, Senate House Library, University of London.

Contemporary assumptions that linked genius and virtue meant the suggestion that Newton suffered a mental collapse had to be refuted. Although conceptions of genius might include madness, this remained controversial, especially when applied to scientific rather than literary genius. In addition, Brewster was disgusted by the suggestion that Newton's religious writings were the product of a weakened intellect: 'That he who among all the individuals of his species possessed the highest intellectual powers, was not only a learned and profound divine, but a firm believer in the great doctrines of religion, is one of the proudest triumphs of the Christian faith' (p. 269). For Brewster, discoverers were 'instruments through which Providence discloses to man the wonders of creative power'.[15] Brewster's writings on Newton are infused with the assumptions of British Newtonian natural theology, demonstrating a fundamental difference between French and British images of Newton. There were those in Britain who were outraged that Newton's works had been 'appropriated by the infidel' across the Channel and Brewster resented the fact that Laplace had set Newton 'at the head of the atheistic sect'.[16] However, before considering Brewster's reinterpretation of Biot's evidence, I will consider Brewster's views on genius and method and place them in the context of the 'Decline of Science' debate. These ideas, together with his personal experiences, encouraged Brewster to form a notion of the 'neglect of science' that he raised within his biography and used as a contributory factor to Newton's illness.

Contradictions: Brewster on Genius and Baconianism

Brewster's view of scientific genius, of which Newton was the highest exemplar, involved contradictions which affect both the reading of his *Life* and his ideas on the support of science. On the one hand he rejected any connection between genius and madness or anti-social behaviour; the Newton he describes is relatively sociable and, in London, lives a life appropriate to his status. Yet, on the other hand, Brewster also believed that minds such as Newton's were not 'of earthly mould'. The tension surrounding the notion of genius is highlighted in his claim that the 'social character of Sir Isaac Newton was such as might have been expected from his intellectual attainments. He was modest, candid, and affable, and *without any of the eccentricities of genius*' (p. 337, my emphasis). Similarly, Brewster cited Newton's words about hard work and patient thought being his route to success but also opined that the 'impatience of genius spurns the restraints of mechanical rules, and never will submit to the plodding drudgery of inductive discipline' (p. 336). He wrote of Newton's use of hypothesis and of his 'fancy' producing the 'wildest conceptions' (p. 337), but elsewhere claimed 'the weakness of his imaginative powers'

(p. 224) was evidence against his having been driven to madness by the shock of losing his papers in a fire.

Brewster rejected the Baconian method as a path to scientific discovery. It is curious, but not necessarily contradictory, that in his own scientific work he demonstrated empirical rigour rather than the theoretical boldness he admired in Newton and Kepler. It is clear, however, that he always saw experimentation as something driven by a deep understanding and believed new experiments and techniques were suggested by speculative thought. But, despite his privileging of genius over method in major discoveries, Brewster did not wish to present the discoverer as purely God-inspired.[17] He emphasized the need for meticulous experimentation to test theories and the amount of time needed to develop ideas. This belief was perhaps behind his rejection of the famous apple anecdote, although his ostensible reason was that it was not well-authenticated.[18] Brewster also played down Newton's abilities as a child and gave the credit for his intellectual flourishing to Cambridge, thus giving a more prominent role to training than would be expected for an advocate of genius (pp. 5–8, 12–13).

In the final chapter of the *Life* Brewster took the opportunity to examine the 'pretensions of the Baconian Philosophy' (p. xv). He objected to those who ascribed Newton's 'method of investigating truth by observation and experiment' to Bacon, so that 'Newton is reported as having owed all his discoveries to the application of the principles of that distinguished writer' (p. 333). Brewster made two claims: first, what was commonly called the 'Baconian' method – an assertion of 'necessity of experimental research, and of advancing gradually from the study of facts to the determination of their cause' – was 'a doctrine which was not only inculcated, but successfully followed, by preceding philosophers' (p. 331); and second, none of 'the philosophers who succeeded [Bacon] acknowledged any obligation to his system, or derived the slightest advantage from his precepts' (p. 334). As an example of the first point, Brewster cited a letter from Tycho Brahe to Kepler, urging him 'to lay a solid foundation for his views by actual observation, and then, by ascending from these, to strive to reach the causes of things'. This advice meant that Kepler 'submitted his wildest fancies to the test of observation, and was conducted to his most splendid discoveries' (pp. 331–2). Brewster also pointed to Nicolas Copernicus, Leonardo da Vinci and Galileo as examples of those who knew 'the proper method of philosophical investigation' (p. 332). Bacon's lack of influence was then 'proved' by the assertion that no discoverer acknowledged him as a benefactor to modern science and that he had not been mentioned by either Newton or 'the amiable and indefatigable Boyle' (p. 334).[19]

While some common-sense techniques, often termed 'Baconianism', might aid discovery, as in the case of Tycho or Galileo, Brewster felt that other parts of Bacon's legacy were more damaging. He advised his protégé, J. D. Forbes, to

'Forget all you have heard of Lord Bacon's Philosophy. Give full reins [*sic*] to your imagination. Form hypotheses without number.'[20] In the *Life* he suggested that Bacon's method would only produce results if you chanced upon the most important facts: 'Nothing even in mathematical science can be more certain than that a collection of scientific facts are of themselves incapable of leading to discovery ... unless they contain the predominating fact or relation in which the discovery mainly resides' (p. 335). In this context Brewster asserted the value of biography for understanding the process of discovery, for the 'history of science does not furnish us with much information on this head'.[21] In other words, progress in science resulted from the efforts of peculiarly gifted individuals, and the process of discovery 'in its generalities at least ... is the very reverse of the method of induction'. Brewster described the 'impatient mind', which 'imagines a thousand consequences', 'forms innumerable theories' and 'exhausts his fancy' before submitting his 'wildest conceptions ... to the rigid test of experiment' (p. 336). This process could lead the discoverer to 'new and fertile paths', 'the invention of methods' and 'new discoveries far more important and general than that by which he began his inquiry'. As proof, Brewster referred his reader to Kepler and, unlike Drinkwater, suggested that Newton worked in a similar manner. Brewster suggested that Newton's process of discovery was not generally understood 'because he kept back his discoveries till they were nearly perfected, and therefore withheld the successive steps of his inquiries' (p. 337).

As Yeo states, 'this severing of Newton from the Baconian tradition had the effect of confirming the image of him as the solitary figure lauded by some of the Romantics'.[22] This endangered both Brewster's view of the man of science as a public figure and his denial of Newton's madness. Brewster attempted to ease the difficulty by referring to the differences between 'poetic' and 'philosophical' genius. He told the reader not to ascribe to Newton, because of his early promise and the quick succession of his discoveries, the 'quickness of penetration' and 'exuberance of invention, which is more characteristic of poetical than of philosophical genius' but to remember the amount of toil involved and the fact that his 'discoveries were ... the fruit of persevering and unbroken study' (p. 329). The distinction is reminiscent of that described Gerard in 1774, although their understanding of 'penetration' clearly differed.[23] However, the contradictions in Brewster's view of genius and his attempt to refute Biot's thesis remain. His view that Newton was 'deeply imbued with a cautious spirit' (p. 329) sits uncomfortably with his description of the 'impatient mind' of the discoverer. Similarly, his denial of Newton's 'imaginative powers' in the context of a breakdown contradicts his emphasis on 'fancy' as a vital element in discovery (p. 224). These contradictions were repeated when Brewster considered how science and its means of support might be reformed.

The Life of Newton and the Reform of Science

As Brewster was composing the *Life of Newton* he was in frequent correspondence with Charles Babbage, who was writing his *Reflections on the Decline of Science* (1830). A significant portion of Babbage's book focused on the belief that the Royal Society had become increasingly corrupt and less scientific, and therefore less capable of encouraging science and its practitioners. Those who styled themselves 'the scientific portion of the Society' and advocated its reform often held up the former President, Joseph Banks, for particular blame. Banks, it was believed, had made the Society a club that saw more value in gentlemanliness and nobility than scientific attainment.[24] This theme was the inspiration for 'Half a Dozen Epigrams on a Pair of Busts', sent to Babbage by the amateur astronomer Thomas John Hussey, which compared the busts of two Royal Society Presidents, Banks and Newton:

> In equal dignity 'tis Newton's lot
> To stand with Banks, par pari, – and why not?
> Small was the difference my muse maintains
> Between the men – except in point of Brains.[25]

In the same year, Babbage, with allies such as Francis Baily, was involved with the mechanics of the campaign to elect John Herschel as a reforming and scientific President of the Royal Society.[26] Brewster likewise supported him enthusiastically. Apart from attacking the current leadership of the Royal Society, the *Reflections on the Decline of Science* also discussed the lack of scientific education and the fact that there was no scientific career structure, particularly for those searching for 'abstract truth'. Babbage called for government money to be spent on rectifying such problems, but thought that it would not be spent effectively while those in power lacked scientific knowledge and men of science lacked influence. The present government, he argued, received poor advice on scientific matters, as evidenced in the case of the recent abolition of the Board of Longitude.[27]

Babbage compared the situation in France, where men of science frequently took on public and political roles, with that in England:

> Newton, was, it is true, more than a century since, appointed Master of the Mint; but let any person suggest an appointment of a similar kind in the present day, and he will gather from the smiles of those to whom he proposes it that the highest knowledge conduces nothing to success, and that political power is almost the only recommendation.

He also compared the Royal Society with foreign societies, noting that its membership was vastly inflated by individuals with little interest in science.[28] Babbage's criticisms of the Royal Society, some of which were personal attacks, revolved around the issues of power being in the hands of the non-scientific, the breaking of the Society's statutes and the quashing of attempts to reform. His suggested remedies began with the thorough reform of the Society but also included educational innovations, the formation of an order of merit to reward scientific achievement and a union of scientific societies to provide a platform for men of science. He appended an account of the meeting of the Deutscher Naturforscher at Berlin, first published in the *Edinburgh Journal of Science* (1829), as a model of the kind of organization that might be created.

Brewster sent Babbage a number of ideas he thought might be used in the book. These, which partly anticipate Babbage's arguments, were to reappear in his review of *Reflections on the Decline of Science* in the *Quarterly Review*.[29] Living in Edinburgh and without steady employment, Brewster, who at this period lived mainly on the proceeds of his literary endeavours, had different targets in mind. First, he asserted that there was a 'total want of situations to which [men of science] can aspire' and that those which they should fill, for example on boards for the encouragement of arts, were given to non-scientific individuals. Second, he felt a 'great evil in this country is the want of some order of Civil Merit, which w^d honour the successful labours of Science'. Third, he believed that the costly and inefficient patent system 'is another great check to Science' and should be reformed along the lines of copyright law. He mentioned 'a very striking example of the state of Science in G. Britain connected with myself', the negative reception in Britain of his new lighthouse lens compared to the French acceptance of 'the same idea', proposed by Fresnel. Brewster claimed that the advantages of the lens were obvious and thought that, 'if these Lighthouse boards were partly or wholly managed by men of Science well paid for their Labour, there would be an annual saving of many thousand pounds'.[30]

In this letter Brewster stated his belief that 'it w^d be necessary to obtain the co-operation of political & influential persons in order to produce a practical result'. Two weeks later, he asked: 'Would it not be useful to organise an Association for ye purpose of protecting and promoting the *secular* interests of Science?'. Again and again he wrote of the need to 'do something effectual', hoping for a unified front from men of science, headed by 'you great men in the Metropolis'.[31] Babbage seemed less enthusiastic about the proposition than he had been in 1829, when Brewster informed Brougham that Babbage had suggested 'the establishment of a great scientific association or Society embracing all Europe' and that they 'would take the oar if you w^d touch the helm'.[32] This appeal for patronage was, for Brewster, an important part of the project. He did not see membership of the association as being based on scientific ability, thinking 'many of our Nobility, tho' not scientific, would willingly promote such a great object'. It was them, rather than the men of science, who 'would have influence enough to direct the existing Government to a system of measures which would put England on a level with other nations'.³³ Despite a lack of encouragement, Brewster put the wheels in motion for what became the first meeting of the BAAS, held in York in September 1831. Since the 1980s, the roles of Vernon Harcourt, Roderick Murchison and the Cambridge 'gentlemen of science' in the formation of the BAAS have been emphasized over that of Brewster, and the role of the BAAS in the professionalization of science has been accordingly sidelined.³⁴ However, for a full understanding of Brewster's *Life*, it is necessary to bear in mind his conception of the BAAS. It would perhaps be misleading to say that Brewster wished for the professionalization of science, for maintaining independence was fundamental, but he did hope for paid scientific work, increased state funding and more scientifically-informed individuals within government.³⁵ Brewster reviewed the Reports of the BAAS in 1835 and, although unreliable in attributing his own views to all the founders of the Association, proclaimed loudly his belief that it was moving in a direction different to that which he had envisioned.

Brewster and the BAAS

Brewster's beliefs that the government should support science and that it could be actively advanced by the BAAS can appear difficult to reconcile with his view that scientific genius was unpredictable and unmethodical. How could minds that 'will not run either on railways or in grooves' be directed by others or be beholden to public money and positions?³⁶ His statements about how science might best be supported are unclear or, in fact, deliberately unsystematized. In his view, an 'association of our nobility, clergy, gentry, and philosophers' would somehow alert 'the sovereign and the nation' to the neglect of science and, once called, the 'aristocracy will not decline to resume their proud station as the patrons of genius'.³⁷ Like Babbage, his main concern was for individuals pursuing research that was not directly profitable. Such men should be allowed time and sufficient funds to prosecute their chosen studies. Teaching, at universities or elsewhere, was incompatible with research and Brewster wrote pathetically of those forced to:

> squeeze out a miserable sustenance as teachers of elementary mathematics at our military academies, where they submit to mortifications not easily borne by an enlightened mind. More waste their hours in the drudgery of private lecturing, while not a few are torn from the fascination of original research, and compelled to waste their strength in the composition of treatises for periodical works and popular compilations.³⁸

Brewster believed there was a danger that the need for money would distract researchers from their original course: 'The mammon of knowledge has beguiled many of her most ardent votaries, and some of our proudest intellects have fallen in their attempts to explore the Eldorado of science'.[39]

It appears that Brewster saw men of science in a number of different roles.[40] Those who invented useful machinery would be rewarded by the financial success of their invention, once the patent laws were amended.[41] Some scientifically-informed men were expected to have time to impart their knowledge to students, to sit on the Lighthouse Board or even to enter government. Original researchers, however, were different creatures and were always limited in number: 'Original and inventive minds are not of earthly mould. Divinely formed, they are the instruments through which Providence discloses to man the wonders of creative power, and the laws of his material universe'. One of his criticisms of the BAAS was that it wished to give 'systematic direction' to science, something he considered 'wholly beyond the power of any body of men to accomplish.'[42] 'Where', he asked, 'is the great intellectual captain who is able to compose, and daring enough to issue such general orders to original talent, – or if such a sage does exist, will he devote his powers to the drill-sergeantry of science?':

> Those high powers, indeed, which are worthy of being directed, will not run either on railways or in grooves, however wisely they may be prescribed. Let true genius be fostered and cheered. Give it a generous impulse. Give it leisure and a fair field, and it will speedily find a starting point – a direction – and a prize.[43]

Such statements make it clear that Brewster's understanding of scientific genius was incompatible with the Baconianism projected by other leaders of the BAAS.[44] He differed too from Babbage, who downplayed the role of genius in favour of a division of labour that echoed economic and mechanical models.[45] Rather, he conformed to what Simon Schaffer has described as the typical early nineteenth-century view, where 'discoveries are viewed as unproblematic mental events with obvious marks of identity' and which coincided with 'the reification of heroic discoverers'.[46]

The Neglect of Science and of Brewster

Rather than the 'decline' of science asserted in Babbage's book, Brewster claimed that Britain was guilty of neglecting her men of science. Several commentators have suggested a correlation between Brewster's position regarding the patronage of science and his career disappointments.[47] Although honoured for his experimental work in optics, it was only in 1838, at the age of fifty-six, that Brewster enjoyed a regular and sufficient source of income, on becoming Principal of the United College of St Salvator and St Leonard in St Andrews. Until that date he supported himself first as a private tutor and, after starting a family, principally as a writer and editor.[48] Although he derived some income from inventions, they are illustrative of his difficulties. The controversy over his lighthouse lens, referred to above, caused him 'half a century' of 'overwhelming anxiety and distress'.[49] He also invented the kaleidoscope, which proved to be immensely popular and 'spread over Europe and America with a *furor* which is now scarcely credible', but his patent did not protect him from entrepreneurs producing cheaper and less 'scientific' copies.[50] These experiences fed directly into his journalism on the decline of science and his *Life of Newton*.

Brewster's correspondence with Brougham demonstrates his financial insecurity. He wrote 'as a last resource to ask your Lordship if I could entertain the hope of obtaining any situation for which my habits of business and of hard work might qualify me'. He claimed that his editorship of the *Edinburgh Journal of Science* gave him no income and the *Edinburgh Encyclopaedia* had left him and the other proprietors in debt.[51] Brewster even considered Brougham's offer of a Church of England living, but shortly after wrote of his decision to avoid both the Church and writing, 'when many others can do better, while I abandon important investigations on which no other person but myself has entered'.[52] However, in 1835 Brewster wrote that the man of science must 'choose a profession, unless he ventures on the dangerous expedient of living by his wits and perhaps losing his wits in attempting to live by them'.[53] Brewster's brothers entered the Church and this was the path he had intended to follow. His daughter tried to dismiss the general belief that Brewster's dread of public speaking meant he failed in this career, but preaching was evidently difficult for him and perhaps prompted his decision to become a tutor instead.[54] Even if the problem decreased in later years, the reputation he had earned counted against him in his attempts to gain employment at the universities. Brewster's ambivalence about such employment, despite applying for the Chairs of Mathematics at St Andrews (1807) and of Natural Philosophy at Edinburgh (1832), was probably also a hindrance. He decided against applying to the Chair of Practical Astronomy at Edinburgh in 1828 because the salary was insufficient.[55] He also advised Forbes that a legal career was a wiser choice:

> There is no profession so incompatible with original enquiry as a Scotch Professorship, where one's income depends on the number of pupils. Is there one Professor in Edinburgh pursuing science with zeal? Are they not all occupied as showmen whose principal object is to attract pupils and make money?[56]

It was, however, probably the state of his finances in 1832 that led Brewster to contest the Edinburgh Chair of Natural Philosophy, a position which he also thought would leave him ample time for research.[57] In the election he was beaten by the young Forbes. His failure on this occasion was, according to his daughter, 'perhaps the most severe disappointment of [his] life'.[58] Although Brewster thought Forbes talented, he had clearly not taken him as a serious rival and considered this 'the most scandalous Job that the History of Science records'.

In a letter to Babbage Brewster blamed the situation on the partisanship of the Edinburgh Council, in whose gift the position lay.[59] This assessment has credibility, for the Council was largely Tory and had close links with Forbes's father, an Edinburgh bank owner.[60] However, other factors obviously came into play, including Brewster's notorious dread of public speaking and his difficult personality.[61] *The Scotsman*, generally sympathetic to the Whig cause, reported his 'great infirmities of temper' and that he would 'make the chair a sinecure, and leave its duties to some raw inferior substitute'. It reported too that Brewster was 'but a slender mathematician'.[62] In comparison to Forbes this was a fair assessment and it is something that his supporters seem to have considered. If this was a factor in Forbes's appointment it indicates a change in the status of mathematics at Scottish universities, where science was traditionally associated with philosophy rather than mathematics, which was rarely taught beyond an elementary level. After his election, Forbes attempted to remodel Edinburgh scientific teaching on that in Cambridge, becoming the first Scottish professor to teach the wave theory of light. Brewster later described these changes as 'treachery to the whole Northern tradition of experimental science'.[63]

Brewster clearly had a talent for making life difficult for himself and those around him. He maintained his opinions consistently and strongly, lacking the ability to view opposition as anything but hostility to the truth. In 1844, after he had made himself unpopular in St Andrews over the issue of university reform and by his attachment to the Free Church of Scotland, Lord Cockburn, a zealous Whig, described his position with some sympathy:

> With a beautiful taste for science, he has a stronger taste for making enemies of friends. Amiable and agreeable in society, try him with a piece of business, or with opposition, and he is instantly, and obstinately, fractious to the extent of something like insanity. With all arms extended to receive a man of whom they were proud a few years ago, there is scarcely a hand that he can now shake.[64]

This portrait is confirmed by Maria Gordon's description of Brewster's 'naturally irritable temper and finely strung nerves'. She noted, however, that 'if he caused distress and trouble to others, it was but a tithe of what he caused to himself' for, although his difficulties were 'often entirely of his own creating' or 'made more difficult by a power of magnifying them as by the lens of one of his own powerful microscopes', they were not the less real to him.[65]

Brewster's position in the 'decline' debate was thus clearly influenced by personal experience, and the issue of the neglect of science was also prominent in his *Life of Newton*. He claimed that Newton, even after he had become famous, was 'left in comparative poverty' and that this was partly to blame for whatever illness he had suffered in 1692–3:

> Such disregard of the highest genius, dignified by the highest virtue, could have taken place only in England, and we should have ascribed it to the turbulence of the age in which he lived, had we not seen, in the history of another century, that the successive governments which preside over the destinies of our country have never been able either to feel or to recognize the true nobility of genius. (p. 247)

Newton's subsequent patron, Lord Halifax, was therefore lauded as an example to the modern nobility. This 'accomplished nobleman' and 'liberal patron of genius' was said to be 'the first and last English minister who honoured genius by his friendship and rewarded it by his patronage' (p. 250). Brewster's claim that English science had not been patronized since Newton's time had also been a fundamental argument in his review of Babbage's *Reflections on the Decline of Science*. This theme was so strong in the *Life of Newton* that reviewers were obliged to comment on a subject 'on which the mind of Sir D. Brewster appears to be most strangely warped'.[66] It reappeared in later works, most notably his 1841 *The Martyrs of Science* which consisted of brief biographies of Galileo, Tycho and Kepler. As the title suggests, this work outlined instances of the neglect or persecution of science in history and, conversely, celebrated the enlightenment of monarchs and princes who supported its practitioners.[67] As with Halifax, these three 'martyrs' were a recurring theme in Brewster's writing and appeared, with similar intent, in his reviews of Babbage and the BAAS reports, and in his *Life of Newton*.

Newton's Breakdown Reinterpreted

To the reviewers, the most interesting parts of Brewster's *Life* were those in which he opposed Biot's account of Newton's breakdown and its consequences. The seventeenth-century diagnosis of melancholy or derangement had been reinterpreted by Biot as evidence of Romantic genius, but for Brewster it was essential to undermine this by unearthing a physical malady. His correspondence evidences his search for a variety of sources to enrich his account, but he was most interested in those which would 'throw a great and an agreeable light over the supposed insanity of Sir Isaac Newton'.[68] His anxiety to present the case well was also demonstrated by his sending Lord Braybrooke – who had provided some of the new sources – the proof sheets, with a request that he and Whewell 'would communicate to me any suggestion which may present themselves, or any alteration which may appear necessary, as I shall be sorry to have treated this delicate part of Newtons life in a way which might be disagreeable to the most sensitive of his admirers'.[69]

Brewster's refutation of Biot began by stating that what was 'an event' or 'incident' had been 'magnified' into 'a temporary aberration of mind' (p. 222). He felt he could dismiss the loss of papers in a fire as something that 'could never have disturbed the equilibrium of a mind like' Newton's (p. 224). Brewster firmly rejected Biot's suggestion that the 'incident' had a lasting effect and 'was the cause of his abandoning the sciences'. He could not deny its occurrence but he did everything he could to play down its seriousness. He based his case on what was fairly described by a reviewer as an 'over strict and erroneous interpretation' of the evidence of de la Pryme and Huygens, which he claimed were the only sources to point to a period of madness.[70] Because the former related to an event before January 1692 and the latter to one in November 1692, he questioned the reliability and compatibility of the reports. He further dismissed the import of de la Pryme's account by asserting that the statement 'every one *thought* [Newton] would have run mad' (pp. 228–9, my emphasis), meant he had not in fact done so. Brewster attempted to refute Huygens's evidence by showing that the longer period of indisposition it indicated coincided with that in which Newton wrote his letters to Bentley on the theological implications of his natural philosophy.[71] These letters were deemed to 'evince a power of thought and a serenity of mind absolutely incompatible even with the slightest obscuration of his faculties' (p. 230).

His trump card was the correspondence acquired from Braybrooke, between Newton, Samuel Pepys and John Millington. Although this proved that there had been a period of illness in September 1693, Brewster believed they showed the cause and nature of the illness to be physical, for Newton wrote that he was not eating or sleeping well. Newton did indicate that he had not had his 'former consistency of mind' for twelve months (p. 232) but, because this period again coincided with that in which he wrote his letters to Bentley, the mental effects could not, Brewster claimed, be considered serious. Pepys mentioned a strange letter from Newton that might indicate 'a discomposure in head, or mind, or both' (p. 233), but Brewster laid more weight on the reassuring reply of Millington, a medical doctor, which sought to quash rumours that Newton's understanding had been fatally compromised:

> He is now very well, and, though I fear he is under some small degree of melancholy, yet I think there is no reason to suspect it hath at all touched his understanding, and I hope never will; and so I am sure all ought to wish that love learning or the honour of our nation, *which it is a sign how much it is looked after, when such a person as Mr Newton lyes so neglected by those in power.* (pp. 234–5)

In this quotation, italicized by Brewster, the nature of Newton's ailment is not only disassociated from madness but is even partly blamed on 'those in power'.[72] Brewster went on to discuss the anxiety Newton experienced in his search for employment, mentioning also that the Royal Society had waived his subscription 'on account of his low circumstances, as he represented' (p. 236). Thus Newton is shown to be another victim of the neglect of science and Brewster was able to use the otherwise unfortunate 'incident' to add more weight to his favourite theme.[73]

Newton's anxiety, now reflected back onto 'those in power', could then be used by Brewster to explain at least part of the letter to Locke that had recently been published in Lord King's *Life of Locke* (1829). Brewster admitted that the letter was written during an illness (p. 238) and evinced 'the existence of a nervous irritability', but insisted that it was a state 'which could not fail to arise from want of appetite and of rest' and that 'it is obvious that its author was in the full possession of his mental powers' at the time he wrote (pp. 240–1). This was, he claimed, Locke's assumption in his reply, which was 'nobly distinguished by philosophical magnanimity and Christian charity' (p. 238). Brewster also felt 'it deserves to be remarked, that Mr Dugald Stewart, who published a portion of these letters, never imagines for a moment that Newton was labouring under any mental alienation' (p. 241) because he had not been aware of the dubious evidence on which Biot relied.[74] However, despite his decided opinion on this subject, Brewster did present all the relevant texts to his readers and reviewers who, if they were not persuaded by his version of events, could attempt their own judgment.

Brewster's other aim was to refute the suggestion that Newton's theological interests had developed and his scientific work had ceased after the breakdown. He bridled at the 'foreign philosophers' who thus 'indirectly questioned the sincerity of [Newton's] religious views' (p. 225). Worst was Laplace, who had investigated the event 'as if it concerned the interests of truth and justice to show that Newton became a Christian and a theological writer, only after the decay of his strength and the eclipse of his reason' (p. 227). Biot's stance was never as dogmatic as his mentor's but he too had dated most of Newton's religious writings to the later period. In countering both men Brewster relied in part on a paper by J. C. Gregory, based on the manuscripts of his ancestor David Gregory, to show Newton's pre-1692 religious interests and post-1692 scientific work. Early drafts of the General Scholium, which, because of its religious content, Laplace had eagerly noted only appeared in later editions of the *Principia*, showed that Newton had considered these themes much earlier.[75] Because the drafts also contained a section on light and ether, 'respecting which he maintains an opinion diametrically opposite to that which he afterwards published at the end of his Optics' (i.e. it supported the materiality of light), Brewster felt convinced that, even if temporarily lost, Newton's faculties were fully recovered (p. 241). He also showed, using evidence from King's *Life of Locke*, that Biot had erroneously dated Newton's essay *An Historical Account of Two Notable Corruptions in Scripture* to 1712–13 (pp. 271–2). Correspondence between Newton and John Flamsteed further revealed that the former was engaged in theoretical work regarding the motion of the moon between 1694 and 1698 (pp. 243–4). Brewster, who believed in the suitability of public roles for men of science, likewise saw no reason to belittle the importance of Newton's positions at the Mint and Royal Society.

Responses to Brewster's *Life of Newton*

The Journal des savans: J. B. Biot

In Biot's view, Brewster's work was missing the narrative and empathetic excitement that should accompany historical writing. Biot compared Brewster to Dr Dryasdust, the fictional antiquarian to whom Walter Scott addressed the prefaces of novels such as *Ivanhoe* (1819). Scott used these prefaces to defend his blending of fact and fiction as a means of recreating the past. The literary foil Dryasdust cannot condone anything that is not verified fact, but Biot claimed that Brewster exercised even greater scruples, for example in dismissing the apple story and, Biot probably intended to suggest, in his interpretation of the evidence for Newton's breakdown.[76] The French, according to Biot, followed Plutarch in attempting to bring their subjects to life, while the English practised a type of history that was 'moins difficile à pratiquer, et plus minutieusement'.[77] This criticism of Brewster's historical style shows sympathy with the approach of Augustin Thierry. In his *Lettres sur l'histoire de France* (1820), Thierry advocated a radical reform of historiography which insisted on the primacy of historical narrative: 'The task is to find a way across the distance of centuries to men, to represent them before us alive, and acting upon the country in which even the dust of their bones could not be found today ...'.[78] However, although Biot claimed that the Dryasdust style was particularly English, Macaulay likewise recognized his debt to Scott and declared that the historian's task was to 'make the past present' by using the tools appropriated by the historical novelist.[79]

Despite Biot's criticism, Brewster was to enjoy Macaulay's *History*, noting that his 'figures stand out before us in three dimensions, in all their loveliness, or in all their deformity, living and breathing, and acting'.[80] Biot was, however, writing in defence of a Romantic historiography and his use of anecdotes to reach an imaginative sympathy with the past. He argued that the apple story was not only well documented but allowed an insight into the development of Newton's ideas. In rejecting accounts of the falling apple and Newton's madness, Brewster had failed to understand the prodigious effort and inspiration required to produce the *Principia*. Biot also defended his more Romanticized account of Newton's education. He quoted Brewster's opinion that Cambridge was 'the real birth-place of Newton's genius' with disapproval, claiming, 'Le compliment est flatteur pour Cambridge; mais il est en opposition complète avec la vérité du caractère, et avec les détails même que le docteur Brewster, comme tous les biographes, rapporte de Newton enfant'.[81] For Biot, the tales of Newton's youthful activities demonstrated the promise that was fulfilled after only a relatively brief time at Cambridge, and 'tout atteste que la naissance de ce génie solitaire fut don de la nature et non pas un ouvrage de l'art'.[82]

Like other reviewers of the *Life*, Biot generally applauded Brewster's account of the history of science. Because it was aimed at a popular audience, his clarity was especially commended, as was the fact that he took the opportunity to put Newton's work in its historical and scientific context. However, he claimed that Brewster did not understand the connections between Newton's discoveries, for his narrative 'rompt si complétement [*sic*] l'ordre logique dans lequel les conceptions de Newton durent successivement se développer'.[83] Brewster's chief weakness was in failing to understand the significance of the development of fluxions and its impact on the rest of Newton's work. The *Life* gave accounts of Newton's optics, astronomy and fluxions, in that order, the reverse of the plan advocated by Biot, based on chronology of discovery rather than publication. Biot also cast doubt on the sections on optics by demonstrating Brewster's mistake on one of the occasions that he charged Newton with having made an error.

Biot devoted considerable space to the topic of the quarrel with Leibniz and, in lengthy footnotes, demonstrated the research that ultimately led him to edit and republish the *Commercium Epistolicum* (1856). Brewster had suggested that Leibniz should properly appear as 'the disciple and follower of Newton' and it was only because Newton desired tranquillity, and so had not published his discovery, that Leibniz 'stood forth with all the dignity of a rival' (p. 196). This was an account Biot could not accept:

> La suit des inventions de ces deux grands hommes, et leurs communications par correspondance, sont racontées avec une adresse si habile, les caractères de leurs méthodes sont présentés comme si analogues, les différences de leurs procédés de calcul comme si légères, et l'irritation de l'un comme si vive comparativement à la mansuétude de l'autre, que tous les torts, toutes les injustices semblent être du côté de Leibnitz, si même on ne doit lui reprocher quelque chose de plus.[84]

Biot pointed out that Brewster did not refer to Leibniz's frank response to Newton's circumspect letter containing the anagram that hid his method of fluxions. He implied that this omission was suspicious but wrote disingenuously that he hoped it was not 'une intention d'infidélité'.[85] However, perhaps because of the expected difference of opinion between Continental and British writers on this topic, Brewster was only pushed to re-examine the subject by the writings of De Morgan, which were tackled in his 1855 *Memoirs of Newton*.

The third section of Biot's review was devoted to the issue of Newton's illness and its aftermath. Biot was clearly on the defensive regarding both his interpretation of the evidence and 'la mémoire de Laplace sur-tout'. Brewster had tried to demonstrate Biot's prejudice against Newton's religious belief, for example by charging him with suppression of de la Pryme's statement that Newton had been in chapel when the fire occurred. Biot, of course, could hardly have suppressed a source which he would not have known before Elphinstone's translation was published.[86] Similarly, Biot vigorously defended himself from the accusation that he had deliberately misdated Newton's *Two Notable Corruptions*, the evidence regarding which was again only published after he had written his 1822 article.[87] However, he also denied Brewster's claim, verified in a letter from Laplace to Herschel, that Laplace had commissioned a search for evidence regarding Newton's illness and theology.[88] Biot scorned the idea that Laplace would have charged anyone with this 'mission antireligieuse', stating that he had merely been struck by Huygens's evidence and with 'un intérêt très-philosophique' wanted to see if it was related to the fact that he had all but abandoned natural philosophy in later life.[89]

Biot defended the validity of his interpretation by claiming that Brewster's dating of the sources was erroneous. In particular, he believed that de la Pryme's diary entry should be dated February 1693 (new-style), placing the fire toward the end of 1692, which could bring it into line with the evidence of Huygens and closer to that of the letters to Locke, Pepys and Millington. In addition, Biot thought that Newton's first and most important letter to Bentley might have predated the incident but, even if written after, did not necessarily indicate soundness of mind. He pointed to Pascal's ability to write his *Pensées* despite a disturbed mental condition and claimed that, in any case, the 'lettres de Newton à Bentley ne sont pas de cet ordre de philosophie'. It was, he asserted, only Brewster's desire to find religious truths in Newton's writings that led him to respect such mediocre ideas. As a Catholic, Biot could not agree with Brewster's assessment of Newton's work on biblical prophecy, including as it did an identification of one of the beast's horns with the Church of Rome (p. 272). He concluded with this point, and added sarcastically, 'Voilà un genre d'argument qui éclaire beaucoup les questions littéraires, et les savans du XIXe siècle savoir gré au docteur Brewster de les avoir ramenés à s'en servir.'[90] Biot's tone of superiority can perhaps be attributed to the fact that he was as much an authority on Newton as Brewster. Other reviewers of the *Life*, however, took a similar stance without this foundation.

The Edinburgh Review: B. H. Malkin

The *Edinburgh Review*'s assessment of Brewster's *Life* 'consisted principally of dissent', although its author claimed to 'think highly of its general value'.[91] The author has been identified as Benjamin Heath Malkin (1769–1842), elected Professor of History at UCL in 1829. However, the article was in fact written by his son, Sir Benjamin Heath Malkin (1797–1837), who, with his brothers Frederick and Arthur, was closely linked to the SDUK.[92] Heath, as he was known, wrote 'Astronomy' (1834) and Frederick a 'History of Greece' (1829) for the Library of Useful Knowledge, while Arthur edited the *Gallery of Portraits* (1833–7) (which included a short biography of Newton, probably by Heath) and wrote 'Historical Parallels' (1831) for the Library of Entertaining Knowledge. Both Heath and Arthur were SDUK council members. Heath Malkin was born in Hackney and attended the King Edward VI grammar school at Bury St Edmunds, where his father was headmaster. He subsequently went to Trinity College, Cambridge, from which he graduated as third Wrangler (1818). He entered Lincoln's Inn and was admitted to the Bar in 1823. In 1832 he left Britain to become first Recorder of Penang and then a judge of the Supreme Court in Calcutta. He died in Calcutta, in October 1837, and a memorial to him bears an inscription composed by Macaulay.[93]

Malkin's connections to the SDUK may explain why he reviewed the *Life*. He pointed out the SDUK's role in publicizing Biot's discovery and, because the Society had already 'incurred the misfortune of exciting in no ordinary degree the alarm of many very excellent, and the enmity of some very well-meaning persons', it appeared that this 'added to the violence of the outcry'. The SDUK and by extension the readers of the *Edinburgh Review* – who were frequently the audience for Brougham's writings on the Society and on political, legal and educational reform – were thus dissociated from these 'national and religious feelings' and those who 'would allow of no other doubt, than whether the statement proceeded principally from enmity to England or to Christianity'.[94] Brewster was seen as a spokesman for these over-anxious individuals and his 'enthusiastic admiration for his subject' as a handicap to 'impartiality', for 'enthusiasm sometimes leads to error and incorrectness, and excessive attachment to the fame of one may occasionally produce injustice to others'. Brewster, Malkin wrote, had 'certain theories which he is evidently anxious to support; and we cannot entertain any very high opinion of the accuracy with which he examines any facts which appear to bear on them'. Although he cleared him of 'any intention to mislead', Malkin pointed out that the reader must exercise caution when reading Brewster on the subjects of Newton's neglect, the question of his temporary insanity and his opponents, especially Leibniz. Malkin therefore felt he must examine each case, for they 'deserve correction in a work likely to continue for some time the standard Life of Newton'.[95]

Malkin gave his opinion of the *Life* and described his critical method in a letter to Macvey Napier, editor of the *Edinburgh Review*:

> I fear you will find the general tone of the article more culpatory than you expected or wished. I could not however make it otherwise. I wish to do full justice to the merits of the work, but this can only be done by some general praise. Censure necessarily consists of detail, & [illegible] much more paper: & many of the misrepresentations of the Life of Newton seemed to me to deserve & even to require to be refuted.[96]

First, Malkin showed that 'for the sake of treating Newton as neglected, he is represented as in a state of privation; and that this is entirely untrue'. Using Brewster's own evidence, Malkin extracted a very different interpretation of Newton's position; pointing out that he cannot have been as anxious for a situation in the early 1690s as Brewster suggested because he declined a lucrative position at Charterhouse late in 1691. On the issue of Newton's illness, he showed that Brewster 'manifested the same eagerness to force evidence into conformity with a foregone conclusion'.[97] Brewster's partiality was mitigated by the fact that Malkin believed, or knew, it to be 'unquestionable that La Place did attach much importance to the question; and was anxious to establish the fact that Newton's theological works were written at a late period of his life, after his intellect had received a shock'. However, in opposition to Brewster's attempt to demonstrate the novelty of Biot's claims, Malkin stated that the 'notion that Newton's theological writings were composed in the decline of his life, is not new; and that period has often been represented as one of mental inaction and comparative imbecility'. While accepting Brewster's findings regarding Newton's writings on theology, Malkin reminded his readers that it was immaterial whether or not they were written in a weakened state of mind, for religious faith should not depend on the authority of any one individual.[98]

Seeing Brewster as 'the advocate for Newton's uninterrupted soundness of mind', Malkin commended him for the 'considerable dexterity' with which he arranged his evidence. Adopting what he suggested was a more transparent technique, Malkin presented events 'simply in the order of their occurrence'. He agreed with Brewster that the suggestion that Newton's intellect was permanently affected could be disproved but read the evidence on the illness differently.[99] Brewster, he claimed, attached 'far too great weight to the evidence of particular writings', especially the anecdotal evidence of de la Pryme and Huygens. Perhaps following Brougham's review of King's *Life of Locke*, Malkin pointed to an earlier letter to Locke which could indicate that Newton's indisposition began early in 1692 but he thought that Newton's mind was only affected after his physical illness in 1693, which caused 'a short paroxysm, during which neither his memory nor his reason were proof against the assault to which they were exposed'.[100] The mental illness was due to the combined 'effects of long continued exertion, and the additional pressure of immediate bodily illness'. Although Malkin thought Newton had recovered towards the end of 1693, he suggested a novel interpretation for his diminished scientific activity. He thought it 'a conclusion hardly to be avoided' that Newton 'might naturally be very careful not again to expose himself to the danger of the like suffering'. Malkin believed that Newton's 'unwilling abstinence' was enforced, as he 'learned from experience to fear the effect of overstrained exertion continued for a long period'.[101]

On the subject of the calculus controversy, Malkin again claimed the ability to judge more fairly than Brewster. In his opinion a 'dispassionate review of the real facts will perhaps leave neither party completely free from blame' but would 'vindicate each from much obloquy which has been cast on him by the supporters of the other, and show that there was very little to condemn in either'. Malkin clearly thought that Brewster was over-critical of Leibniz's behaviour and that Biot was similarly unfair to Newton, although Newton, if responsible, was unjustified in removing his acknowledgment to Leibniz from the third edition of the *Principia*. Malkin considered this

> both undignified and unfair; and we would readily suppose, either that Newton had no part in it, or yielded, in almost the extremity of old age, to the persuasions of those about him, equally zealous with himself for his reputation, but less scrupulous as to the means of asserting it.[102]

Malkin preferred to find excuses for Newton but, unlike Brewster, did not wish to deny all culpability.

Lastly, the review addresses Brewster's 'very disparaging view, both of the value and the effects of the Baconian Philosophy', which is ascribed to his 'zeal for Newton's glory'.[103] Malkin undoubtedly respected Bacon more than Brewster did. His 'Astronomy', which leads readers along an inductive path, has been used to illustrate the SDUK's typical Baconianism.[104] This approach, Malkin believed, would teach correct reasoning and give satisfaction to minds unused to the 'hypothetical system'.[105] In addition, Malkin's account of Newton in the *Gallery of Portraits* suggested the sobriety of his method. Although the apple story was included, Malkin mentioned that, as Newton could not immediately make his calculations work, he laid aside his speculations because it 'was not Newton's habit to force the results of experiments into conformity with hypothesis'.[106] However, this point is not dissimilar to Brewster's attitude and there is nothing to suggest that Malkin dismissed the role of hypothesis. A letter from Malkin to Napier demonstrates that, in fact, the section on Bacon was added by the editor:

You expressed a wish to add something to the article in reference to Brewster's attack on Bacon, if it did not contain any notice of it. I have mentioned it: but only very slightly. If you have time & inclination to add or substitute any thing on the question I should be very glad to see the article so much improved.[107]

There are similarities between this part of the review and Napier's 1818 paper, which insisted that, while 'the Inductive Method had been happily *exemplified* in the discoveries of some of [Bacon's] contemporaries', no previous writer had attempted to 'systematize the true method of discovery; or to prove, that the *Inductive*, is the *only* method by which the genuine office of philosophy can be exercised'. This section may thus be read as part of what Yeo terms 'The Scottish Debate on Bacon', belonging to an earlier period, rather than a continued defence of Bacon by Englishmen in the 1830s.[108]

Regarding the depiction of Bacon, Napier wrote that Brewster's ideas were 'eminently unsound and illogical' and 'expressed in a tone of arrogance, and of confident assertion contrary to fact, not a little calculated to lessen our respect for his judgement, and our belief of his competency as a historian of science'. Bacon, in Napier's opinion, was the first to give a 'deliberate and detailed exposition and enforcement of the Inductive, as the only method of legitimate enquiry'.[109] Brewster's statement that no one subsequently acknowledged a debt to Bacon was 'pregnant with error, and inconsistent with fact'. Napier could not doubt Newton's acquaintance with the *Novum Organum*, 'or his obligations to those logical instructions which it had diffused throughout that school in which his mind was formed'. In addition, Napier emphasized Newton's 'patient and laborious application of his whole mind to the gradual evolution of a theory, to which, and not to sudden conjecture or intuition, he uniformly attributed the success of his researches'.[110] Finally, directly contradicting Brewster, Boyle was shown to have frequently mentioned Bacon.

The appreciation of Bacon in this section is as obvious to the reader as Brewster's of Newton. This in itself might point to the authorship of Napier, for Malkin's review condemns Brewster for playing the advocate.[111] He presented himself as enlightened and dispassionate enough to condemn the hero unflinchingly, if the evidence required it. By claiming that the evidence could be rearranged or better reasoning used to uncover a more genuine version of events, Malkin also defined himself as a better historian than Brewster. There is a hint that Brewster had not done his work fully, for Malkin noted with regret that the collection of papers in the hands of the Portsmouth Family had not been examined.[112] Malkin did not himself carry out any original research, basing his critique on Brewster's use of evidence, but his letters to Napier attest to a desire for scrupulous accuracy. He apologized to his editor for the delay in producing the review, which was due to waiting for information from Cambridge 'about one or two dates of minute circumstances'.[113] It is interesting to note, however, where Malkin remained wedded to an idealized image of Newton. It was not until the publication of Francis Baily's *Account of Flamsteed* in 1835 that long-held opinions about Newton's temperament, excepting the period of his illness, were challenged. Thus Malkin referred to 'the general calmness of Newton's temper' and claimed that it 'would be difficult to find a more admirable combination of temper, simplicity, humility, benevolence, and perseverance' than in Newton.[114]

An 'SDUK' Response?

It is tempting to see a negative response to Brewster's work from a largely London-based group that can be loosely identified with the SDUK and, as the following chapter on Baily's *Account* demonstrates, the RAS. Like Brewster, they backed political reform but appear to have been more radical than him in their opinion of authority, whether political, religious or intellectual. Like Malkin, they tended to object to uncritical hero-worship and to the appearance of 'partiality' in historical writing. Biot may have mocked Brewster for his antiquarian approach to biography, but these commentators reproached him for a lack of fidelity to his sources. The author of the article on Newton in the SDUK's *Penny Cyclopaedia*, De Morgan, stated that if we

> judge of Newton from the life of him recently published by Sir David Brewster, we could only infer that his moral character had suffered from no one instance of human infirmity, and that every action had been dictated by feelings of benevolence and the love of truth.[115]

Again, like Malkin, De Morgan used Brewster's own sources to give an apparently more honest portrait of Newton. It might be suspected that they highlighted Brewster's interpretation of evidence in order to reassert the worth of the SDUK translation of Biot; this was the Society for which they had acted as joint referees with Elphinstone, and for which all three had written or translated biographies of Newton.[116] In one sense, then, Malkin and De Morgan can appear to be spokesmen for the Society. But they also spoke to a wider set of values. Another reviewer of Brewster, who was only later a close friend of De Morgan and Baily, a Fellow of the RAS and writer for the SDUK, brought forward similar criticisms. Thomas Galloway, formerly teacher of mathematics at the Royal Military College at Sandhurst and, since 1832, an actuary in London, wrote an article for the *Foreign Quarterly Review*, designed to acquaint British readers with Biot's review. Galloway claimed that Brewster had an advantage over Biot in, theoretically, having access to a variety of archival sources, but that nevertheless his 'work is far more remarkable for the manner in which the ingenious author has contrived to mix up his own idiosyncrasies with the narration'.[117]

Galloway closely followed the arguments in Biot's review, but his intolerance of Brewster's advocacy of Newton is expressed in stronger language. Brewster 'appears to have been animated by the spirit of a zealous partisan' in his account of the Leibniz affair, and this 'unhappy spirit of prejudice and intolerance, so alien to philosophy, and so incompatible with the impartial investigation of historical truth, betrays itself in almost every page of the work ... and, indeed, forms one of its most prominent features'.[118] Interestingly, Galloway was 'disposed to agree' with Brewster on the subject of the Baconian philosophy, since 'Lord Bacon never performed an original experiment, or discovered a new truth', or explained '*how* nature can be best interrogated'. However, Galloway disagreed with Brewster's means of asserting this fact in a revealing manner. Brewster, he said, did not support 'his argument by general reasoning' but 'has recourse to authority'. One of the authorities to whom Brewster referred was, of course, Boyle and Galloway referred his reader to the *Edinburgh Review* article to see Brewster's error 'triumphantly exposed'.[119] It appears that in the 1830s support of Bacon was not part of the disagreement with Brewster. In the initial stages of writing at least the topic was not of great importance to Malkin. Galloway positively agreed with Brewster on the topic of Bacon's method and De Morgan, too, did not see Bacon's method as a description of the process of discovery that Newton had followed.[120] Just as it was shown in the previous chapter that the men of the SDUK did not all agree with Brougham on the subject of natural theology, this suggests that many did not advocate a strict Baconianism.

These men can, however, be linked to those SDUK members who saw virtue in the attempt to write impartial history or biography. It was a stance set against the belief that the more negative side of a subject, being both a poor example for readers and irrelevant to an understanding of the subject's public achievements, should be suppressed. The belief they opposed was clearly stated by John Davy, who wrote in his biography of his brother that 'to hold up the infirmities of a man of genius to observation is neither necessary nor useful; on the contrary, injurious, as tending to lower him as an example in the minds of posterity, and diminish the influence of his name'.[121] Many writers cited Diogenes Laertius's maxim 'de mortuis nil nisi bonum' ('nothing but good of the dead'), and Brewster was not alone in being shocked that Newton 'should be attacked more than a hundred years after his death ... when the hand and the tongue of the accused and his contemporaries were safely mouldering in the grave'.[122] Their stance against this position was a criticism of hypocrisy and of the alliance between Newtonianism and the established order.

Conclusion

The debate over the 'decline of science' and Brewster's own experiences in the scientific community fed into his portrait of Newton. Equally, his sometimes contradictory view of genius, explored through the figure of Newton, informed his arguments about how the nation should support science. These contradictions were nowhere more evident than in his refutation of the suggestion that Newton, even briefly, suffered a breakdown that affected his mind. Despite having described his genius as 'wild fancy' and not conformable to Baconian methodology, Brewster claimed that 'the weakness of his imaginative powers' meant that he was unlikely to suffer this kind of mental alienation. However, in the process of attempting to rebut Biot, Brewster accumulated most of the known evidence regarding Newton's character as well as his illness. Although imbued with theological imagery and references to Newton as the High Priest of Science, the *Life of Newton* therefore also recorded details that fought against Brewster's preferred interpretation.

The evidence for Newton's breakdown, and his overwhelming dislike of publicizing his discoveries, was only reinforced by the interpretation placed on Brewster's facts by the reviewers. Brewster had denied that genius was inevitably accompanied by 'peculiarities' but the general impression of his book went against him. Thus, in his popular *History of Natural Philosophy*, Baden Powell referred to the 'peculiarities of [Newton's] character', which 'have, till of late years, been little known or remarked'. These, which he thought 'were unquestionably connected, in the closest manner, with the powers of that pre-eminent genius with which he was endowed', included:

> The extreme repugnance of Newton's mind to the publication of his researches, the weariness and disgust which he, more than once, speaks of feeling towards scientific subjects, and the strong revulsion of his mind towards those 'mystical fancies,' as he himself calls them, in which he delighted to lose himself; the 'refreshment' he found in the driest details of ancient chronology; his excessive sensibility to the annoyances of controversy; his preference of tranquillity to every other consideration; his positive determination, on more than one occasion, to give up all scientific labours; [and] his constant refusal, during the later years of his life, to answer enquiries put to him on mathematical subjects ...[123]

Thus, although Baily's *Account* was described as 'the first English work in which the weak side of Newton's character was made known', if we consider the reviews of Galloway and Malkin, and the popular history of Powell, it would appear that the campaign, if it may be called that, against the uncritical hero-worship of Newton had already begun.[124]

3 FRANCIS BAILY'S *ACCOUNT OF THE REVD. JOHN FLAMSTEED* (1835)

> ... our dear illused Flamsteed ...
>
> Caroline Herschel[1]

Although Newton's reputation was used to emphasize the 'decline of science' agenda, one of the reformers adopted a different hero. Francis Baily's 1835 *Account of the Revd. John Flamsteed* appeared to some as much an attack on Newton and Edmund Halley as it was a vindication of the first Astronomer Royal and, although Baily always claimed neutrality, it is clear where his sympathies lay. Because of the controversy surrounding the publication and the importance of its contents to Newtonian biography, Baily's work is an integral part of the story that relates the increase in knowledge about Newton and the sources that described his life. This chapter first concentrates on the production of the work, drawing particular attention to tactics used in the presentation of controversial material. Baily's motivations in publishing can be seen to reflect his scientific interests and the sphere of the scientific community with which he was identified. Only 250 copies of the *Account* were printed and the chosen audience largely overlaps with the scientific constituency of the RAS, of which Baily was a key member. Baily used the *Account* to advertise a particular set of values that he thought fundamental to both practical astronomy and documentary history.

The *Account* was extensively commented on in the press and in private correspondence. Most notoriously, some, particularly the Cambridge-based William Whewell and the Oxford Savilian Professor Stephen Rigaud, reacted with alarm to the apparent attack on Newton's character. This hostile reception has been discussed by a number of historians. Richard Yeo has highlighted the debate's 'appeal to assessments of character, both intellectual and moral, as a means of attributing blame' and its implications for notions of the relationship between observers and theorists and the nature of private and public scientific property.[2] William Ashworth has developed these themes, asserting that the debate illuminates the notions of accountability, methodology and discovery which were being vigorously debated in the intellectual world of 1830s and 1840s Britain. Central to the debate were the scientific labour process and the place of the practical observer and philosopher in the manufacture of knowledge.[3]

Ashworth's main interest lies in the themes of intellectual property and labour, discussed in the context of the ownership of Flamsteed's observations and their value relative to Newton's theoretical work. Adrian Johns has also pointed to the contemporary significance of Baily's book, for the defence of Flamsteed was seen to be supportive of the London-based RAS – to which Baily was both 'parent and protector'.[4] I reassert and amplify these interpretations but, by including additional evidence of positive responses sent privately to Baily, I also identify a group that welcomed the *Account* for political, social and religious, as well as scientific, reasons. The discussion over the negative aspects of Newton's character must be understood within the context of other biographical studies of Newton produced at this time and as part of the critique of hero-worship, stimulated by the SDUK, in which both style of history and political commitments are implicated.

The Flamsteed/Newton Controversy Revisited

On becoming Astronomer Royal in 1674, Flamsteed had been given the task of setting up the new observatory and producing a more accurate catalogue of stars.[5] Before he began to print the catalogue, some thirty years later, he supplied Newton with a large number of observations to develop his lunar theory in the second edition of the *Principia*. The two men had initially enjoyed as cordial a relationship as could be expected between two bad-tempered individuals, but difficulties arose when Flamsteed decided that Newton was asking for too many observations and showing too little gratitude. To make matters worse there were general complaints that Flamsteed had produced no publication to warrant his salary. Flamsteed, however, believed the observations to be his personal property and wished to withhold them until he could produce as complete a catalogue as possible. In 1704 Prince George of Denmark was persuaded to pay for the publication, and production of the catalogue was taken out of Flamsteed's hands, decisions effectively being made by a Royal Society committee that included Newton and Halley. To bind his unwilling co-operation, Flamsteed was asked to present a sealed copy of his observations and manuscript catalogue to Newton, President of the Royal Society. However, the printing was frequently delayed and slowed to a halt after the Prince's death in 1708. Matters remained thus until 1711 when Flamsteed discovered to his outrage that his observations were being edited and printed by Halley, and that the sealed documents had been opened without his knowledge.

Halley was the chief object of Flamsteed's wrath, his implacable dislike focusing on his supposed atheism and 'loose and irreligious conduct', but also his alleged lack of expertise in practical astronomy.[6] Flamsteed blamed Halley for turning Newton against him, and spoke of the latter as 'Halley's dupe'. However, he also claimed he was convinced that Newton 'was no friend to [my] work; and ... that, whatever he pretended, his design was either to gain the honour of all my pains to himself, to make me come under him ... or to spoil or sink it' (p. 76). Flamsteed was horrified by Halley's 'garbled and incorrect' version of the catalogue, *Historia Coelestis* (1712) (p. xli). The difference between this catalogue and Flamsteed's, published posthumously in 1725, is marked.[7] The former was a bare record of the catalogue, followed by a summary of the observations on which it was based. In his preface, Halley, who had also been editor of the first edition of the *Principia* (1687), suggested that with this work the reader might 'weigh all the hypotheses of celestial motions' and confirm that Newton's 'theory alone should be embraced'.[8] The later version consisted of three volumes and placed a history of astronomy, an account of Flamsteed's methods and all of the observations before the catalogue itself. Thus, while the earlier version showed 'practice subservient to philosophy', the later one constructed a complete history that reflected the manner in which the knowledge was constructed, where 'practical astronomy was not simply a constituent of natural philosophy' but 'an end in itself'.[9] According to De Morgan's later article on Flamsteed, which was largely a restatement of Baily's views, the 1725 volume 'seems to us to occupy the place in practical astronomy which the Principia of Newton holds in the theoretical part'.[10]

The *Account of Flamsteed* begins with a long Preface by Baily which describes his enterprise and outlines Flamsteed's biography and disputes with Newton. This is followed by 'Flamsteed's *History of his own Life and Labors*, compiled from Original Manuscripts in his own handwriting', which includes autobiographical writings from the suppressed preface to Flamsteed's *Historia Coelestis* that climax in a series of complaints relating to the establishment of the Royal Observatory, Newton's demands for observations and the publication of the catalogue. To this is attached a long appendix 'containing a variety of *Original Documents*, confirming and illustrating the several facts therein recorded; and extending that history beyond the period narrated by himself'. This consists of letters to and from Flamsteed, his assistant Abraham Sharp, Newton, Halley and others. The second volume contains the *British Catalogue*, a 'corrected and enlarged' version of Flamsteed's star catalogue, with an introduction and copious notes by Baily. The debates were to focus on the first part of the work, and especially on Flamsteed's own account of events.

The controversial part of Baily's work appeared almost by accident. In 1832 he was shown a collection of letters from Flamsteed to Sharp by his neighbour

Edmund Giles, a descendant of the latter. Baily recorded in his Preface that seeing these papers put him in mind of looking at the 'vast mass of MS books, papers and letters belonging to Flamsteed, which had been lying on the shelves of the library [at Greenwich] for the last sixty years, unnoticed and unknown' (p. xiv). These papers contained the other half of this correspondence, further letters, autobiographical sketches and, most importantly,

> the *original entries* not only of Flamsteed's astronomical observations made at the Royal Observatory, but also those which he previously made at Derby and at the Tower ... [and] a great variety of computations connected with his astronomical labors and researches, more especially those from which the *British Catalogue* has been deduced. (pp. xv–xvi)

These allowed Baily to appreciate Flamsteed's working methods and revise his catalogue.[11] In the suppressed preface and in correspondence Baily constantly asserted the importance of this section of the *Account*. It is essential to keep this in mind in order to understand why Baily undertook the project, why he took it upon himself to rescue Flamsteed's reputation and, especially, why the work was published and distributed at the cost of the Admiralty.

The *British Catalogue* has precedents in Baily's other work, both historical and astronomical. His previous revisions of star catalogues by deceased astronomers, which John Herschel termed the 'archaeology of practical astronomy', included those by Ptolemy, Ulugh Beg, Tycho, Halley, Johannes Hevelius and Tobias Mayer.[12] Baily also examined recently produced catalogues and observations that were under suspicion for inaccuracy or the manner in which they were produced, and his expertise in this area is well attested.[13] As Ashworth has shown, Baily felt that he and Flamsteed fitted into the same tradition of English practical astronomy. Baily had been a stockbroker before accumulating a fortune sufficient to allow his retirement, and his was a bookkeeping style of science that valued accuracy, patience and repetition.[14] Herschel's description of him, read to the Society to which he was devoted, was typical:

> To term Mr. Baily a man of brilliant genius or great invention, would, in effect be doing him wrong. His talents *were* great, but rather solid and sober than brilliant, and such as seized their subject rather with a tenacious grasp than with a sudden pounce. His mind, though, perhaps, not excursive, was yet always in progress, and by industry, activity, and using to advantage every ray of light as it broke upon his path, he often accomplished what is denied to the desultory efforts of more imaginative men.[15]

In his article on Baily in the *Penny Cyclopaedia*, De Morgan cited this passage with approval but suggested that only the audience of Fellows of the RAS would have understood its impact. He claimed that 'there was in his habits of execution something unique: to us it seems right to say that in this respect he was a genius of an uncommon order'. Denying that Baily worked in the manner of a 'retired accountant', De Morgan claimed that 'instinct seemed to point out to him the proper mode of undertaking things to which he had never been accustomed', and thus reclaimed for him some of the heroic qualities of the scientific enterprise.[16]

That contemporaries understood the debate as a conflict between the rival advocates of primarily theoretical or practical approaches to astronomy is nicely illustrated with a cartoon by De Morgan, entitled 'Discordance between theory and practice' (Figure 2). De Morgan was a close friend of Baily's and, with a developing interest in the bibliography and history of science, no doubt spent many evenings with him discussing Flamsteed and Newton. De Morgan's later writings on Newton show the extent to which he was influenced by Baily's findings. Aware of his friend's foibles, however, he commented, two years after publication, that Baily finally felt 'Flamsteed is canonized'.[17] De Morgan, who reported that he was 'much amused' by the *Account*, showed in this cartoon his humorous attitude to a subject that was perhaps more serious to Baily.[18] Here he imaginatively recreated an argument of 1711 during which, Flamsteed claimed, 'Puppy' was the least of the insults that Newton addressed to him (p. 294). St Newton and St Flamsteed spar as Newton puffs out the reported insult. Suitably, Flamsteed has a halo of observable stars, while Newton's is a mathematically-calculated elliptical planetary orbit.

Figure 2. 'Discordance between Theory and Practice', inserted in A. De Morgan, 'Mathematical Biography extracted from the Gallery of Portraits', Royal Astronomical Society, De Morgan MSS 3, f. 70. Permission Royal Astronomical Society.

The theory-practice element of the debates, clearly apparent in the texts and correspondence discussed by Ashworth, is also confirmed by the correspondence analysed below. This is likewise the case with regard to the issue of intellectual property, raised over the question of whether the unfinished catalogue, sealed by Flamsteed and presented to the Royal Society, was Flamsteed's or public property. Champions of Newton preferred the latter possibility, for it gave him the right to break the seal in the event of Flamsteed refusing to co-operate, while those favouring Flamsteed pointed to the fact that he had paid for his instruments, assistants and the initial printing from his own pocket and that his official salary of £100 was hardly a sufficient remuneration.[19] One of the British Library copies of the *Account* appears to have been used by Baily in preparation for his 1837 *Supplement to the Account* and reveals points that were of particular concern. In a number of places Baily wrote 'Copyright' in the margin and, while this was not the word used in the *Account* or its *Supplement*, it is perhaps significant in the context of debates on copyright and patent law during this period.[20]

A dispute had forced the question of intellectual property to Baily's attention not long before his discovery of Flamsteed's papers. In June 1830 he reported to the Royal Society on the observations made at Paramatta Observatory, New South Wales, recently presented to the Royal Society by Charles Rumker, one of the assistants at the observatory.[21] It was only by accident that the observatory's founder, Sir Thomas Brisbane, discovered that these observations were not the original records, kept in bound volumes, but copies, made on loose sheets of paper.[22] Rumker was a more proficient observer than Brisbane; he had made most of the observations and claimed that he had provided the original records. However, the committee, consisting of Baily, Francis Beaufort and Davies Gilbert, found against him. They stated that the original bound volumes, 'unquestionably the property of Sir Thomas Brisbane', had been illicitly copied. The observations they contained were made by Rumker '*as the paid assistant*' to Brisbane and according to his instructions.[23] Rumker had taken these volumes to make copies and, despite repeated requests from Brisbane, they were 'still withheld, and are in fact unwarrantably mutilated and partially destroyed'. Brisbane, like Flamsteed before him, considered legal action in order to recover his property.[24] These findings suggest that Baily supported Flamsteed's claim because of his financial investment rather than because of his expertise as an observer.

The Rumker incident also allowed Baily to make a point about good astronomical practice, highlighting 'the great importance, in every Observatory of preserving the *original* observations that are made; inasmuch as it tends to prevent errors, and insures the confidence of the Public in the observer'.[25] It is possible that Baily found himself prejudiced by Rumker's methods when he examined the copied observations, which were 'on separate & rough pieces of paper in his usual way, scarcely legible, & evidently extracts from some other work; since none of them seemed to be *original* observations'.[26] Conversely, Flamsteed's record-keeping was admirable and one of the chief reasons for Baily's admiration. Flamsteed's original observations were recorded with sufficient clarity to enable Baily 'not only to detect many errors in the catalogue, but also to discover the source of them, and thus correct them with more confidence' (p. xvii), and in addition throw light 'on many stars supposed to be lost'.[27] Like Priestley, who had been an 'intimate acquaintance' of Baily's youth, he believed that Newton's accounts of discovery set a poorer example than Flamsteed's *Historia Coelestis*, which laid down a process of induction that might be followed by others.[28]

It is interesting to note the similarity between Baily's approach to historical documents and to scientific data. In both cases his aim was to print as much raw data as possible, to make it clear where he had added notes, interpretation or alterations, and to allow the reader to use the material to draw their own conclusions. Even if new discoveries were made and old methods discredited, because his methods were fully described subsequent researchers would be able to make use of his research. His scientific reports were renowned for their fullness of explanation and provision of data, usually in the form of tables. Herschel described one of his papers as 'very elaborate and masterly', excellent 'as a specimen of delicate experimental inquiry and induction' and having 'the further merit of bringing into distinct notice a number of minute circumstances'. Baily's first historical productions were an 1812 extension of Priestley's *Historical Chart* accompanied by a work entitled *An Epitome of Universal History* (1813). These, again described by Herschel, provided 'an easy and useful work of reference, in which the number and accuracy of the dates, and the utility of the appended tables, are especially valuable'.[29] Baily argued that, unless history was pursued systematically and non-judgmentally, 'the whole mass of historical records become confused; the transactions of various countries are divested of their two most essential qualities, *time* and *place*'.[30] In both his scientific and his historical work Baily was thanked for his patient labour rather than his insight.

The *Account* was a different kind of work, but it had direct similarities to his scientific approach. In an account of Henry Foster's pendulum experiments, Baily noted that, 'in order to distinguish every additional figure that has been inserted on the pages, since the manuscript left Captain FOSTER's hands, they are written in *red* ink'.[31] Echoing this, the introduction to the RAS copy of the Flamsteed/Sharp correspondence described how doubtful passages were underlined and missed words added in red ink, so that the reader could 'distinguish what has been actually added by myself'.[32] Interestingly, both this transcription and much of Baily's experimental work relied on the labour of others and Baily was the director rather than the prosecutor of such projects, providing instructions intended to cover all contingencies.[33] As I will show in this chapter, Baily's approach was influential and his was probably the first life of a British scientific figure to utilize the new standards of historical research. He was, however, working in a similar manner to several of those to whom he sent copies of the *Account* in thanks for their aid and advice. These included men such as Robert Lemon, Keeper of State Papers, Joseph Hunter, and the county historian John Britton, who saw history as the 'systematic collection of a body of factual knowledge ... the corollary of which was a hostility to theory'.[34]

The Account of Flamsteed

Baily brought Flamsteed's manuscripts to the official attention of the RAS on 8 November 1833. After describing the manuscripts and their contents, he noted in the final line that his find 'divulges this lamentable fact, that even amongst men of the most powerful minds, science is no protection against the infirmities of human nature'.[35] The next day he wrote to Babbage for help in deciphering Sharp's shorthand, saying that he was thinking of publishing much of the material.[36] On 10 January 1834, Baily presented the Society with a transcript of the Flamsteed/Sharp correspondence and was thanked for the addition to the library of 'one of the landmarks of Astronomical History'.[37] In January 1834, Baily was making enquiries about further manuscripts relating to Flamsteed. He received information and encouragement, notably, and in light of subsequent developments ironically, from Brewster: 'If you have not already resolved upon it I would venture to urge you to prefix a Life of Flamsteed to your Edition of his British Catalogue; and this would afford you an excellent opportunity of giving an account of the difference between him and Newton'.[38] On 12 February Baily told Herschel that he was 'decided' on publishing both the catalogue and a life of Flamsteed.[39] Whether or not he initially intended to publish at his own expense, as he had done on previous occasions, on 3 June he wrote a letter requesting the Board of Visitors to apply to the Lords of the Admiralty for payment of the costs of printing the work.[40] The Admiralty had paid for his last major work in the same manner.[41]

In his undertaking Baily assumed the role of revisionist, challenging previous accounts of both Flamsteed's work and character: 'Having minutely examined the whole of the manuscripts, I soon found that the character of Flamsteed had not been fully developed by his biographers' (p. xvi). In contrast, his presentation was intended to look as uncontroversial as possible and, as Ashworth has written, 'History, like Baily's (and Flamsteed's) astronomy, was to be about documented facts as certain as stars passing across the wires of a micrometer'.[42] Baily's scientific persona, as described by Herschel, was also projected into his historical work: 'Far-sighted, clear-judging, and active; true, sterling, and equally unbiassed [*sic*] by partiality and by fear; upright, undeviating, and candid, ardently attached to

Figure 3. Francis Baily, by Thomas Goff Lupton after Thomas Phillips. The *Account of Flamsteed* lies horizontally on the table; the paper is headed 'Pendulum Experiments'. National Portrait Gallery, London (NPG D7466).

truth, and deeming no sacrifice too great for its attainment'.[43] Extensive use of manuscript materials was becoming increasingly common in biographical writing, but Baily's *Account* barely included anything else. He emphasized the fact that Flamsteed's life was presented *'in his own words'*; the contents of the papers were undeniably contentious and Baily wished to minimize the presence of the author.[44]

Baily also asserted his impartiality by claiming that he had, unsuccessfully, 'sought documents which might tend to extenuate and explain the conduct of Newton and Halley in these proceedings' at the British Museum, Trinity College in Cambridge, Corpus Christi and the Bodleian Library in Oxford (p. xx). As he neared completion of the work, he was also granted permission to see the Portsmouth Papers.[45] Baily believed that new research methods were beginning to improve historical knowledge. The Newton manuscripts had long been dismissed but Baily found much 'that is *now* highly interesting, and not generally known' (p. xxi). Like Malkin and Galloway, Baily felt that his historical sophistication allowed him to attack the methods of other writers who had previously discussed Flamsteed, including Brewster. This attack was inevitable, since earlier accounts so disagreed with his own, but it also allowed him to imply the superiority of his own methods. He declared that Brewster, 'by a singular error', showed Flamsteed 'in a character which he by no means deserves' and, like earlier writers, did damage through 'partial statements and unfounded remarks' (p. xvii). He asserted that 'a proper regard for truth and justice prevents any suppression, at the present day, of the many curious and important (though at the same time lamentable) facts which these manuscripts have, for the first time, now brought to light' (p. xx). This claim – that modern history was able to tackle painful subjects fairly – placed Baily's approach with that of the SDUK men.

However, Baily's apparently neutral presentation of the manuscripts in fact ensured that his *Account* was inevitably biased. Although it was presumably quickly apparent to most readers, Baily did not specifically warn them that most of his evidence belonged to one side of the argument. It seems certain that, for him, previous bias meant the balance had to be redressed. Baily's desire to restore Flamsteed's reputation was, in part, an attempt to show that his work would achieve its reward, even if only a century later. An important aspect of both Baily's work and that of the RAS was the promotion of astronomy in order to attract more individuals to its pursuit.[46] The Society hoped that existing theories might 'be supported or refuted by the slow accumulation of a mass of facts' and that their 'associated labour' would carry out this task.[47] It was, therefore, important that accessible role models for the practical astronomer were brought out and dusted off and that the prospect of some degree of fame was held out. In this he succeeded, for if Robert Grant's *History of Physical Astronomy* is to be believed, by the 1850s Flamsteed was 'universally admitted to have been one of the most eminent practical astronomers of the age in which he lived'.[48]

Baily declared his regret that Newton was cast in a bad light, but was convinced that Newton's reputation, unlike Flamsteed's, would recover unaided: the 'passing cloud will soon blow away, & Newton's failings will be lost in his general merits'.[49] Flamsteed's autobiographical writings, which showed a selfish, spiteful Newton who was under the sway of 'those that were worse than himself' (p. 66), were, Baily admitted, written in anger but, together with the correspondence, demonstrated that Newton and others had pursued 'a line of conduct towards Flamsteed, which tends to make them appear less amiable in our eyes' (p. xx). Baily felt that Flamsteed's depiction of Newton as 'insidious, ambitious, and excessively covetous of praise, and impatient of contradiction' was

> so much at variance with that mild and modest behaviour which most of [Newton's] biographers have attributed to him, that it might seem like the excess of spleen and malice on the part of Flamsteed to dwell so much on these topics, were not his opinions strengthened by that of some of his contemporaries. Whiston, who knew him well, says he was impatient of contradiction, and that he was of the most fearful, cautious, and suspicious temper that he ever knew.

Unpleasant though it was to repeat the condemnation by William Whiston and others, Baily felt that 'justice to Flamsteed's memory requires that he should be defended even from the suspicion of misrepresentation' (p. 74).

There was little in the *Account* written in Newton's hand that backed Flamsteed's story but, in a letter of 6 January 1699, Newton expressed his anger towards Flamsteed for having mentioned in a paper that he had given observations to him: 'I do not love to be printed upon every occasion, much less to be dunned and teased by foreigners about mathematical things; or to be thought by our own people to be *trifling* away my time about them, when I should be about the King's business' (p. xxxiii). This was shocking both for its abrupt treatment of Flamsteed, who replied in a most conciliatory manner, and because Newton called his work on the lunar theory a 'trifling' matter compared to his work at the Mint. Baily noted that, in his *Life of Newton*, Brewster had mistakenly claimed that this letter was written by Flamsteed, saying that it was 'characteristic of Flamsteed's manner' and demonstrated his lack of awareness about the importance of Newton's theoretical work.[50] With the tables turned, the comments had to be explained away, as both Brewster and Whewell were to attempt to do.

It was Halley, rather than Newton, who received most criticism from both Baily and Flamsteed, especially for his alterations to Flamsteed's catalogue. Flamsteed's reputation was implicated by these charges but, Baily claimed, 'if he has expressed his opinion of Halley's conduct (in his confidential letters) in terms which sound, at the present day, extremely harsh to our ears, it must be confessed that he had much to irritate and excite him' (p. xlii). Flamsteed criticized Halley's character and atheism, and for Baily his *'garbled and incorrect'* edition of the *Historia Coelestis* suggested his dishonesty and duplicity (p. xli). Baily had earlier written a paper on Halley's unpublished observations, made as Astronomer Royal, which were 'very badly, and sometimes rather confusedly written'. Baily also noted that the non-publication of Halley's observations 'had probably been the subject of public complaint', for Newton was forced to bring it to the attention of the Royal Society Council. Halley's excuse was that he had kept his observations to himself until he had worked on a theory that might aid the calculation of longitude *'that he might take the advantage of reaping the benefit of his labours'* and win the reward offered by Parliament.[51] This was rich indeed, in Baily's view, coming from the man who had stolen and published Flamsteed's catalogue.

A Select Audience

Baily applied to the Admiralty for payment of the estimated £250 cost of publication. He stressed the catalogue as the major part of the publication and submitted

> the expediency and propriety of causing this first production of the Royal Observatory to be given to the world in a more perfect and enlarged shape ... cleared of all those errors which it is now known to contain; whereby it will be rendered more worthy of the nation and more deserving of that Royal patronage which has uniformly been directed towards the establishment, that produced the original work.

This letter was read to the Board of Visitors at the annual visitation on 7 June 1834 and they resolved to recommend Baily's project. They agreed that the amendment and publication of Flamsteed's catalogue was 'not only an object of great interest and utility to Astronomy, but necessary to the history of the Royal Observatory and highly creditable to the scientific character of the country'. Four days later the report was annotated by one of the Admiralty Secretaries, John Barrow, with the message that 'their Lordships duly appreciate the labours of Mr. Baillie [*sic*] and most willingly accept the proposal to print the amended British Catalogue'.[52]

The Admiralty printed 250 copies of the *Account*, which were distributed to a selected group of individuals and institutions.[53] This was contentious: it was felt that such material should be made available to the general public, especially as public money was involved, and once it had become the subject of public debate. If it seems at odds with Baily's historical style, which made a virtue of accessibility of facts, that the intended audience should have been so small, there is evidence that the arrangement was dictated by the Admiralty. De Morgan, who thought at least some copies should have been for sale, stated that the

Account's 'presentation was the act of the Lords themselves' and mentioned the specific nature of the directions 'under which Baily acted'.[54] However, both parties considered the revised catalogue, which was unlikely to be of public interest, to be the most important part of the book. The manuscripts, being held at the Royal Observatory, were government property and arguably it was they who should pay for publication; Baily was doing a considerable favour by taking on so much gratuitous work. In addition, the choice of recipients was chiefly Baily's.[55] Baily's list divided the 250 copies of the *Account* between 37 observatories, 72 other institutions, 36 foreign men of science and 105 British individuals (see Table 1).[56] Given the small print-run, the geographical distribution of the copies was remarkably wide, reflecting the international character of astronomy, with its connections to navigation and geodesy, at this period. The reach of Baily's contacts meant that 93 copies were sent to the Continent and beyond. The most remote destinations were often British-run institutions, such as Paramatta Observatory, the East India Company's observatories at St Helena and Madras, or that of the British Government at the Cape of Good Hope. Six copies were sent to North America, testament to Baily's early travels on that continent.[57] All the foreigners listed were practitioners of astronomy and related disciplines and all but one were associates of the RAS.

The institutional affiliations and occupations of the 'Individuals' listed are revealing of Baily's scientific world and confirm Herschel's comment that 'in a narrative of [Baily's] life it becomes impossible ... to separate the Astronomical *Society* from astronomical *science*, in our estimate of his views and motives'.[58] Of these individuals, 80 per cent were or would become Fellows of the RAS and 32 per cent had been members since the Society's foundation in 1820.[59] By comparison, 65 per cent were Fellows of the Royal Society, usually becoming so only after their election to the RAS. This was a community of active scientific practitioners or cultivators. By and large this group did not consist of the leisured gentleman amateurs who have rightly been acknowledged as a large constituency of scientific societies at this period. Of those for which there is sufficient information, only 16 per cent had never needed paid occupation. This constituency was typical of the RAS, as was the fact that thirteen were 'Scientific Servicemen' who had made their names through scientific work, such as surveying, undertaken in the army or navy. Another nine individuals taught mathematics at naval and military colleges and schools and the school of the East India Company at Addiscombe, who can, along with the eight individuals who were employed as observers in private or state-owned observatories, be counted among another important group, the 'Mathematical Practitioners'. Twelve men taught at the Universities, the majority of whom were part of the 'Cambridge Network', the last group forming the triple alliance that led to the foundation of the Astronomical Society.[60]

Observatories

Åbo/Turku (Finland)
Altona (Denmark)
Armagh (Northern Ireland)
Berlin (Prussia)
Bologna (Italy)
Brussels (Belgium)
Buda (Hungary)
Cadiz (Spain)
Cambridge (England)
Coimbra (Portugal)
Copenhagen (Denmark)
Cracow (Poland)
Dorpat (Estonia)
Dublin (Ireland)
Edinburgh (Scotland)
Geneva (Switzerland)
Göttingen (Lower Saxony)
Greenwich (England)
Good Hope (South Africa)
Königsberg (East Prussia)
Madras (India)
Manheim (Lower Rhine)
Marseilles (France)
Milan (Italy)
Naples (Italy)
Oxford (England)
Padua (Italy)
Palermo (Italy)
Paramatta (New South Wales)
Paris (France)
Portsmouth (England)
St Helena (South Atlantic)
Seeberg (Saschsen-Gotha)

Institutions

Academy of Science Boston
Admiralty (10)
Antiquarian Society
Astronomical Society (2)
Athenaeum
Bologna University Library
British Museum
Bureau de Longitude Paris
Cambridge Philosophical Society
East India Company
Geographical Society
Geological Society
Harvard College
Imperial Academy Petersburg
Imperial Academy Vienna
Kings College
London Institution
London University
Mathematical Society
Mechanics Institute
Nautical Almanac
New York Society
Newcastle Society
Philadelphia Society
Royal Academy Berlin
Royal Academy Bologna
Royal Academy Brussels
Royal Academy Copenhagen
Royal Academy Dijon
Royal Academy Geneva
Royal Academy Göttingen
Royal Academy Modena
Royal Academy Munich

Individuals

Adare, Lord
Airy, G. B.
Allen, William
Ashley, Lord
Babbage, Charles
Baily, Francis (10)
Barlow, Peter
Beaufort, Francis
Beaufoy, H. B. H.
Beaumont, Ed B.
Becher, Alex B
Beechey, F. W.
Best, Richard
Bishop, George
Bostock, John
Brewster, David
Brisbane, Thos
Britton, John
Cape, Jonathan
Catton, Thomas
Christie, S. H.
Cloyne, Bishop
Colby, Thomas
Cooper, E. J.
Dawes, W. Rutter
De Morgan, A.
Dolland, George
Donkin, Bryan
Drinkwater, J. E.
Drummond, T.
Dunlop, James
Ellis, Henry
Epps, James

Individuals cont.

Herschel, Sir J.
Holford, Charles
Hudson, J.
Hunter, Joseph
Hussey, T. J.
Hutton, Miss
Innes, George
Ivory, James
Jardine, David
Johnson, M. J.
Lardner, Dionysius
Lee, John
Lemon, Robert
Leybourn, Thos
Lubbock, J. W. (2)
Macclesfield
Maclear, Thos
Melville, Lord
Miller, W. H.
Nicolas, Sir N. H.
Oxmantown, Ld
Peacock, George
Pearson, William
Pond, John
Powell, Baden
Raper, Henry
Rickman, John
Riddle, Edward
Rigaud, S. P.
Robinson, T. R.
Rothman, R. W.
Sedgwick, Adam
Sheepshanks, R.

Foreign Individuals

Amici, Giovan-Battista
Arago, D. F. J.
Argelander, F. W. A.
Bessel, F. W.
Biot, Jean-Baptiste
Bouvard, Alexis
Bowdich, Nathanial
Brioschi, Carlo
Cacciatori, Niccolò
Carlini, Francisco
Cerquero, Jose Sanchez
Damoiseau, M. C. T.
Encke, Johann F.
Francoeur, Louis B.
Gambart, J. F. A.
Gauss, Carl F.
Gautier, Jean Alfred
Hansen, Peter A.
Hassler, Ferdinand
Inghirami, Giovani
Lindenau, Bernard von
Littrow, Joseph
Moll, Gerard
Mossotti, Octtaviano F.
Olbers, Heinrich
Plana, Giovanni
Poisson, Simon Denis
Pontécoulant, P. G. D.
Quetelet, L. A. J.
Santini, Giovanni
Schubert, Theodor F.
Schumacher, H. C.
Slavinski, Piotr

Spires/Speyers (Rhine)	Royal Academy Naples	Everest, George	Skirrow, Walker	Soldner, I. G. von
Turin (Italy)	Royal Academy Paris	Fellowes, Henry	Smyth, W. H.	Struve, Fredrich
Vienna (Austria)	Royal Academy Stockholm	Fellowes, Newton	Snow, Robert	Visconti, Ferdinando
Wilna/Vilnus (Lithuania)	Royal Academy Turin	Fisher, George	Somerville, Mary	
	Royal Academy Uppsala	Frend, William	South, James	
	Royal Asiatic Society	Galloway, Thos	Squire, Thomas	
	Royal Institution	Gibbes, Thomas	Stratford, W. S.	
	Royal Irish Academy	Giles, Mrs	Sussex, Duke of	
	Royal Society	Gompertz, Ben.	Tiarks, John L.	
	Royal Society, Edinburgh	Gregory, Olinthus	Turner, Dawson	
	Sion College	Hall, Basil	Wallace, William	
	Trigonometrical Survey	Hamilton, W. R.	Whewell, William	
	Trigonometrical Survey, Dublin	Harcourt, Vernon	Woollgar, J. W.	
	United Service Museum	Henderson, Thos	Wrottesley, John	
	University of Aberdeen	Herschel, Caroline		
	University of Cambridge (5)			
	University of Edinburgh			
	University of Glasgow			
	University of Oxford (5)			
	University of St Andrews			
	York Society			

Table 1. List of recipients of the *Account of Flamsteed* (source Cambridge University Library, Royal Greenwich Observatory Archives, Baily Papers, RGO 60/4, list dated 18 July 1835). Numbers in brackets indicate number of copies sent, where more than one.

With the exception of the Cambridge Network, these individuals were those for whom Flamsteed might be a suitable role model, achieving prominence through scientific work that was more concerned with data than theory. In addition, Baily's *Account* stressed the problems of interference by the Royal Society and its President in the affairs of the Astronomer Royal. Many Fellows of the RAS remembered the autocratic ways of Banks's Royal Society, in opposition to which the Astronomical Society was founded. Baily had not sought election to the Royal Society until after Banks's death, claiming that, 'To those who know the secret history of the last administration, it is unnecessary to state that such an admission was at least but an ambiguous honour'.[61] Like Babbage he complained that the Royal Society, even after Banks's death, was run largely by non-scientific men. The 'Mathematical Practitioners' had, from the end of the previous century, defined themselves as an 'underground resistance' to the Banksian regime and 'acquired the self image of a persecuted minority'.[62] These men were the heirs of the mathematicians and practical astronomers who rebelled against Banks in the 1784 'Mathematicians' Mutiny', standing for 'professional skill' and 'that accuracy of science which arises from having been employed only on one subject'.[63] They threatened to secede from the Royal Society, leaving Banks with his 'train of feeble *amateurs*'.[64] However, there were other nuances to the mutiny, for the mutineers were typically of a lower social class and one opponent claimed that they caused disruption 'merely from that levelling spirit and impatience of all governments which infects the present age'.[65]

David Miller has shown that the Astronomical Society was 'a bastion of the reform movement in British science'.[66] By the 1830s Baily and other members of the RAS were successfully claiming a voice in the organization of government-funded science, gaining full or partial control of the Board of Visitors to the Greenwich Observatory, the Board of Longitude, the production of the *Nautical Almanac* and the calibration of national standards in weights and measures. George Airy saw Baily, Babbage, Francis Beaufort and James South as leaders of 'a reforming party' and the 1820s and 1830s was a time of 'redistribution of power' between the Royal Society and the RAS.[67] Their successes caused consternation among some and Thomas Young, superintendent of the *Nautical Almanac* and recipient of criticism at the hands of Baily and South, declared that 'Mr. Baily will never rest satisfied until the Astronomical Society not content with the humiliation of the Royal Society shall succeed in dictating to the *Admiralty* and the *British Parliament*'.[68] Baily's growing prominence within the scientific community is symbolized by an 1835 honorary degree awarded in Dublin. De Morgan commented to his father-in-law, the radical rationalist Dissenter William Frend, that this indicated changes within the establishment: 'I think you can remember when a retired stockbroker would as soon have been an Archangel as a Doctor in one of our orthodoxy shops, if he had been Newton three times over. This is no bad sign of the times.'[69] However, the memory of persecution and 'tradition of dissent' seems to have remained. Flamsteed's difficulties with the seventeenth-century Board of Visitors, composed entirely of Royal Society gentlemen, must have been a reminder of the system they replaced. So too were his complaints about the poor salary and inadequate provision of funds, instruments and assistance.[70]

Initial Responses

The individuals to whom Baily sent copies of the *Account* replied with thanks and comments. Happily, many of these letters survive in the Greenwich Observatory Archives, placed there by Baily's sister in 1852. De Morgan, who was given the task of sorting Baily's literary remains, informed Herschel, 'I have retained ... the letters of thanks for Flamsteed. Offhand opinions given on the first look of a book ought not to be preserved – And yet some of these letters give additional information. They must be further considered.' He added, 'Baily must have weeded his letters carefully soon after they were rec^d – of anything which would give offence to others – In some cases, pieces are cut out'.[71] Keeping these warnings in mind – that the letters were probably censored, that initial comments might not reflect mature judgment, and also that there was probably a tendency to praise Flamsteed as a politeness to the author – this correspondence remains a fascinating source.

Those who disagreed most violently would perhaps say least to Baily, or not write at all. This was the case with William Whewell and Stephen Rigaud, both of whom provided Baily with information respecting Flamsteed, received copies of the *Account* and published critiques, which are discussed below. Herschel, a closer friend, expressed great interest to Baily regarding the corrected edition of the *British Catalogue* but seemed to avoid the subject of the other manuscripts that Baily published. To Beaufort, however, he apparently confided that he 'had a much higher idea of Flamsteed before, than since this publication' but much regretted 'that the matter seems to have been taken up as "Newton versus Flamsteed"'.[72] The opinion of the Cambridge-educated Herschel is an interesting contrast to that of his aunt, Caroline Herschel, who thanked Baily for 'your valuable Catalogue and Biography of our dear illused Flamsteed' which 'must be my companion to the last'.[73] John Herschel confirmed that Flamsteed 'was always her magnus Apollo in Astronomy' and that she was nearly as delighted at Baily's undertaking as she was at having been elected an honorary Fellow of the RAS.[74] These comments from two generations of Herschels reflect the main split among Baily's correspondents, which relates to the values – scientific and moral – attached to theoretical and practical science.

Commenting on the Preface prior to publication, Airy, who left Cambridge in 1835, felt that no passage was 'objectionable on the ground of *too strong expressions*' but thought Baily should 'remove any appearance of partisan spirit' by explaining Newton's behaviour. He thought it would be easy enough 'to shew that Newton was not necessarily a rascal at the beginning, though he might be an irritated man at the end'. Airy pointed to the impatience Newton must have felt at thirty years' delay in the publication of the observations, for he and those who could comprehend his theory understood the value of the lunar observations, 'the importance of which Flamsteed (a despiser of theory) could not estimate'.[75] Airy certainly understood the importance of large numbers of accurate observations but he saved his chief admiration for 'the results of combined theory and observation'. Privately, Airy had written that 'the work of a mere observer is the most completely "horse-in-a-mill" work that can be conceived'.[76] Similarly, Adam Sedgwick, another Cambridge-based correspondent, asserted that 'Flamsteed *did not see* the infinite importance of Newton's theory. He had not the most distant notion of it', while 'Newton *did see* the great importance of Flamsteed's observations to the completion of the lunar theory'. He concluded, 'I think you lean too much, in your estimate, on Flamsteed's side – or perhaps I ought to say you seem to press too hard upon Newton'. The belief in the pre-eminence of Newton's theory seems generally to have dictated the reader's estimation of his behaviour. Sedgwick thought that, '*For years* Newton wrote not only with courtesy but with *kindness* to Flamsteed' and, like Herschel, felt that these documents in fact ensured that 'so far from falling' Newton's character 'rises in my estimation above the place where my imagination had placed it', for he 'knew before that he was capable of losing his temper'.[77]

However, many more of Baily's correspondents replied in the manner of John Rickman, a House of Commons clerk and statistician, who declared that Baily had 'gratified me much with Flamsteed's Autobiography and the justice (tardy but decisive) now done to his character'. He went on: 'that the Demigods of an Idolatory ... should have behaved so infamously, – treacherous to the fame & fortune of no unworthy associate, – an assistant to his own great work, – is shocking to human nature'.[78] Others, while accepting Baily's estimation of Flamsteed's worth and his account of the ill-treatment he had received, nevertheless sighed for the loss of their former hero. John Bostock, Professor of Chemistry at Guy's Hospital, assured Baily that 'it is truly gratifying to see justice done to desert, especially such as appears to have been that of Flamsteed', but added 'yet it is, at the same time, not without considerable mortification to lower one['s] opinion of the estimate of characters that we have been accustomed to regard as almost superior to the weakness of human nature'.[79]

Some of those who approved of Baily's rescue of Flamsteed responded with hints about other scientific forbears with whom they identified. The mathematician William Wallace wrote at great length about his chance visit to the house of the 'Great Calcul^r' Abraham Sharp. After being refused a souvenir of Sharp's handwriting, Wallace was eventually allowed to make off with one of his wigs, which had 'adorned the heads of all our Mathematical friends who have visited Edinburgh since, in particular Messrs Whewell, Airy, South [and] Sedgwick'.[80] George Innes, an Aberdeen clock-maker and astronomer, was led to consider the 'excellent youth [Jeremiah] Horrox'. Although Innes claimed 'I am too far North to venture my humble opinion as to a comparison of Horrox and Newton', he suggested that had the former 'been spared he would have done all that Newton did, and, with less noise among the most eminent in science'.[81] Innes was also interested by Whiston, whose outspokenness about his unorthodox religious beliefs cost him his professorship at Cambridge. He suggested that Whiston had been more honest than Newton who, according to Whiston, shared these beliefs but remained silent. Thus, although Flamsteed was an Anglican, his stance against Newton was, as in Baily's Preface, linked to that of the Dissenting Whiston.

A number of Baily's correspondents made explicit reference to contemporary affairs and figures. Flamsteed's manner and troubles with the Royal Society made Thomas Maclear think of the former Astronomer Royal John Pond.[82] The antiquary John Britton claimed that the book reflected 'not only honor on its author & editor but on the Government body which has thus laudably applied *some* of its funds to the advancement of Science & Literature'. This statement relates to the 'Decline of Science' debate, but Britton went further:

> When Governments thus devote some of the many millions they extort from a hard working population & spend the other with equal discretion we shall all be loyal … but whilst Silly, reckless kings & queens – murderous & debauched Princes – & Stupid Dukes, Marquesses &c. fare sumptuously every day on the taxes wrung from a half starving people – dissatisfaction – radicalism, & other similar isms & schisms, must be the consequence – We shall all be delighted with Cobbatts [sic] Legacies – Howitts Priestcraft – Paynes [sic] Age of Reason, & other similar books.[83]

In a subsequent letter, probably written in response to Whewell's criticisms, Britton urged Baily on in his task: 'Surely the whole blame – the whole of illtemper, jealousy and injustice – is not attributable to Flamsteed?'. He thought it essential that '*Justice* should be done to all the sons & daughters of Genius & Talent' and claimed '*I* should be ashamed to assert that Shakespeare, Stukeley or Camden were immaculate – that their whole lives & writings were without error & blame'.[84]

The radical publisher Sir Richard Phillips, who had been a friend of Priestley, also wrote to Baily about the *Account*, asking if he might borrow a copy.[85] Phillips, according to De Morgan, 'was not only an anti-Newtonian, but carried to a fearful excess the notion that statesmen and Newtonians were in league to deceive the world'.[86] It is therefore unsurprising that he told Baily, 'If I respect you for your high attainments during many past years; I almost worshipped you for your manly independence in tearing the mask from the character of Newton'. He commended Baily for not suppressing statements, as many other editors would have done, that verified those of 'Whiston, Hutchinson[,] Hooke & others'. He also declared that Newton 'was more indebted to the fascinations of a pretty niece than the Schools of Philosophy are willing to allow'.[87] Baily had referred to this gossip regarding the possible relationship between Newton's niece and Lord Halifax in a note in the *Account* (p. 72). It interested him enough to provoke a correspondence with several individuals on the subject, and both Rigaud and De Morgan subsequently wrote essays on the topic, which are discussed in Chapter 6. For the first time this long-existing rumour was put to the test of archival research.

Broadly, the correspondence on the *Account* gives the impression that Baily received approval for his work from the 'Mathematical Practitioners', like Wallace, and the 'Scientific Servicemen', like Brisbane, who believed the *Account* 'must induce every one to change that opinion he had been led to form of the Dignity of our Great Philosopher'.[88] That is, those who were most likely to applaud Baily's scientific work, to be involved with similar enterprises and, according to Miller, to be 'defensively suspicious of high-flying, abstract mathematical gyrations'.[89] They tended also to have political views ranging from Whig to Radical and to have Nonconformist sympathies.[90] More cautious responses, in general, came from the Universities, although J. Wood of St John's College, Cambridge, gave an opinion with which Baily would have entirely agreed:

> Justice has now been done to the merits and exertions of that great Founder of the modern system of Astronomy, amid the most unexpected, unmerited, unjust and oppressive opposition. A melancholy proof is indeed necessarily disclosed of the effects of jealousy on minds otherwise supposed to have been of the most amiable and exalted character, and actuated solely by the pure love of science.[91]

In general, however, the importance of theory and the legacy of the *Principia* to the Cambridge men coloured their view of Flamsteed.

Published Responses

The publication and distribution of the *Account* marks it out from the cheap publications discussed in the previous chapters. It was, however, to become something of a 'sensation'.[92] Indeed, Galloway, who read the book before publication, told Baily that this 'extraordinary piece of literary history' would 'certainly *produce a sensation*'. When Powell looked back on the publication of these 'racy records' he remembered the 'singular warmth' of the debates and 'how fiercely the controversy raged anew, as if it had been a personal affair of the present day'. De Morgan recalled that the book had been 'talked over, in every scientific company'.[93] Interest had quickly spread beyond the original constituency and enquiries arrived from individuals hoping to see a copy.[94] Rumour, newspapers and reviews all played a role in diffusing the story, generating fear that a distorted version of events would be transmitted to undiscerning readers. This fear was fed by the manner of distribution and a letter to *The Times* took the 'private distribution' of the publicly-funded *Account* as its principal complaint. Considering the 'existing circumstances of the country, and the general system of retrenchment', it was particularly strange that the work – which the correspondent called a 'magnificent drawing-room ornament' – should have been published in an expensive format. To add insult to injury, the writer had heard that nearly half of the printed copies were given to foreign institutions and individuals: 'A stigma upon Newton's character, as a man of common honesty, and resting upon *ex parte* statements, thus circulated amongst the continental men of science at the expense and by the authority of the British Government! It is monstrous, and without parallel!' The whole transaction is deemed 'a general grievance and a public injury'.[95] These comments are typical in that they ignore the *British Catalogue*, although this was the chief reason for the Admiralty's sponsorship, for the length of the volumes, and for the limited audience. In addition, the writer apparently based his opinion of the biographical sections on hearsay.

Some of those closest to Baily questioned his decision not to publish a second edition. Basil Hall told him 'it seems to me but fair that you should furnish the Scientific world at large, & not merely 250 favoured individuals – with an opportunity of judging of the question for themselves'. Unfortunately the reply to this letter, in which Baily explained his decision, does not survive, but it appears that he wished to treat the work as a catalogue of stars and not the controversial piece of history that many readers saw. The Admiralty probably would not finance further copies and Baily would not reprint the biographical sections without the catalogue. His response to criticism was to publish a *Supplement to the Account*, which was only distributed to those individuals and institutions who received the original book. He seems not to have been swayed by Hall's fear that 'the opinion against Newton which has gone abroad', which was worse than that actually contained in the *Account*, 'will be much more likely to fix itself & become permanent in the public mind, than if the work were in general circulation'.[96] A similar concern was expressed by W. H. Smyth, who found the book 'painfully interesting',

> but except in all that related to the personal character of Flamsteed, I could almost have wished the documents had been destroyed. People of judgement well know that men without faults are monsters, but vulgar minds delight in seeing the standard of human excellence lowered.[97]

The Quarterly Review: Sir John Barrow

The first review of the *Account* appeared in the *Quarterly Review* for December 1835. It was by the *Quarterly*'s regular author of reviews 'on anything connected to China and travel, fisheries, ship-building, history of inventions, "quackery in general", canals, railroads, geography, nations and their inhabitants', John Barrow.[98] Because of his position as Secretary to the Admiralty his review was positive and he defended the means of publication, applauding the Admiralty for bringing out an 'expensive work ... in a limited impression', and distributing it as they did. The review is generous to Baily and, especially, to Flamsteed. It gives all credence to his complaints and condemns Newton for being determined 'to harass and annoy' this 'great and good man'.[99] His conduct is described as 'quite inexplicable', although Barrow did tentatively link it to 'that distressing malady' of 1693; 'a malady rashly ascribed by some to mental aberration, but which was clearly occasioned by want of sleep, want of appetite, excessive restlessness, and a great nervous irritability', but, 'making all allowances ... it would still be difficult to find any excuse for the overt acts of meanness, injustice, and ingratitude, of which Flamsteed had but too much reason to accuse Newton'. The only excuse for Newton was that he had responded to the influence of Halley, who Barrow claimed 'was undoubtedly in all respects the very reverse of Flamsteed. Low and loose in his moral conduct – an avowed and shameless infidel.'[100]

However, Barrow differed significantly from Baily when he discussed Flamsteed's relationship with the Royal Society. He stated, incidentally, that the Society had experienced a decline 'more especially since the presidency of Sir Joseph Banks', and agreed with Banks's prediction that the 'numerous offsets', such as the RAS, would precipitate this decline.[101] Barrow was never a Fellow of the RAS and had become a Fellow of the Royal Society not for his scientific attainments, but as a 'clubbable person' and friend of Banks. For Barrow, Banks, a man 'conversant in general knowledge, especially in the knowledge of the world, courteous and agreeable in his manners and conversation, ready to oblige and to forward to the best of his power the objects brought to the consideration of the Society', was a more suitable President than someone 'elevated in any one particular department of science'.[102] In his discussion of this review, Ashworth misrepresents Barrow's comment on the decline of the Royal Society, claiming that he thought the Society had declined 'especially *under* the presidency of ... Banks' and thus explaining his pro-Flamsteed position by placing him amongst the RAS/Declinist reformers.[103] Miller, however, links Barrow to the 'Banksian Learned Empire' and a 'Royal Society-Admiralty coterie'.[104] Together they had represented the enemy to Baily in the arguments over the reform of the *Nautical Almanac* and the Board of Longitude in the 1820s.

Ashworth is, however, correct to suggest that Barrow's own virtues of reliability, punctuality and hard work meant he admired Baily's historical approach.[105] Barrow had been central in the reorganization of the Admiralty, saving money, removing abuses and taking responsibility for the new classification and filing of the Admiralty papers.[106] Barrow's lack of scientific expertise would, in addition, mean he could better appreciate careful labour than mathematical theory. However, Barrow also lacked expertise as a historian of science and his entry into the debate was not appreciated by writers on either side. Because it was less well informed and also because it was less circumspect in its manner of expression, and with a wider readership, Barrow's review was in fact more controversial than Baily's book. It proved, therefore, a convenient target for those wishing to counter Flamsteed's statements, especially those who wished to avoid offending Baily.

William Whewell and Stephen Rigaud

Whewell chose to respond to Baily publicly, and the dispute was thus identified as one between London and Oxbridge, or between scientific practitioners and academics. However, Whewell's pamphlet, *Newton and Flamsteed* (January 1836), nominally aimed at Barrow's article, appeared to accept Baily's impartial stance.[107] He praised the *Account* and accepted that its contents must lead to a revised understanding of the characters of Newton and Flamsteed.[108] However, he felt that it was not made sufficiently clear to 'the great body of Review Readers' that Barrow – he does not say Baily, who was almost equally culpable – took 'for his sole guide the statements of one of the parties, written in the warmth of the moment'. This unmodified, partisan account, Whewell suggested, might induce these readers to 'cast away all their reverence for the most revered name of our nation', leading to who knew what disastrous consequences.[109]

Defence of Newton's character was Whewell's chief priority. He questioned the *Quarterly* reviewer's acceptance of Flamsteed's condemnation of 'almost all the eminent literary and scientific men of the day'. This position, Whewell claimed, had no backing except that of Whiston, 'whose judgement is perfectly worthless, for he was a prejudiced, passionate, inaccurate and shallow man', while all other accounts demonstrated that the 'mildness of Newton's character shewed itself rather in his horror of disputes, than in his skill in conducting them'.[110] Whewell further undermined Flamsteed by claiming that he did not appreciate the importance of Newton's work and overestimated that of his own. His culpability was compounded by his creation of needless delays in the publication of his catalogue. Whewell deemed the observations 'national property', and therefore considered the Royal Society referees justified in taking control of the publication process out of his hands.[111] In the *Philosophical Magazine*, Rigaud also focused on Barrow's account and played the role of someone outside Baily's selected group, falsely claiming that he had not had the chance to compare the review with the book. In general, Rigaud reiterated the points made by Whewell, and judged that 'Newton's philosophical and moral character come out from this examination blameless and admirable, as they have always been esteemed by thinking men'.[112] Rigaud suggested that Flamsteed's temper and his 'passionate and wrong-headed statements' should be held culpable for the affair.[113]

The most offensive piece of writing to appear in the whole debate was a reply to Whewell's pamphlet in the February 1836 issue of the *Quarterly Review*, presumably also by Barrow. As a correspondent informed Baily, this short note gave Whewell 'a dressing down ... with no very good style'.[114] It accused him of 'dogmatic assertions' and suggested that 'college tutors are apt to conceive rather an overweening idea of their own authority'.[115] This 'Note' provoked a second article by Rigaud and a letter from Whewell in the *Philosophical Magazine*, although Rigaud doubted whether Whewell 'would condescend to soil [his] shoes by kicking the dirty blackguard'.[116] In his article Rigaud declared, 'I have much regretted the line which has been taken by the Reviews. The public mind will be made up on the differences between Newton and Flamsteed'. He regretted that, as a result, 'the British Catalogue, as republished by Baily, has been noticed with merely transient praise'.[117]

While Rigaud disagreed emphatically with Baily's conclusions, he had been supportive of the publication of the book: both he and Whewell had written to Baily about relevant manuscripts. Rigaud was in regular correspondence with him about matters historical, astronomical and personal and they met regularly at the RAS, Royal Society and the Board of Longitude. Baily's letters to both men remained friendly and he assured Rigaud that he respected their difference of opinion while trusting Whewell to distribute his reply to *Newton and Flamsteed* within Cambridge.[118] There seems to have been agreement that Baily was sincere and convinced of his position. As Beaufort wrote to Whewell, 'we all know how prone the human mind is to fall in love with its hero, and to glide from the biographer into the apologist'.[119] However, relations were undoubtedly strained. In February 1836, while Baily was encouraging Rigaud to publish his article on Halley, which in part refuted the *Account*, Rigaud was informing Whewell, 'If the Flamsteedians will not be quiet, I am ready to fight knee-deep for Newton'.[120] The following month, however, Rigaud told Baily that he was 'very sorry that we come to different conclusions about Newton & Flamsteed. It is to me a painful discussion; because the one cannot be defended without blame being imparted to the other.'[121] Baily seemed oblivious of the depth of Rigaud's convictions, for he hoped an hour's discussion would convince him that Flamsteed was neither ignorant of the significance of Newton's theory nor unwilling to supply observations for its improvement.[122]

The responses of Rigaud and Whewell have been discussed by both Ashworth and Yeo as attempts to protect the links between Newtonian science and Oxbridge, Church and State, by maintaining the 'union of intellectual and moral excellence'.[123] The validity of this approach is underlined by Rigaud's comment that 'if Newton's character is lowered, the character of England is lowered and the cause of religion is injured'.[124] Rigaud and Whewell saw themselves as a 'Newtonian confederacy' and their response as a joint campaign for a righteous cause.[125] In addition, Rigaud felt required to defend Halley, his predecessor as Savilian Professor of Astronomy in Oxford.[126] He did not live to complete the biography or the defence of Halley against charges of atheism that he was working on in 1839.[127] At the same time, Rigaud was investigating another scandal that threatened Newton, the relationship between Catherine Barton and Lord Halifax, in order to answer 'those ill-natured individuals, who assume that he was so eager for preferment that he could sacrifice the fair fame of his favourite niece for the attainment of it'.[128] Again he admitted that in these endeavours 'Newton's character is my real object'.[129]

The Edinburgh Review: Thomas Galloway

Galloway, who reviewed the *Account* for the *Edinburgh Review*, was a friend of Baily's who had corresponded with him about the book both before and after its publication.[130] Ashworth presents Galloway's review as entirely favourable, as would be expected from one who fitted the profile of the 'Mathematical Practitioner' so well, but, although Galloway appreciated Flamsteed's achievement and understood why Baily admired him, his position was far more balanced than that of Barrow. For Rigaud the review was 'much superior to the Quarterly on the question', although it still did not 'meet my view'.[131] Galloway in fact largely blamed Flamsteed for the initial confrontation with Newton and told Baily that it was only with 'a *bad* grace' that Flamsteed had provided Newton with lunar observations.[132] As with Rigaud and Whewell, Galloway's sense that Flamsteed gave only grudging aid to Newton seems to have been informed by his greater appreciation of the role of scientific theory. Galloway's attitude is demonstrated in an 1830 review of the RAS's *Memoirs*. This was complimentary of the Society's aims but suggested there was a 'neglect of theory'. He wrote of observation as 'only the recreation or amusement of the astronomer' compared to mathematical deduction from theory, 'of which the human genius has infinitely more reason to be proud'.[133]

Galloway allowed that Flamsteed was 'probably the best practical astronomer in England' and had played an invaluable role in the foundation of the Royal Observatory, but his general estimate was less complimentary:

> The character of his mind is more remarkable for activity, and that sort of sagacity which leads to practical skill, than for any of the higher endowments. In point of genius, his name is not to be mentioned with that of Newton; he was immeasurably inferior even to his rival Halley. His mathematical knowledge, even for the time, appears to have been extremely limited. He set no value on the physical speculations of Newton and evidently never understood them.[134]

Galloway did, however, back Flamsteed over the question of the publication of his observations. Newton could be blamed for allowing the seal on Flamsteed's observations to be broken and, in removing his acknowledgments to Flamsteed from the second edition of the *Principia*, for allowing 'his conduct to be influenced by vindictive feelings'.[135]

Galloway considered 'Flamsteed's representations to be on the whole true, though exaggerated'. This demonstrated the falsity of most biographies of Newton, including Brewster's *Life of Newton* and, as in his review of this work, Galloway showed his suspicion of eulogistic writing. He described Newton's temper as 'always misgiving and suspicious' and, he asked Baily, 'why should Newton not be allowed to be human?'.[136] While Brewster was criticised for his inadequate use of original sources, Galloway approved of Baily's aim of bringing all available material to the attention of the public. However, like Whewell, he felt that the reader was inadequately warned of the one-sided nature of the published texts and he could not unquestioningly endorse an attempt to make Flamsteed the hero of a book in which Newton was all but silent.[137]

Baily's Reply

For Baily and his supporters, only one review 'has appeared worthy of the subject', which 'has properly appreciated your design & the manner of its execution'.[138] This was by Biot, despite his being 'an Admirer of Halley' who 'evidently idolizes Newton'.[139] This judgment may have been due to the relief that Biot did not focus, as the British did, on issues such as the opening of the seal and showed less concern for questions of morality.[140] Baily's 1837 *Supplement to the Account* therefore tackled the other approach, exemplified by Whewell's pamphlet. Baily professed surprise at the furore that the book provoked and claimed, like Galloway, that his avowed purpose was merely the elucidation of certain difficult points, 'to enable the general reader to arrive at an unbiased and correct opinion'.[141] However, he consistently found in favour of Flamsteed. Baily was particularly aggrieved that one tactic of his critics had been to link character to intellect: 'The mere fact of mental superiority, which no one is disposed to deny, ought not to weigh one feather in the scale of justice'. Seeking to answer the four 'totally unexpected' charges raised by Whewell, Baily admitted that Flamsteed's abilities were different from Newton's but did not agree that his hesitation over the lunar theory demonstrated his lack of understanding.¹⁴² The theory was a startling novelty that was still undergoing refinement, and Baily felt it might even 'be fairly questioned whether Newton himself was fully aware of the immense benefit and advantages that were destined eventually to flow from his own researches'.¹⁴³ However, he argued that Flamsteed supplied all the observations that Newton required, with occasional delays on account of illness, or when he ran out of data. Flamsteed's subsequent bitterness about Newton's lack of gratitude should not, Baily argued, be confused with the actual record of their exchange.¹⁴⁴ He concluded that Flamsteed exhibited 'an earnest desire to promote the great object that Newton had in view' and that 'the perfecting of it was ... in a great measure owing to Flamsteed's voluntary services'.¹⁴⁵

Baily produced evidence that suggested many of the delays in the printing of the catalogue could be attributed to Newton, but he considered Newton's decision to open the sealed copy of Flamsteed's observations the only serious allegation against him. That such observations were the property of the Astronomer Royal Baily thought proven by the fact that, when the Royal Society and government had attempted to claim ownership of the observations made by James Bradley, his executors succeeded in keeping them despite a decade-long lawsuit.¹⁴⁶ Maintaining the moral high ground, Baily did not descend to charges against Newton's character and continued to express his regret at finding 'Newton's name mixed up with a transaction of this kind'. While the initial 'sudden ebullitions of temper and apparent perversity of conduct' were 'mere venial offences of our common nature', the opening of the sealed observations was behaviour he likened to that of illicit financial dealers:

> I suspect that it was in that day, as at the present hour, that individuals of high and honourable character (when acting in concert with others having interested objects in view, and not quite so scrupulously austere in their conduct as themselves,) may oftentimes be led to countenance and sanction certain acts, which as private persons, and on their sole responsibility, they could cautiously avoid.¹⁴⁷

Baily's *Supplement* was a convincing, and reasonably conciliatory, statement that was not publicly rebutted until the 1850s. Baily had 'contradicted Whewell's historical and philosophical authority – and survived'.¹⁴⁸ From Cambridge George Peacock told Baily that he considered the *Supplement* 'one of the ablest pieces of criticism upon a most obscure & difficult event in scientific history which I have ever read: your case about the Lunar Theory is unanswerable: Whewell confesses it'.¹⁴⁹ Many others wrote to congratulate Baily, finding the *Supplement* 'quite conclusive' and Flamsteed 'completely vindicated'.¹⁵⁰ Rigaud was still plotting his response, but his *Essay on the Principia* did not concern itself with Flamsteed, his defence of Halley only appeared posthumously and his essay on Catherine Barton was never published. Whewell's 1837 *History of the Inductive Sciences* stepped back from this particular controversy, and graciously accepted much of Baily's *Supplement*.[151]

A Proposed Publication of the Portsmouth Papers?

Baily's emphasis on original documents won him the praise of many commentators. Joseph Hunter felt the *Account* was 'what we so rarely have, a book of Biography formed out of new and authentic materials, of a celebrated name in English Science'.[152] Baily's call to publish more such material (p. xxi) was met with universal approval. Our reading of the *Account* should be tempered by the knowledge that Baily wished to make the writings of others, including Newton, available in print. He corresponded with Rigaud about the publication of Lord Macclesfield's papers and, in October 1836, accompanied him on a visit to see them at Shirburn Castle.[153] It also appears that Baily may have had plans to publish a selection of the Portsmouth Papers. He had been given permission to see these in April 1835 by Henry Fellowes, nephew of the third Earl of Portsmouth. After publication, however, he received a rather strange letter from Newton Fellowes, Henry's father and subsequently fourth Earl of Portsmouth:

> I have always felt a jealousy with respect to the manuscripts being seen & in any way used without the full authority & concurrence of those whose property they are. I am not unwilling that useful matter sh^d be made public yet I will always reserve to my Family the right & choice of persons to whom I may delegate or grant the power.[154]

Walker Skirrow, who forwarded this letter, felt the need to apologize for its tone: 'Although Mr. Newton Fellowes is a person of singular temper & disposition, I did not expect he would write the note which I send for your perusal'. Fellowes wrote in response to an 'application' by Baily, presumably to view or publish the manuscripts a second time. Fellowes was clearly upset by the *Account* and felt a duty to protect the reputation of his namesake.[155]

The possibility that Baily intended to publish from the Portsmouth Papers is borne out by a letter from W. S. Stratford which comments that there was 'plenty of work cut out for you if the Portsmouth family should surrender the Papers!'.[156] In addition, Galloway's review claimed:

> We have good reason to believe that the government would be disposed to print them at the public expense; and indeed we have been informed that an offer of this nature has been actually made to the Portsmouth family; some members of which have signified their assent, whilst others have objections which cannot for the present be removed ... The character of Newton may be considered a species of national property – it is in fact the nation's glory – we therefore trust that what we may call this *national* appeal to the Portsmouth family, will not be made in vain.[157]

It was probably no accident of Newton Fellowes's temper that it was Brewster, rather than Baily, who was allowed to view the manuscripts in May 1837. On his arrival in London from Hurtsbourne, Brewster felt able to declare, 'Every thing that I have seen among the Portsmouth MSS has contributed to exalt Newtons Character in my estimation, high as my impression of it had previously been'.[158] Indeed, because *The Times* mistakenly reported that the Royal Society had recently purchased the papers, someone, presumably Brewster, wrote a correction stating publicly that he had recently seen the papers at Hurtsbourne,

> which not only throw much light on the early life and studies of our immortal countryman, but tend to refute the groundless rumours respecting a temporary derangement of his mind in 1692, and to exalt in the highest degree his moral and intellectual character.[159]

His second attempt to restore Newton's character had begun, although it was only completed in 1855 and Baily's *Supplement* appeared, in the short-term, triumphant. Newton's dislike of communicating his ideas and acknowledging the aid of others had, for many readers and reviewers of the *Account*, become acknowledged fact and contributed to the disintegration of the connection between intellect and morals.

Conclusion

The evidence presented in this chapter adds to the accounts of this dispute provided by Ashworth and Yeo by revealing Baily's immediate constituency and their comments on the book, as well as by placing it fully within the context of contemporary writings on Newton. While both the published and unpublished responses to the *Account* demonstrate the importance of attitudes to theory and practice, there were other factors that the supporters of Flamsteed had in common. Some of the letters that Baily received in relation to the *Account* demonstrate a relationship between radical political views and a rejection of the idealized image of Newton. More generally, Baily's middle-class constituency of 'Mathematical Practitioners' and 'Scientific Servicemen', who were paid for scientific work and desired the removal of social and religious barriers, had a self-identity formed through a sense of injustice against establishment scientific interests. This created an audience ready to read of a practical astronomer whose heroic efforts were thwarted by an autocratic President of the Royal Society. The picture is, however, complicated by recognizing that Barrow, the most outspoken defender of Flamsteed, was a Tory and staunch supporter of Banks, who praised the *Account* because of his connection to the Admiralty.

In general, however, like Brewster and the men of the SDUK, both Baily's supporters and critics can be linked to calls for the reform of science in the

1830s. This indicates the diversity of this group and a division between those based in the Anglican strongholds of Oxford and Cambridge and those living and working in the metropolis. The London-based Fellows of the RAS approved Baily's and Flamsteed's criticism of Newton's character and scientific approach, just as London-based writers for the SDUK promoted Biot's article and criticized Brewster's defensive biography. This attitude can be related to Baily's historical style. It provided a critique of traditional accounts that relied on the 'impartial' presentation of historical 'facts', drawing authority from its similarity to his bookkeeping style of scientific research. Baily's reputation within astronomical circles for meticulousness and trustworthiness could be transferred to a contentious historical dispute. The following chapter will focus on the tactics of other writers who made similar claims to 'impartiality' either to defend or criticize Newton. One who chose the latter path, and is of particular importance to this story, was De Morgan: friend of Baily, Fellow of the RAS, writer for the SDUK and Professor of Mathematics at the 'godless' UCL.

4 NEWTONIAN STUDIES AND THE HISTORY OF SCIENCE 1835–1855

> The history of Newton is in a great measure the history of science ...
> Baden Powell[1]
>
> ... there is much more curiosity among you now than formerly, & greater diligence of research ...
> John Lee to Stephen Rigaud[2]

In 1843 Baden Powell, who had succeeded Stephen Rigaud to the Savilian Professorship of Geometry, reviewed his predecessor's *Historical Essay on the Principia* and *Correspondence of Scientific Men*. He noted with pleasure that 'the attention of several eminent persons has been more closely than heretofore directed to the details of our scientific history in general, and more especially to the eventful period of which Newton formed the brightest ornament'.[3] In an essay published in the same year, De Morgan also commented on this phenomenon. After listing Brewster's *Life of Newton*, Rigaud's *Works and Correspondence of Bradley* (1832), Mark Napier's *Memoirs of John Napier* (1834), Baily's *Account* and Rigaud's *Historical Essay*, he noted that these 'coming so close together, make a remarkable epoch in biographical writing'.[4] The succeeding decade saw the brief existence of the Historical Society of Science (HSS), a number of important essays by De Morgan, and C. R. Weld's *History of the Royal Society*. Looking on to the 1850s we note works such as Joseph Edleston's *Correspondence of Newton and Cotes*, Brewster's *Memoirs of Newton* and Robert Grant's *History of Astronomy*.[5] What the HSS and these works had in common was a focus on manuscript sources; it was their evidential detail and precision that commentators found noteworthy. However, this period is now known better for the publication of works of broader scope, based on secondary sources, including Powell's own *History of Natural Philosophy* (1834) and Whewell's *History of the Inductive Sciences* (1837).

These narrative or philosophical works have received great, and often exclusive, attention from those historians who have examined the nineteenth-century historiography of science.[6] H. Floris Cohen believes that Whewell's *History* should have guided other historians of science but this was the 'follow-up that failed to occur'. He claims that, 'to become really historical, the historiography of science had to liberate itself from its foster father, the philosophy of science', and that this could have happened if Whewell's historicist hints had been combined with the new historical techniques developing on the Continent. However, Cohen believes that instead, the 'crude philosophy' of Comte exercised an inhibiting influence on the discipline.[7] This, I suggest, demonstrates an ignorance of the range of work being done in the history of science at this time. While much of this work was far less ambitious than that of Whewell or Comte, there were those who sought to write a history free of philosophy that attempted to develop 'refined methods of criticism of sources'. The pioneering work of De Morgan, Baily, Rigaud and Edleston in Britain and Guglielmo Libri and Michel Chasles on the Continent has been less studied, despite the fact that their focus on original sources is recognizably connected to an increasingly rigorous historical discipline.[8] In many ways they set themselves up in opposition to works such as Whewell's; making more modest claims about the scope of their publications, they asserted the superiority of their methodology.[9] It is the purpose of this chapter to consider the flowering of historical works on science in the 1830s–50s. My analysis draws particular attention to the different forms of historical writing, the moral claims that were attached to them and the important role of works relating to the life of Newton in creating expertise in the history of science.

A. N. L. Munby's short *History and Bibliography of Science* (1968) is one of the few publications to link the historical work of the individuals discussed in this chapter. He saw the 1830s and 1840s as the period in which 'documents and sources were being got into print for the first time' and queries why astronomers were so prominent in this process: 'Worship of Newton and Galileo', he thought, provided the clue.[10] Certainly the peculiar interest in Newton, as the archetypical scientific hero, whose cause had been linked to the religious and political establishment, ensured that the research into his life was carried out with unusual vigour. The unique interest in his moral as well as scientific character meant that an unprecedented amount of probing was done, both by those questioning the traditional perception and by those defending it. It therefore does seem possible to argue that the particular debate over Newton's reputation, sparked by Biot and continued by Baily, was responsible for a new era in the history of science. However, these writers were not interested only in Newton but wrote with aims beyond his exposure or rescue. In particular there was a sense that other names and traditions should be highlighted, for they were in danger of being eclipsed by Newton's reputation.

Stephen Rigaud's Historical Writings

Rigaud's historical work was strictly based on analysis of primary evidence. A contemporary claimed that, when writing histories and biographies 'his perseverance in seeking for materials was exceeded only by the discrimination, and impartiality which accompanied his researches & rendered them of permanent value'.[11] Modern opinion agrees that his knowledge of the sources for the history of seventeenth-century science was unrivalled.[12] Rigaud is probably now best known for his role in the debate over Baily's *Account* and for his *Historical Essay* and *Correspondence of Scientific Men*. In addition to these, however, he also produced memoirs of James Bradley, Thomas Harriot, John Hadley, Jeremiah Horrox, Halley and Pappus of Alexandria.[13] All relied heavily on primary sources and several, like Baily's *Account*, were essentially collections of manuscript material with a biographical introduction. Although Rigaud did not become a Fellow of the RAS until 1838, the year before he died, during the 1830s he was in close contact with Baily, with whom he served on the Council of the Royal Society. They corresponded over matters such as star catalogues, standard measures and pendulum experiments, but also on the history of astronomy. In 1830 Rigaud sent Baily parts of his first important historical work, on Bradley, and Baily replied, 'I cannot suggest anything for their improvement; except that I hope you will give us *every scrap* of information in your possession: for Bradley belongs to the *world*, and not to Oxford alone'.[14] While the two had much in common, the fact that politically 'he was strictly a Conservative' and 'a member of the Church of England of the old school', helps account for their differences over Flamsteed and Newton.[15]

Philip Bliss, an antiquary and the keeper of archives at Oxford, wrote that Rigaud's love of Oxford was 'that of a most affectionate son to a beloved and revered mother. He was most jealous for her honour, & resented with great warmth, every attack upon her fame'.[16] Although this readiness to defend Oxford and her sons might seem at variance with the impartial history that Bliss also credited to him, it is intimately related to his work. Not only did Rigaud research Oxford figures such as Bradley and Halley, but his original move towards history can be seen as a result of his desire to maintain Oxford's reputation by underlining its role in pre-Newtonian science and reviving its activity in the history of mathematics. His interest was first sparked by learning of a failed attempt at the end of the eighteenth century to publish Thomas Harriot's papers that would have continued the Oxford tradition of publishing texts important to the history of mathematics.[17] As Savilian Professor, Rigaud took on this responsibility, and his notebooks testify to the amount of research he carried out among these papers. His work on Bradley was also the result of a similar sense of duty, for he wrote privately that he was not much concerned by its reception: the 'thing was a task, and having executed it the rest may take its chance'. However, despite this apparent indifference, it was Rigaud who had traced the papers and persuaded the holders to return them to the university.[18]

Rigaud does not, in fact, seem to have taken up historical studies until his late fifties, his interest first being signalled by the part he played in the foundation of the Ashmolean Society (1828). Rigaud published both scientific and historical papers through them in the 1830s.[19] In what may have been less than a decade, he acquired ample knowledge of the history of science. Both his manuscripts and his publications testify to a very meticulous style of research.[20] In his memoir of Bradley, Rigaud described his approach:

> I have in every instance in my power derived my information from original authorities, and have done my best to verify the facts and dates which are taken from printed accounts: wherever I thought any thing was to be learned, I did not hesitate in making inquiries ...[21]

His correspondence demonstrates the truth of this claim and his unwillingness to accept information that he could not trace back to its original source. As with his biographical works, his *Historical Essay* (1838) was based on primary sources, several of which were printed in appendices.[22] This work was innovative in treating a particular episode in the history of science in such detail and in investigating the production of a single book. This testifies to the unique importance of the *Principia* but is also an indication of the importance of bibliography to the field at this period. As will be discussed below, this was an approach that Rigaud shared with De Morgan and others.

The *Historical Essay* was important for telling the full story of Halley's involvement in the publication of the *Principia*. Rigaud also promoted his understanding of the nature of Newton's genius and character. For example, he found evidence to explode the myth that Newton had only been able to verify his theory of gravitation after 1682 when he belatedly acquired Jean Picard's measurement of the earth. According to this story, Newton returned to his calculations, laid aside since 1666, but was so overcome by nervous anticipation that he had asked a friend to finish it for him.[23] Rigaud demonstrated that it was likely that Newton had heard of Picard's measurement before 1682 and that his 1666 calculations were sufficiently accurate to convince him that he was on the right track. Rigaud thus succeeded in both removing the image of the agitated Newton and reiterating the importance of the 1666 moment of discovery. The centrality of that moment was also reinforced by Rigaud's decision to open his account with the apple story. He thought the anecdote 'neither devoid of interest, nor improbable in itself', for Newton's 'powerful mind was to be able to follow out a chain of reasoning, in which the connecting links would escape the notice of a common observer'.[24] The decision to prioritize this story was controversial enough to provoke De Morgan to a refutation, in which he insisted that Newton's idea owed more to the work of others than happy chance.[25]

The *Historical Essay* discussed the development of Newton's fluxions and, inevitably, the question of priority over Leibniz. Rigaud was able to offer important evidence regarding Newton's early works on fluxions and their relationship to the composition of the *Principia*. In addition, a paper in the Macclesfield Collection provided a means of explaining away an error noted in the *Commercium Epistolicum*, which threatened to cause the case against Leibniz to 'fall to the ground'.[26] Rigaud also took the opportunity provided by his discussion of precursors to Newton's theory to quash the pretensions of Hooke to the discovery of gravitation. Hooke, who lacked Newton's 'clear and steady view of his own first principles' and had not verified his theory experimentally, was charged with possessing the same 'jealous temper' as Flamsteed. Hooke's claims had been backed by John Aubrey but, with access to the relevant letter, Rigaud was able to demonstrate that these passages had been composed, or at least annotated, by Hooke. This fact not only removed testimony in Hooke's favour but also revealed him employing underhand tactics.[27] Here and elsewhere Rigaud demonstrated that his instinct was to find statements of Newton inherently more trustworthy than those of Hooke or other rivals. However, the work of Biot, Brewster, Baily and their reviewers had combined to ensure that some negative aspects of Newton's personality were an essential component of his biography. Rigaud accepted the evidence for a breakdown, though thinking 'the duration, as well as the nature of the disease were much exaggerated', and acknowledged that a 'nervous habit made [Newton] quick in feeling any disturbance or injury', even if he were 'always anxious to make amends'.[28]

The *Correspondence of Scientific Men* (1841) consists largely of letters, sent to William Jones by various philosophers, held in the Macclesfield Collection. They helped enrich knowledge of Newton but also indicated the strength and variety of the British scientific tradition. Rigaud died before completing the work and his input as editor, apart from the laborious task of copying out the letters, was solely in the form of footnotes.[29] Compared to the *Historical Essay* these are not voluminous – if he had lived to complete the work he might have added more – but Powell suggested that they were uniquely valuable.

> His discharge of the duty of Editor has been by no means confined to the mere exactness in presenting the materials – though this was a matter demanding some attention from the inaccuracies in many of these extracts already before the public; – but he has throughout been careful to add any requisite illustration of persons, books and circumstances referred to, which are often necessary for rendering the text intelligible. It is curious and often amusing to observe how in his hands, circumstances apparently insignificant are brought together from the most remote sources to bear on some question of personal biography, or on the progress of discovery – how from

> a memorandum book, a tombstone, a parish register, a postmark, testimony in point is ingeniously extorted.[30]

The use of prominent footnotes can be an indicator of both the extent of research carried out by historians and the values that they attached to their work – 'the outward and visible signs of this kind of history's inward grace' – and historians increasingly referenced their sources rather than their authorities.[31] Rigaud's earliest works pre-dated Ranke's fame in Britain, and there is no evidence that he even admired Niebuhr and the Göttingen School, but he, like Baily, was working in the same manner. It has been said that Ranke's distinctive contribution to historiography was to create a radical distinction between primary and secondary sources, but it is clear that this was already acknowledged by some historians in 1830s Britain.[32]

Although Rigaud's desire to defend both Newton and Halley from the attacks of Flamsteed suggests his overt concern for the character of his scientific heroes, like Baily his historical style also carried a moral message. Bliss made an explicit connection between Rigaud's morals, observational astronomy and archive-based history. Rigaud was said to have been

> constantly applying to his own moral improvement, the accuracy of observation, and correctness of judgement, which qualified him for mathematical pursuits, and enabled him to recover and ascertain so many particulars respecting Bradley, Hamilton, Hadley, and other eminent scientific men, the biography of whom had been previously neglected.[33]

His use of original sources also gave authority to his case; the *Historical Essay* was extremely persuasive in its presentation of Halley as the hero of the *Principia*. Its exhaustiveness made its conclusions, apparently derived inductively from 'the evidence', seem judicial and impartial. However, Rigaud's unpublished essay on Newton's niece demonstrated his marshalling of sources in support of a particular case. When asked for his opinion, Whewell admitted that he sensed 'a somewhat vehement and advocate-like tone'. For Whewell, it was a question of tempering the tone rather than changing either the intention (the defence of Barton from the charge of being the mistress of Halifax) or the conclusion (that no impropriety had taken place). He advised Rigaud to 'leave out all the more vehement part of the argumentation' and, indeed, to 'avoid putting in to words the worst of the insinuated charges'.[34] Rigaud was still to be an advocate, only a more skilful one. Likewise, his concern for Halley's character led to some suppression of evidence: after reading the reviews of the *Account*, he admitted that he 'could tell them some more scandalous stories about Halley than any, which they seem to know', but because the source was doubtful he would 'think myself much to blame if I revived them'. Of Halley, he added candidly, 'He was a harsh overbearing man, & I see no possibility of getting rid of the prophaness [*sic*] of his conversation: but [as] I am not aware of his having been guilty of the same faults in print, more harm therefore than good is now done by dwelling on his faults in this respect'.³⁵

Despite his interest in Newton and his reputation, Rigaud should be remembered for the breadth as well as the depth of his research. Because of Newton's fame, Rigaud's writings on other natural philosophers tended to be sidelined.³⁶ Powell's review attempted to correct this balance by preceding the discussion with an outline of the 'history and progressive stages' that led to Newton's discovery. His chief interest was the 'small band of philosophers struggling against every disadvantage' – including William Milbourne, William Gascoigne, Horrox and William Oughtred – who prepared the way for Newton by bringing the ideas of Copernicus, Kepler and Galileo to England.³⁷ Those who commented on these individuals, such as Rigaud, Powell, De Morgan and J. O. Halliwell, claimed that they had been unjustly neglected and detrimentally compared the poverty of knowledge regarding English science before Newton with the greater understanding of early Continental science. An attempt was being made to educate the nation about the less well-known British scientific tradition.

Rigaud's reputation as a historian of science was high. His care and accuracy were particularly applauded, as was his wide knowledge of the period about which he wrote. However, his star was, in the opinion of some, eclipsed in the decades after his death. The obituary that appeared in *Knight's Cyclopaedia of Biography* (1858) opined that there was

> probably no other person of his age who was equally learned on all subjects connected with the history of literature of astronomy; as a mathematical antiquary and bibliographer he was unrivalled, at least in this country, until the gradual adoption of similar pursuits by Professor De Morgan.³⁸

However, Rigaud was perhaps the first writer to assert himself as an expert in the field of the history of science. This is suggested by the sheer number of individuals who addressed queries to him and by the fact that he felt himself to be in a position to judge problems in the field. When it became necessary to ascertain whether a certain William Wallis was a descendant of the mathematician John Wallis, it was Rigaud to whom the authorities wrote, and when the Royal Society decided to reject the catalogue of their library prepared by Anthony Panizzi, it was Rigaud who published a response to his complaints. His knowledge of the history of science gave him the ability and the credibility to point to errors in the bibliography of the chief keeper of the printed books at the British Museum.³⁹ This authority was derived from the increasingly wide-spread appreciation of archive-based knowledge.

Antiquarians, Archivists, Librarians and Historians of Science

The historical interest in primary sources overlapped with that generated by genealogical, bibliographical and antiquarian studies. Those who pursued such researches moved in the same circles and inhabited the same societies and archives. The correspondents of Rigaud and Baily included antiquaries such as Bliss, T. J. Hussey, Joseph Hunter and J. O. Halliwell (later Halliwell-Phillipps), and several were recipients of the *Account of Flamsteed*, including Hussey, Hunter, John Britton and Nicholas Harris Nicolas. Their interest was often not, or not only, amateur: Halliwell had been librarian at Jesus College, Cambridge, Hunter was assistant keeper of Public Records, and Baily's recipients also included Robert Lemon, keeper of the State Papers, and Henry Ellis, principal librarian of the British Museum.[40] Just as Baily's astronomical friends were 'proto-professional', members of this group have been identified as among 'the earliest of professional historians'.[41] They were, like Rigaud and Baily, concerned with cataloguing and publishing manuscripts, and called for public access to archives. Nicolas believed that the publication of manuscripts was the principal task of the Society of Antiquaries; his criticisms of that Society for neglecting this responsibility were compared to the campaign against the Royal Society and led to his being called 'the Babbage of the antiquarian and literary world'.[42]

As manuscripts came to be seen as indispensable to historical writing, and, increasingly, as the nation's heritage, there were calls for government involvement in making them accessible. There had been a series of Record Commissions since the beginning of the century but these, like the Society of Antiquaries, were criticized, by Nicolas and others, for their expense and incompetence. The many privately-established printing societies, beginning with the Surtees Society (1834) and the Camden Society (1838), can be seen as filling the gap.[43] The intention was to make texts available, either by improving catalogues and physical access or by publishing them. Publications appeared with a minimum of scholarly apparatus, being intended to serve future historians rather than to display the erudition of their editors. The government's response was, however, felt by the 1850s when it established the new Public Record Office and the Historic Manuscripts Commission and sponsored cataloguing and publications such as the Calendars of State Papers. Such enterprises were 'laying the groundwork of future expertise', but many archivists and librarians developed expertise enough to publish monographs as well as catalogues.[44] Hunter, for example, was known as a historian of Yorkshire. We have seen that Rigaud's professional commitments led him to historical research and the same was true of Joseph Edleston at Trinity College, Cambridge, and C. R. Weld, the Royal Society's librarian and historian.[45] These individuals make up an alternative story of nineteenth-century historiography to that usually associated with famous historians such as

Carlyle and Macaulay.[46] Although both groups frequently spoke of antiquarianism as the handmaid of history, there were those who had high claims for their work. Thomas Wright was one such, claiming that 'Antiquarianism as a science allied to history, belongs to a more advanced state of intellectual refinement', and that antiquarians assemble materials that the historian would 'arrange and make ... intelligible'.[47] His description of antiquarianism as a science, amassing the raw data from which the historian could form theories, is typical, as is his implication that resistance to premature theorizing is laudable. The moral implications of this ethos, perhaps especially applicable to men of science, are clear.

The strictly antiquarian, rather than historical, impulse was to play its part in adding to the biography of Newton. Dawson Turner of Pembroke College, Cambridge, was a botanist and antiquary whose collection of manuscripts included several letters from Newton.[48] Turner's *Thirteen Letters from Newton to John Covel* (1848) was one among several publications based on his collections. These letters relate to Newton's business as MP for Cambridge and shed some light on an otherwise under-represented facet of Newton's career. They give some flavour of his political opinions, one letter arguing that 'when K. James ceased to protect us, we ceased to owe him allegiance by ye law of ye land'.[49] However, Turner considered the information they contained regarding the political climate of Cambridge and England of secondary interest. The primary importance 'derived from their illustrious author', for 'whatever can be collected, even by tradition, touching him who was the glory of his country and his age, deserves to be recorded'. Turner captured the peculiar interest of autograph manuscripts, which had long enthralled him, in his description of these Newtonian letters as preserving the 'sparks from his mind and lines from his pen'. Turner also admitted to a motivation that has perhaps played a not unimportant part in Newtonian studies over several centuries: 'I feel a pride, that I hope is honest, in the opportunity thus afforded me of associating my name in any manner, however humble, with Newton's'.[50]

The collector's instinct lay behind other historical-scientific works. De Morgan was an avid collector of early scientific books, which were then extremely cheap. The same is true of his friend Guglielmo Libri and acquaintance J. O. Halliwell, who introduced himself to De Morgan by sending him a copy of Edward Cocker's *Arithmetick*.[51] Libri and Halliwell had an equal interest in manuscripts, and the precocious Halliwell had begun collecting as early as 1835, the year he matriculated at Trinity College as a tutee of George Peacock. He had assembled an impressive collection by 1840, when, for financial reasons, he auctioned the books at Sotheby's.[52] Chasles, another contact of De Morgan's, also accumulated a collection of thousands of manuscripts, quotations from which adorned his writings. The temptations of this world are amply demonstrated by the fact that Libri, Halliwell and Chasles were caught up in scandals of forgery and theft.

The case of the unfortunate Chasles is considered in Chapter 6, but the other two, probably more culpable, can be dealt with briefly. Libri was charged with having stolen books from the public libraries of France, to which he had access as an inspector. He denied the charge, claiming it was trumped up by political opponents, and, in exile in Britain, was backed by defenders such as De Morgan and Brougham. Many years after his death the charges, extended to include the forgery of provenances, were substantiated.[53] Halliwell was accused in 1844 of having stolen from Trinity College some manuscripts which he claimed to have bought in London. An investigation was begun but ultimately the matter was dropped and the manuscripts remained at the British Museum, to which they had been sold by Halliwell. Friends and *The Times* backed Halliwell's story and he was further aided by the fact that, because the legal representatives of Trinity and the Museum quarrelled, the case never came to court.[54] These stories demonstrate the relative newness of this field, showing both the temptations held out to those with the greatest expertise in archival identification, verification, cataloguing and storage – and the possibility of detection by other experts.[55]

Halliwell's scientific-historical publications, which focused on the early period of English science, were 'on the whole, descriptive rather than analytical'.[56] On its foundation in 1838, Halliwell began to be published by the Camden Society, the perfect vehicle for his style of scholarship. These publications involved even less analysis than his early essays, usually having extremely short introductions and a minimum of footnotes.[57] In 1840 he proposed the formation of the HSS, on the same principles as the Camden Society, 'to render materials for the history of the Sciences accessible to the general reader, by the publication of manuscripts, or the reprinting of very rare works connected with their origin and progress in this country and abroad'.[58] The council members of the society were drawn from the antiquarian and scientific worlds. There was an overlap with councils of the Society of Antiquaries and the Camden Society, including names such as Sir Francis Palgrave, Charles Purton Cooper, Thomas Pettigrew, Thomas Wright and Sir Frederick Madden. Also on the council were Halliwell's close friend Thomas Stephens Davies, Professor of Mathematics at the Royal Military Academy at Woolwich, Robert Willis, Jacksonian Professor of Natural Philosophy at Cambridge, Hunter, Powell and De Morgan. The membership of the Society, which peaked at 179, included Baily, Chasles, Libri, Peacock and Dawson Turner. It is likely that, had he not died the year before, Halliwell's correspondent Rigaud would also have joined, for his sons Stephen, John and Gibbes are listed.[59] This Society therefore linked some of the most important names in the history of science.

There were also some notable absences. Whewell excused himself, telling Halliwell that he feared he would 'be of no use' to the HSS and wished to avoid involving himself in additional societies, 'though I admire your objects and wish you all success'. He also acknowledged some corrections to his *History* suggested by Halliwell, saying he expected 'that additional researches would supply additions and corrections to my general sketch' but indicating that he was not about to embark on this himself.[60] The HSS did not produce Whewell's kind of history. He and other absentees like John Lubbock and Herschel may have been wary of the young Halliwell's ambition. This caution was justified, for the Society only published two books.[61] Although the Society nominally existed until the end of 1846, it seems not to have functioned after 1841, and in July 1842 Powell asked to have his name removed from the council; neither his nor Davies's names appeared in the second HSS publication, edited by Halliwell's collaborator Wright.[62] The Society's failure has been ascribed to lack of interest, and it is perhaps true that 'only a few scattered persons regarded outmoded science as of much value'.[63] In fact the Society never got off the ground: seventy-six members, including Powell, never paid a subscription, and only thirty-nine paid a second year.[64] The active members were Halliwell, Wright, Pettigrew, who helped recruit several noble vice-presidents, and, to a lesser extent, De Morgan, who requested further prospectuses for distribution and reviewed Halliwell's HSS publication.[65]

Although the HSS had avowedly been formed to make texts available to the 'general reader', Halliwell did not appear to either expect or desire the popularization of the field. Nor did he expect to change the 'lamentable apathy towards matters of history which is too frequently characteristic of the lover of demonstration'.[66] Halliwell advocated knowledge for its own sake.[67] There is no hint in his writings that the publication of primary sources might be a 'scientific' way of doing history, or that men of science were likely to learn anything useful from a history of their field. This may have displeased some members of the Society, like De Morgan, or potential members, like Whewell, who saw history as a means of countering the increasing specialization that threatened the unity not just of the sciences but of learned knowledge in general. Like Halliwell, De Morgan hoped that the Society would 'succeed in placing this neglected department of history on its proper footing'. However, he was both more convinced of the genuine need for research in the history of science and less convinced by the adequacy of the type of publication that Halliwell had produced.[68] He suggested that there was something Baconian in the scheme of the Society, and he was no advocate of that methodology. Halliwell's work was potentially useful, but undirected: 'There is a vein which they are to open, and an ore which they are to extract. Whatever may be the percentage of apparently useless matter which they bring up with the metal, it is not their business to make any final selection.' De Morgan approved this lack of direction from writers such as Halliwell, claiming he 'could not fall into any more pernicious error, than publishing upon any definite system of rejection. Whatever there may be to dig up, let us at least have large specimens of everything', but asked, 'When will it happen that a palaeographist is also a mathematician, with enough energy and leisure both to work the ore and the metal?'.[69] The implication is that if this rare beast could be found something of genuine use and interest could be produced.

Joseph Edleston's *Correspondence of Newton and Cotes* (1850)

The Correspondence of Newton and Cotes by Joseph Edleston of Trinity College, Cambridge, was a publication in the spirit of the times, although it contained a great deal more than the texts produced by the printing societies. Its main content was the eponymous correspondence, held at Trinity College Library, supplemented by other documents at Trinity, such as Newton's buttery bills, his payments from college and the list of his exits and redits, or his comings and goings from Cambridge. Edleston also included a 'Synoptical View of Newton's Life' and an extremely detailed set of notes that synthesized all the published manuscripts relating to Newton. It is a book worthy of the most pedantic of antiquarians or librarians, although Edleston was not quite either. It was in many ways the long-overdue official statement on Newton by Trinity and Cambridge.[70] The most obvious person to carry out the work was Whewell, but – another sign of his lack of interest in this kind of history – the task was delegated to a former tutee. Edleston is little known, for *Newton and Cotes* was his only achievement of lasting significance.[71] He was born around 1816 in Halifax, the son of a cloth manufacturer. He entered Trinity College in 1834 and, after graduating as 15th Wrangler (1838) and taking Holy Orders, became a Fellow (1840) and remained to make a career in university and college administration.[72] In 1845 he campaigned for election as University Librarian but, despite the fact that some considered him 'just the man for the office', he was not nominated by the Heads of the University.[73] Instead, by 1847 he had become Steward of Trinity, was Bursar by 1860 and became Senior Bursar in 1861. He left the College in 1863 to marry the daughter of James Cumming, Professor of Chemistry at Cambridge, and received the living of Gainford, County Durham. A clue to his political leanings is found in his obituary of his future wife's father, whom he described as 'thoroughly independent' and 'a Liberal at a time when it was fashionable to be so, but ... a sober and sensible Liberal, and probably on some points more really conservative than many who are called by that name'.[74]

Apart from Edleston's interest in becoming University Librarian in 1845 there is little evidence of his concern for historical matters before he began research for *Newton and Cotes* about 1848. That this publication was not done out of a mere sense of duty is suggested both by the exhaustive research that went into the book and by the fact that the following year he offered to catalogue the manuscripts in the library at Trinity.[75] Although he never became librarian at the College, he seems to have taken an interest in the library's acquisitions, especially after his publication appeared. When, in his seventies, Dawson Turner ran 'off to Gretna Green with a girl he had been keeping', Edleston was informed that Turner's collections were to be auctioned.[76] The Newton-Covel correspondence was duly acquired by Trinity, presumably at Edleston's prompting. To judge from the correspondence he received, Edleston was considered an expert on both Newton and the history of the college. He at least partially took on the mantle of the deceased Rigaud as a source of information on Newton and, especially, his manuscripts. Whewell redirected an enquiry of De Morgan's to him, saying that he was 'deeply skilled both in College lore and in Newtonian letters'.[77] He frequently responded to the queries of Brewster and others and was involved with the identification of the papers of the Gregory family.[78] Edleston also became the local committee chairman for the erection of a statue to Newton in Grantham.[79] Edleston's interests continued after he left Trinity, although he published nothing further. By the time of his death he had formed a collection of rare scientific books, which were left by his daughter to the University of Durham.[80]

Edleston's volume was a thoroughly Cantabrigian production, using Trinity's manuscripts and elucidating both Newton's time in Cambridge, which had hitherto appeared as something of a blank, and the very fabric of seventeenth-century Trinity. In addition, like Rigaud's *Historical Essay*, it is primarily concerned with the production of the *Principia*, since it consisted of correspondence regarding the publication of its second edition. The *Principia* had, of course, played a central role in Cambridge mathematical education since the mid-eighteenth century, its first book becoming 'the absolute pinnacle of elite undergraduate studies'.[81] It was eulogized by Edleston as 'the most remarkable production of the human intellect'.[82] One of his primary aims was to demonstrate that the *Principia* was not, as often suggested, initially ignored but that it was quickly adopted by teachers in both Cambridge and Scotland. On a number of other points the volume was a defence of Newton himself, placing Edleston within the ongoing debates regarding Newton's character. This was evident in the Preface, which specifically blamed Flamsteed for the delay in the appearance of the second edition:

> if Flamsteed ... had cordially co-operated with him in the humble capacity of an observer in the way that Newton pointed out and requested of him, (and for his almost unpardonable omission to do so I know of no better apology that can be offered than that he did not understand the real nature and, consequently, the importance of the researches in which Newton was engaged, his purely empirical and tabular views never having been replaced in his mind by a clear conception of the Principle of Universal Gravitation,) the lunar theory would, if its creator did not overrate his own powers, have been completely investigated, so far as he could do it, in the first few months of 1695, and a second edition of the *Principia* would probably have followed the execution of the task at no long interval.[83]

xxx SYNOPTICAL VIEW OF

1685 Oct. 14. Letter to Flamsteed (acknowledging the receipt of Flamsteed's two letters in answer to the preceding).

Dec. 30. Letter to Flamsteed (with thanks for information about comet of 1680 and Jupiter's satellites. Kepler makes Saturn's orbit too small. Requests the greatest elongations of any of Jupiter's satellites, and of Saturn's satellite)(80).

1686 Jan. 13. Letter to Flamsteed (wishes to know the major axes of the orbits of Jupiter, Saturn and his satellite)(81).

—— 22. Votes for John Laughton, of Trinity, the successful candidate for the Librarianship.

Apr. 28. FIRST BOOK OF THE PRINCIPIA EXHIBITED AT THE ROYAL SOCIETY(82).

May 19. At a meeting of the Society it was ordered "that Mr Newton's *Philosophiæ Naturalis Principia Mathematica* be printed forthwith in 4to. in a fair letter; and that a letter be written to him to signify the Society's resolution, and to desire his opinion as to the print, volume, cuts, &c."(83).

June 2. Halley undertakes the publication of the *Principia* at his own expense(84).

—— 20. Letter to Halley (demolishing the claim set up by Hooke of having communicated to him the law of decrease of gravity according to the inverse square).

—— 30. At a meeting of the Council of the Royal Society, the President was desired to license the *Philosophiæ Naturalis Principia Mathematica.*

July 14. Letter to Halley (approves of the suggestion of having wood-cuts. Conciliatory remarks respecting Hooke).

—— 27. Letter to Halley (further remarks on Hooke's claim).

Aug. 20. Letter to Halley (with Cor. 2 and 3 of Prop. XCI. Lib. I. of *Princip.* on the attraction of a spheroid on a point in its axis produced, and on an internal point)(85).

Sept. 3. Letter to Flamsteed (Cassinian satellites. Cassini's observation of Jupiter's oblateness).

Autumn. Second Book of Principia made ready for the press(86).

Oct. 18. Letter to Halley (corrections of Scholium to Prop. 31. Lib. I.: transformation of a trapezium into a parallelogram).

1687 Feb. 18. Letter to Halley (may have the second book of Principia when he pleases: has the sheets up to M: thanks him for putting forward the press again)(87).

March 1. Tuesday. Letter to Halley, advising him that the 2nd book will arrive on Thursday night or Friday, by coach: obliged to him for pushing on the edition because of

NEWTON'S LIFE. xxxi

people's expectation, though otherwise he could be as well satisfied to let it rest a year or two longer (read to the Soc. March 2).

1687 March 11. Deputed with Billers, the Public Orator, to carry to the Vice-Chancellor the opinions of the Non-Regent House respecting King James's second mandate, requiring the University to confer upon Alban Francis, a Benedictine monk, the degree of M.A. without the usual oaths(88).

Apr. 6. The 3rd book of the Principia "produced and presented" to the Royal Society(89).

—— 11. Appointed one of eight delegates to represent the Senate, in conjunction with the Vice-Chancellor, before the Ecclesiastical Commission(90).

PUBLICATION OF THE PRINCIPIA (about Midsummer)(91).

1688 Spring. Charles Montagu vacates his fellowship(92).

Dec. 15. Votes for Archbishop Sancroft (for Chancellor of the University) who declines the office.

1689 Jan. 15. Elected one of the representatives of the University in the Convention Parliament(93).

First acquaintance with Locke. Furnishes him (March) with an easy proof of elliptic motion about a centre of force in one of the foci(94).

June 12. Huygens and Newton at the Royal Society(95).

Aug. 20. Parliament prorogued.

—— — Contemplated appointment to the Provostship of King's College(96).

Oct. 19. Meeting of Parliament(97).

1690 Jan. 27. Parliament prorogued.

Feb. 6. Parliament dissolved.

—— 21. Sir Robert Sawyer, who had been expelled the House of Commons, Jan. 20, for having been, as Attorney-General, one of the prosecutors of Sir Thomas Armstrong in 1684, re-elected M.P. for the University. Newton votes for him.

Oct. 28. Letter to Locke: (will send, as desired, his "Historical Account of two notable corruptions of scripture." Acknowledgments to Lord and Lady Monmouth for their endeavours to procure him preferment)(98).

Nov. 14. Letter to Locke, with the "Historical Account."

1691 Feb. 7. Letter to Locke (Daniel and Apocalypse).

June 30. Letter to Locke. (Locke's good offices in trying to get him the place of comptroller of the Mint. Effects of looking at the Sun's image in a mirror).

Figure 4. J. Edleston, *Correspondence of Sir Isaac Newton and Professor Cotes* (Cambridge and London: John Deighton and J. W. Parker, 1850), pp. xxx–xxxi of the 'Synoptical View of Newton's Life'. From the personal collection of the author.

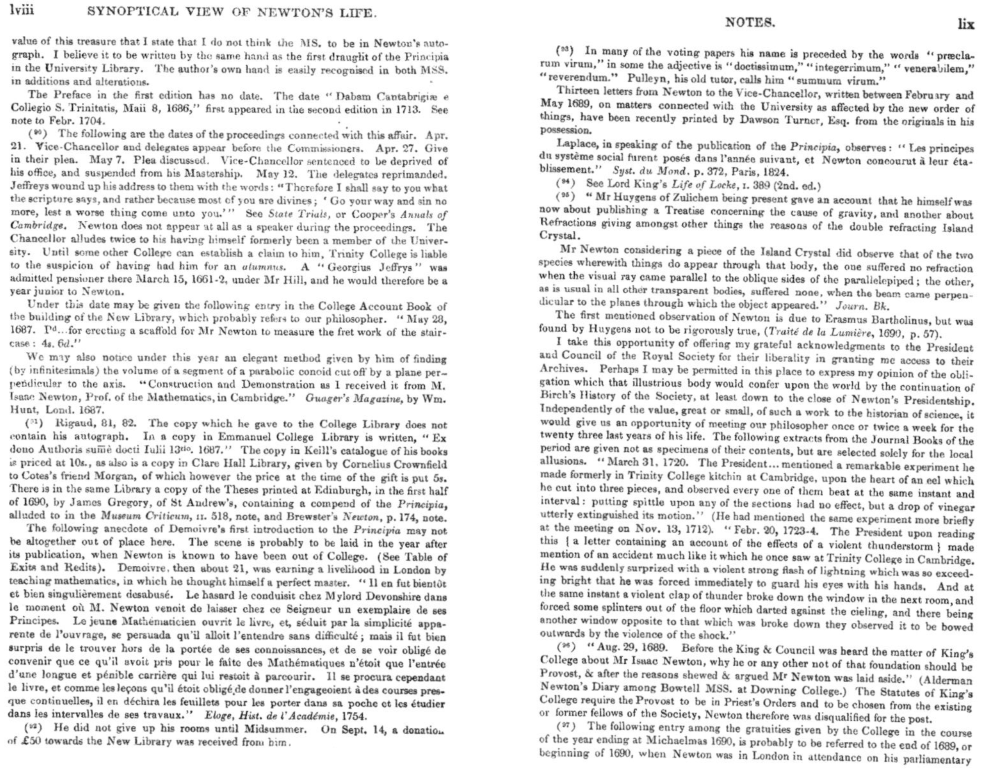

lviii SYNOPTICAL VIEW OF NEWTON'S LIFE.

value of this treasure that I state that I do not think the MS. to be in Newton's autograph. I believe it to be written by the same hand as the first draught of the Principia in the University Library. The author's own hand is easily recognised in both MSS. in additions and alterations.

The Preface in the first edition has no date. The date "Dabam Cantabrigiæ e Collegio S. Trinitatis, Maii 8, 1686," first appeared in the second edition in 1713. See note to Febr. 1704.

(90) The following are the dates of the proceedings connected with this affair. Apr. 21. Vice-Chancellor and delegates appear before the Commissioners. Apr. 27. Give in their plea. May 7. Plea discussed. Vice-Chancellor sentenced to be deprived of his office, and suspended from his Mastership. May 12. The delegates reprimanded. Jeffreys wound up his address to them with the words: "Therefore I shall say to you what the scripture says, and rather because most of you are divines; 'Go your way and sin no more, lest a worse thing come unto you.'" See *State Trials*, or Cooper's *Annals of Cambridge*. Newton does not appear at all as a speaker during the proceedings. The Chancellor alludes twice to his having himself formerly been a member of the University. Until some other College can establish a claim to him, Trinity College is liable to the suspicion of having had him for an *alumnus*. A "Georgius Jeffrys" was admitted pensioner there March 15, 1661-2, under Mr Hill, and he would therefore be a year junior to Newton.

Under this date may be given the following entry in the College Account Book of the building of the New Library, which probably refers to our philosopher. "May 28, 1687. P^{d}...for erecting a scaffold for Mr Newton to measure the fret work of the staircase: 4*s*. 6*d*."

We may also notice under this year an elegant method given by him of finding (by infinitesimals) the volume of a segment of a parabolic conoid cut off by a plane perpendicular to the axis. "Construction and Demonstration as I received it from M. Isaac Newton, Prof. of the Mathematics, in Cambridge." *Guager's Magazine*, by Wm. Hunt, Lond. 1687.

(91) Rigaud, 81, 82. The copy which he gave to the College Library does not contain his autograph. In a copy in Emmanuel College Library is written, "Ex dono Authoris sum̄e docti Iulii 13^{tio}. 1687." The copy in Keill's catalogue of his books is priced at 10*s*., as also is a copy in Clare Hall Library, given by Cornelius Crownfield to Cotes's friend Morgan, of which however the price at the time of the gift is put 5*s*. There is in the same Library a copy of the Theses printed at Edinburgh, in the first half of 1690, by James Gregory, of St Andrew's, containing a compend of the *Principia*, alluded to in the *Museum Criticum*, II. 518, note, and Brewster's *Newton*, p. 174, note.

The following anecdote of Demoivre's first introduction to the *Principia* may not be altogether out of place here. The scene is probably to be laid in the year after its publication, when Newton is known to have been out of College. (See Table of Exits and Redits). Demoivre, then about 21, was earning a livelihood in London by teaching mathematics, in which he thought himself a perfect master. "Il en fut bientôt et bien singulièrement desabusé. Le hasard le conduisit chez Mylord Devonshire dans le moment où M. Newton venoit de laisser chez ce Seigneur un exemplaire de ses Principes. Le jeune Mathématicien ouvrit le livre, et, séduit par la simplicité apparente de l'ouvrage, se persuada qu'il alloit l'entendre sans difficulté; mais il fut bien surpris de le trouver hors de la portée de ses connoissances, et de se voir obligé de convenir que ce qu'il avoit pris pour le faîte des Mathématiques n'étoit que l'entrée d'une longue et pénible carrière qui lui restoit à parcourir. Il se procura cependant le livre, et comme les leçons qu'il étoit obligé de donner l'engageoient à des courses presque continuelles, il en déchira les feuillets pour les porter dans sa poche et les étudier dans les intervalles de ses travaux." *Eloge, Hist. de l'Académie*, 1754.

(92) He did not give up his rooms until Midsummer. On Sept. 14, a donation of £50 towards the New Library was received from him.

NOTES. lix

(93) In many of the voting papers his name is preceded by the words "præclarum virum," in some the adjective is "doctissimum," "integerrimum," "venerabilem," "reverendum." Pulleyn, his old tutor, calls him "summum virum."

Thirteen letters from Newton to the Vice-Chancellor, written between February and May 1689, on matters connected with the University as affected by the new order of things, have been recently printed by Dawson Turner, Esq. from the originals in his possession.

Laplace, in speaking of the publication of the *Principia*, observes: "Les principes du système social furent posés dans l'année suivant, et Newton concourut à leur établissement." *Syst. du Mond*. p. 372, Paris, 1824.

(94) See Lord King's *Life of Locke*, I. 389 (2nd. ed.)

(95) "Mr Huygens of Zulichem being present gave an account that he himself was now about publishing a Treatise concerning the cause of gravity, and another about Refractions giving amongst other things the reasons of the double refracting Island Crystal.

Mr Newton considering a piece of the Island Crystal did observe that of the two species wherewith things do appear through that body, the one suffered no refraction when the visual ray came parallel to the oblique sides of the parallelepiped; the other, as is usual in all other transparent bodies, suffered none, when the beam came perpendicular to the planes through which the object appeared." *Journ. Bk.*

The first mentioned observation of Newton is due to Erasmus Bartholinus, but was found by Huygens not to be rigorously true, (*Traité de la Lumière*, 1690, p. 57).

I take this opportunity of offering my grateful acknowledgments to the President and Council of the Royal Society for their liberality in granting me access to their Archives. Perhaps I may be permitted in this place to express my opinion of the obligation which that illustrious body would confer upon the world by the continuation of Birch's History of the Society, at least down to the close of Newton's Presidentship. Independently of the value, great or small, of such a work to the historian of science, it would give us an opportunity of meeting our philosopher once or twice a week for the twenty three last years of his life. The following extracts from the Journal Books of the period are given not as specimens of their contents, but are selected solely for the local allusions. "March 31, 1720. The President... mentioned a remarkable experiment he made formerly in Trinity College kitchin at Cambridge, upon the heart of an eel which he cut into three pieces, and observed every one of them beat at the same instant and interval: putting spittle upon any of the sections had no effect, but a drop of vinegar utterly extinguished its motion." (He had mentioned the same experiment more briefly at the meeting on Nov. 13, 1712). "Febr. 20, 1723-4. The President upon reading this {a letter containing an account of the effects of a violent thunderstorm} made mention of an accident much like it which he once saw at Trinity College in Cambridge. He was suddenly surprized with a violent strong flash of lightning which was so exceeding bright that he was forced immediately to guard his eyes with his hands. And at the same instant a violent clap of thunder broke down the window in the next room, and forced some splinters out of the floor which darted against the cieling, and there being another window opposite to that which was broke down they observed it to be bowed outwards by the violence of the shock."

(96) "Aug. 29, 1689. Before the King & Council was heard the matter of King's College about Mr Isaac Newton, why he or any other not of that foundation should be Provost, & after the reasons shewed & argued Mr Newton was laid aside." (Alderman Newton's Diary among Bowtell MSS. at Downing College.) The Statutes of King's College require the Provost to be in Priest's Orders and to be chosen from the existing or former fellows of the Society, Newton therefore was disqualified for the post.

(97) The following entry among the gratuities given by the College in the course of the year ending at Michaelmas 1690, is probably to be referred to the end of 1689, or beginning of 1690, when Newton was in London in attendance on his parliamentary

Figure 5. J. Edleston, *Correspondence of Sir Isaac Newton and Professor Cotes* (Cambridge and London: John Deighton and J. W. Parker, 1850), pp. lviii–lix of the 'Notes'. From the personal collection of the author.

Edleston's notes form the most remarkable part of this book. While the synopsis is twenty pages, the notes, in much smaller type, take up forty-one. They are, because of the density of print, an uninviting read and it is questionable how many readers actually bothered to refer to each note in turn as they looked through the 'Synoptical View of Newton's Life' (see Figures 4 and 5). It was clear, for example, that Weld, Edleston's reviewer in the *Athenaeum*, did not. Referring only to the Preface and the letters themselves, Weld blamed Edleston for not printing more material from other collections, considering that 'he does not exhibit that elucidatory research which his subject so eminently deserves'.[84] This elucidatory research is clearly apparent in the 'Notes', which add further references, information and authority to a work already largely based on primary sources. However, the notes also introduce a great deal of interpretation and discuss points of controversy. It is this that makes Edleston's book entirely unlike a production of the HSS and, although presented uniquely, more like Rigaud's *Historical Essay*.

De Morgan did read the notes in detail and thought they contained 'a deal of interesting matter'. He was therefore in a position to commend Edleston as a 'laborious' editor who had put a 'vast means of investigation ... into the hands of the historical inquirer', and in whose research he could have confidence. However, even when writing on the day he received the book and after 'a very eager', though hardly detailed, 'look at it', he was able to spot points on which they differed. Edleston, he felt, had 'made fast his painter to Newton, and must go the rest of the voyage in his wake'. Apparently De Morgan 'expected a strong Anti-Leibnitio-Flamsteedian bias' and would therefore 'always take care to have my gr[ain of] sal[lt] in readiness'. His current interest was the calculus dispute and he saw that note 35 'contains some matter which I shall have more to say upon'.[85] This note made use of the new evidence put forward in the first volume of C. J. Gerhardt's edition of Leibniz's mathematical works (1849). De Morgan had not yet perused this volume but Edleston's use of this material seemed to him 'extraordinary – meaning either that I or the editor is in a fearful state of bias on the matter'. When, after reading Gerhardt, De Morgan published on the dispute, his differences with Edleston remained. He judged Edleston's work highly: 'the synopsis is followed by a body of notes of such research and digestion as make it difficult to give adequate praise to the whole without appearance of exaggeration'. However, he still found that he 'differ[ed] much from the editor as to many matters of opinion and statements the character of which is determined by opinion' and he took 'particular exception' to the account of exactly what information concerning Newton's fluxions Leibniz received before publishing his own calculus.[86]

De Morgan's claim of an 'Anti-Leibnitio-Flamsteedian bias' is further backed by a series of notes relating to the dispute with Flamsteed. One of these took three pages to point to each instance in which Flamsteed's letters demonstrate his 'bad faith'. Edleston wished to counter Baily's attempt to show, 'in opposition to a prevailing opinion', that Flamsteed had freely communicated all his observations with Newton. In addition, he noted the 'torrents of vituperation' poured by Flamsteed on the 'illustrious' Halley.[87] Even in the synopsis, apparently the most neutral of texts, Edleston quoted sections of the correspondence with Flamsteed that put Newton in a favourable light, or described them in an opinionated manner. For example, one letter was designated the 'answer to Flamsteed's childish question respecting a book which Flamsteed, two or three years before, had intended as a present to him' and another, the 'Manly answer to Flamsteed's ungenerous suspicions of his observations having been communicated to Halley'.[88] Another point to which Edleston devoted significant space was Biot's theory about Newton's breakdown. In part he relied on Brewster's refutation in the *Life of Newton*, especially his demonstration that the evidence from Huygens did not fit chronologically with other details. Edleston added a convincing point regarding Abraham de la Pryme's account, which 'an attentive perusal will prove to refer to a period some years antecedent to the epoch under consideration', linking it to accounts of a fire in Newton's rooms before 1683. In addition, having seen de la Pryme's diary, he was able to demonstrate that Biot's suggestions about the use of the old calendar were wrong.[89] Acquaintance with two of de la Pryme's descendants, as well as the evidence of the diary itself, may have affected his conclusion. George Pryme, a Fellow of Trinity, and his son Charles de la Pryme were obviously interested in the reputation of their ancestor. The former offered a memoir of him for a projected Cambridge version of *Athenae Oxonienses* and the latter both published the diary for the Surtees Society and re-adopted the full surname.[90]

Edleston also had the evidence of the Trinity record-books to hand. These indicated that Newton remained only briefly in London once his illness began but, despite his prompt return, the Steward's Book did not record him as an invalid as would have been usual. This was inconclusive, but Edleston added something more significant from the Letter Books of the Royal Society. These included a letter that mentioned the 'Rumor ... concerning Mr Newton as if his House & Books & all his Goods were Burnt, & himself so disturbed in mind thereupon, as to be reduced to very ill circumstances. Which being all false, I thought fit presently to rectify that groundless mistake.'[91] Edleston accepted this as evidence that the story was exaggerated, without stopping to consider its remarkable persistence. His treatment of this topic demonstrated the extent to which Newton's physical illness, which could not be denied once the letters of the autumn of 1693 were known, was successfully divorced from the suggestion that the effects were long-lasting. Yet the issue remained important enough to warrant Edleston's efforts. Other notes took up smaller points and in each case

Edleston strove to defend Newton's reputation. When Locke claimed that Newton was 'a nice man to deal with', Newton's letters apparently showed that 'the groundless suspicions were on the part of Locke'.[92] To De Morgan's charge that Newton had removed references to Flamsteed from the second edition of the *Principia*, Edleston showed that the 'name however will still be found in pages 441, 443, 455, 458, 465, 478 and 479 ...'.[93] He also argued against Biot's old claim that Newton's appearance in Parliament to argue the case for the Longitude Bill was 'presque puérile'. He quoted Whiston's original account, implying that Biot's summary was distorted, although the two versions in fact differed little.[94] These instances demonstrate the extent to which Edleston used a form of writing usually connected with non-partisan scholarship – the reproduction of texts with the addition of scholarly notes – as a vehicle for controversial opinions.

Augustus De Morgan's Historical Writings

The historical writing of De Morgan has already made a number of appearances in the course of this book. Like Brewster's 1830s reviewers Malkin and Galloway, he wrote frequently for the SDUK, cutting his teeth as a historian in many of the hundreds of articles he produced for the *Penny Cyclopaedia* and the *Gallery of Portraits*. He was a close friend of Baily, both supporting him against the attacks of Whewell and Rigaud and protecting his posthumous reputation from Edleston and Brewster. He was involved, however briefly, with the HSS. While his articles for the *Penny Cyclopaedia* encompassed a very large range of subjects, his writings for periodicals – including the *Companion to the Almanac*, another publication edited by the SDUK's publisher, Charles Knight – indicate a number of areas of particular interest. These were the bibliography of historical scientific texts, the Newton-Leibniz calculus controversy and pre-Newtonian science and mathematics.[95] Despite De Morgan suggesting that Halliwell's HSS publication was only of secondary interest to men of science, he saw historical research as of real importance to them and others. While it should be remembered that many of his articles were written as much to amuse as to instruct, and that he only published one short monograph on a historical topic, there was an underlying seriousness to his writing.

Joan Richards has argued convincingly that De Morgan believed 'the twists and turns of mathematical history should be openly examined because they provide crucial clues to understanding mathematics'. I am not concerned here with De Morgan's more technical writings – more especially as Richards and Adrian Rice have dealt fully with De Morgan as a historian of mathematics and arithmetic – but Richards here highlights two important facets of his historical work. First is the insistence that things should be 'openly examined', and second is the suggestion that the history of science has 'twists and turns'.[96] Although De Morgan believed the history of science could teach something about the correct way to proceed with current research, he also warned against any straightforward application of these lessons or belief in the progress of science. For De Morgan the study of 'what actually happened' had to come before any lessons could be extracted, and it could only be meaningful if the historian understood the context: 'In reading an old mathematician you will not read his riddle unless you plough with his heifer; you must see with his light if you want to know how much he saw'.[97] De Morgan differed from Baily, Rigaud and Edleston by not, in general, reprinting manuscripts: his primary sources were chiefly printed books, the key means by which scientific ideas were disseminated. These approaches were complementary, for the 'history of science is almost entirely the history of books and manuscripts'.[98]

De Morgan's review of Whewell's *History* provides a useful overview of what he saw as important in the subject.[99] He was immediately suspicious of the usefulness of Whewell's bold 'Herculean labour'. In his view, the '"History of Physical Science" collectively, cannot be prudently undertaken before each separate department shall have its diligent historian'. Whewell, De Morgan felt, lacked the detailed knowledge necessary to back his bold claims and he addressed himself successfully to neither the specialist nor the general reader.

> The inquiries, *when* and *by whom* each valuable truth was obtained, demand a degree of minute research, to which our author lays no claim, and which he appears to have considered foreign to his purpose. Yet it is of great importance in the history of philosophy to show, that the germs of brilliant discoveries have often been long in the hands of mankind, unappreciated and little thought of, till some accidental association with a fertile principle or abstract truth, developed their nature, and gave them new value. The more we apply ourselves, with antiquarian industry, to examine the history of the human mind, the more apparent it will be, that the present accumulation of science, however massive, has grown particle by particle, and has never really experienced any sudden increase.

In addition, although Whewell has been acclaimed as uniquely aware of the need to contextualize the history of science, De Morgan felt he had failed to do this:

> The necessity of pointing out *how* science has advanced, involves in some degree a history of the human mind in every stage of its culture, and thereby entails a difficulty not easily got the better of by one, who, like our author, avoids in general the discussion of the collateral influence exerted by other branches of knowledge on the philosophy of material nature ...[100]

Whewell's analysis of Greek science has likewise been welcomed by modern commentators as demonstrating his willingness to learn from failures, but De Morgan felt that he neither gave sufficient credit to Greek successes nor adequately explained the perceived failure. He felt, indeed, that 'the question why physical science should at any particular time have advanced so far and no farther, is one rarely admitting of a brief answer'. The project of the Greeks, he pointed out, was not the same as modern science and could not therefore be examined on the same terms. De Morgan thought Whewell's treatment of the Middle Ages somewhat better, as here he 'surveys more widely the condition of the human mind ... and allows us to catch a glimpse of the influence of metaphysic-theology in creating mysticism and dogmatism'.[101] De Morgan was in agreement on such points as the importance of the imagination, hypothesis and guess-work in discovery; that science did not proceed by revolutions; and that history, as well as science, was progressive – but as history he found the work inadequate.[102]

De Morgan's 'References for the History of the Mathematical Sciences' gives further insights into what he considered good history. He admired books that referred to primary materials, had clear, accurate references and that provided a good index. However, to be truly admirable the author required 'the power of choosing and deciding between authorities' and the 'moral qualities of an historian'.[103] But even books that were poor or prejudiced could become useful as primary sources. Whiston was an invaluable source of 'good gossip', while John Hill's *A Review of the Works of the Royal Society* was a 'collection of errors, absurdities, and incredibilities', but still might 'help any one who was desirous of comparing the disposition of different times'.[104] The dedication of De Morgan's 1847 *Arithmetical Books from the Invention of Printing to the Present Time*, addressed to Peacock, whose work on the history of arithmetic he particularly admired, explained the value of old texts. 'The most worthless book of a bygone day is a record worthy of preservation. Like a telescopic star, its obscurity may render it unavailable for most purposes; but it serves, in hands which know how to use it, to determine the places of more important bodies'.[105]

De Morgan's interest in obscure books matched his interest in obscure figures in science. Although Copernicus, Galileo and Newton would always have an important place in the history of science, he objected to the fact that 'they, and their peers, are made to fill *all the space*'. In part this research was necessary to do justice to lesser men, whose discoveries were all too often credited to the more famous, but it was also necessary in order to appreciate what resources were available to each generation. Much of De Morgan's work can be seen as a critique of existing histories or naive popular conceptions. His task was to point out 'myths', such as the belief that Copernicus had immediately been challenged by hostile opponents or that Bacon's writings had produced a revolution in science.[106] His interest was often in the milieu that produced the great ideas as much as in the authors of them. When, in 1859, Anthony Panizzi requested De Morgan's assessment of a seventeenth-century manuscript purportedly by Galileo, De Morgan claimed that he hoped it was by a contemporary:

> As Galileo's it is a curiosity which proves nothing ... But if the tract should not be either by or from Galileo, it illustrates a point of history to an extent which make it, in my eyes, much better worth a hundred pounds than any autograph would be, were it

of the three dialogues themselves. The strong under-current of Copernicanism which was setting away from the old opinions in the period 1616–1633 has, I am satisfied, not been sufficiently brought to light by the historians.

If it were by 'an obscure teacher, of not prominent name or fame', it would be 'the strongest of all the isolated facts which, put together, show what was going on'.[107]

As well as being interested in the reasons why a new hypothesis might be rejected or accepted, De Morgan also believed that new ideas contained elements of older theories, demonstrating the relative nature of 'truth'. Here he was again working with ideas that have been uniquely attributed to Whewell, with the exception that his examples remained particular where Whewell attempted to generalize. It was to Whewell that he wrote 'When I have time & opportunity I intend to work out the thesis, "That Newton was more indebted to the schoolmen than to Bacon, and probably better acquainted with them"'.[108] The two men had very different loyalties in religion and politics and disagreed over many points regarding the history and philosophy of science, education and logic. Yet they remained friendly, and corresponded regularly on topics over which they disagreed. De Morgan's letters reveal his openness in criticizing Whewell's concept of induction.[109] Although the two were essentially on opposite 'sides' in debates such as that provoked by Baily's *Account*, this did not mean that they ceased contact. It is testament also to De Morgan's continuing debt to his Cambridge education and to Whewell, his former teacher, despite his career in a secular university and attempts, in both mathematics and history, to separate rational enquiry from religion.[110] However, De Morgan placed himself within a different tradition and pointed to Rigaud, Peacock, Baily, Joshua Milne and, to a lesser extent, J. E. Drinkwater as having 'set the example of delving into one thing at a time, and presenting both the results and a full account of the sources from which they had been obtained'. He suggested that both in Britain and on the Continent 'in the second quarter of the nineteenth century the examination of special points of scientific history became epidemic among us'.[111] The advantage of this approach, he argued, was that it allowed specialists fully to understand the significance of new sources. This, he believed, allowed the truth to surface through the myths.

Richards describes De Morgan's presentation of himself 'as a self-consciously historical figure who was willing to accept present ambiguities in the hope of future rewards'.[112] The 'future rewards' are necessarily vague, for overt claims to utility would have spoiled the impartiality of the work and muddied whatever lessons history might be able to teach. It was only through 'impartial' research that the historian had a chance of revealing the conditions that led to past discovery. However, De Morgan did have strong claims for the role of science in society. He wrote of 'the liberalizing effect of scientific study' on medieval Oxford and the 'independence of thought' that accompanies scientific investigation.[113] Remembering that much of his historical writing came under the umbrella of the SDUK, we can therefore see it as a means of disseminating knowledge of science's 'liberalizing' power. However, he wished to spread historical knowledge amongst the scientifically literate as much as scientific knowledge amongst general readers:

> Too large a proportion of modern physical inquirers are what the French call *hommes de metier*: they have made a workshop of the natural sciences, from which they turn out admirable results, and greatly improve the arts of life, so far as such arts equally concern the honest man and the knave, the intelligent man and the fool. But of their great pursuit as 'a thing to be desired to make men wise' they take no cognizance: they know it only as a thing to be desired to make men comfortable. And they have their reward: for men would rather be made comfortable than wise.[114]

De Morgan admired men like Libri, Chasles and Peacock who were both scientifically and historically literate. He felt that if such a combination of knowledge could become more general, society would be improved.

A strong message in De Morgan's work is that suppression of historical documents is always unwise. His writings imply that openness was necessary for historians to do justice to all parties, and not just their heroes. For example, in 1855 he published in *Notes and Queries* some petty and vindictive marginalia, written by the mathematician Ruben Burrow, as a balance to the published excerpts of Burrow's diary, whose editor, he suggested, had fallen 'into the error of biographers' in omitting 'anything which may show him unfavourably'. The published parts of the diary contained some harsh judgments of fellow mathematicians but, in De Morgan's view, suppressed the more outrageous claims and therefore gave more credibility to the former opinions than they deserved.[115] Much of De Morgan's work can be read as commentary on this point of view. A number of times he pointed out to those guilty of suppression that their approach could not only harm the reputation of innocent parties but, ironically, also damage their own interests. In the case of his research on Newton's niece, Catherine Barton, and his errant nephew, Benjamin Smith, De Morgan argued that 'reserve of biographers about the illustrious dead is a very unwise proceeding'. An admirer of Newton had gone so far as to burn some letters in which Newton had used 'vulgar phraseology' in remonstrating with his reprobate nephew. De Morgan knew the family of the clergyman who had committed this act of historical vandalism and was able to discover the justifiable cause of Newton's anger.[116] Fear of honest enquiry allowed gossip to develop and nearly caused posterity to lose 'the argument that an uncle, who reproved his nephew's vices in strong and irritating terms, could hardly have been an uncle ... [who was] a consenting party to the dishonour of his own niece'.[117]

In an essay on the calculus controversy he made a similar argument regarding a full investigation of the Committee that judged the rival claims of Newton and Leibniz. In 1846 he sent a paper to the Royal Society, which was published in the *Philosophical Transactions*. This 'defended' Newton, and the Royal Society, against a possible charge that the jury had been packed, for De Morgan showed that it was never intended to be an impartial body. A second paper, which 'intended to point out and repair a small portion of the wrong which *actually was done under the name of the Society to Leibnitz*', the Royal Society did not, to his indignation, publish.[118] De Morgan believed that the Society's decision was evidence of bias:

> It is then the duty and pleasure of the Society to guard the fame of Newton, not only from what has been, but what might be, said against it; but it is affirmed to be either not its duty or not its pleasure to repair the effect of falsifications made in a publication issued under its name, when the sufferer, if any, must be Leibnitz.[119]

This bias had existed since Newton's time and had been propagated to the nation: 'We were taught, even in boyhood, that the Royal Society had made it clear that Leibniz stole his method from Newton'.[120]

This evidence of the Royal Society's refusal to right past wrongs done in its name only strengthened De Morgan's opinion against a Society he refused to join. In the same year that he published this essay he told his friend W. H. Smyth 'I cannot enter the Royal Society – Quant à la physique, want of time: Quant à la morale, difference of principle'. This letter contained a drawing of a 'swann in his owne lake' and a 'goos' with 'Ye mace' lording it above him on a hillock, presumably representing the scholar and the aristocratic, pretentious Royal Society. 'Now', he told Smyth, 'you will guess why I don't FRS my tail'.[121] A similar drawing appeared in an ironic coat-of-arms for the Royal Society that included a copy of Debrett's and subverted the Society's motto, 'Nullius in Verba', to 'Nisi Nobilis Nullius in Verba Jurare Magistri'– 'I will not swear to the words of any master unless he is noble' (Figure 6).[122] De Morgan's opinion of such authority is clear in a letter to his father-in-law. He noted that everyone seemed to have left London for the country, and said that if the same happened the next year, 'They must let me carry on the Government; and if they find a House of Lords and a church when they come back, they will find more than I intend they shall find!', adding that, although he was not yet certain if he would abolish the monarchy, he would certainly reduce the civil list.[123]

Figure 6. 'Coat of Arms of the Royal Society', inserted in inserted in A. De Morgan, 'Mathematical Biography extracted from the Gallery of Portraits', Royal Astronomical Society, De Morgan MSS 3, f. 22. Permission Royal Astronomical Society.

De Morgan's Image of Newton, 1840–1855

De Morgan's publications on Newton included a biographical article for the *Penny Cyclopaedia* in 1840, and a longer account in 1846 for Charles Knight's *Cabinet Portrait Gallery of British Worthies*, which devoted significant space to the calculus dispute and Newton's religious beliefs. In 1852 De Morgan published three articles on the subject of the calculus dispute, which built on his entries on Leibniz, the calculus and similar topics for the *Penny Cyclopaedia* and on the articles he sent to the Royal Society.[124] In 1855 he wrote some 'Notes on the History of the English Coinage', which included details of Newton's role at the Mint and reviewed Brewster's *Memoirs of Newton*, revisiting his favourite Newtonian themes: religion, the calculus dispute and the biased treatment of Newton's enemies by his biographers.[125] These writings can be seen as a series of attempts to investigate cases of suppression – of Newton's role in the calculus dispute, of the true nature of Catherine Barton's relationship with Halifax, of Newton's religious opinions and of his behaviour towards Flamsteed and Whiston. De Morgan believed a 'mythical' Newton had been created by his supporters, to the detriment of his adversaries and of historical truth. This rewriting of events, begun by Newton and his immediate circle, had been maintained by the Royal Society in particular and the establishment in general.[126] His purpose was to write 'impartial' histories of events and persons that had been misconceived by previous writers and traditions. He encouraged historians to be fearless in their investigations and not to be content to assume, or hope, that moral and intellectual greatness always coexisted. It was in itself a moral position, which almost necessarily went hand in hand with the advocacy of the critical use of primary sources.

De Morgan's wife Sophia wrote that his 1846 biography of Newton was,

> after Baily's *Life of Flamsteed*, the first English work in which the weak side of Newton's character was made known. Justice to Leibnitz, to Flamsteed, even to Whiston, called for this exposure; and the belief that it was necessary did not lower the biographer's estimate of Newton's scientific greatness, and of the simplicity and purity of his moral character.[127]

This is a believable summary of De Morgan's views. It seems that it was his friend Baily's *Account of Flamsteed* that first turned his thoughts to how Newton's character had traditionally been represented, whether this allowed a fair treatment of Newton's contemporaries and what were the wider implications of this interest in the moral character of the man of science. De Morgan's admiration for Newton's mind and the *Principia* remained constant but was frequently put into the shade by his other concerns. It is true, too, that De Morgan always believed that Newton was basically honest, that, in regard to sexual matters, he was pure and that he was a deeply religious man. However, he also claimed that 'a morbid fear of opposition from others ruled his whole life', coming to believe that 'the *moral intellect* of Newton – not his moral *intention*, but his power of judging – underwent a gradual deterioration'.[128]

De Morgan's biographies of Newton leave the reader in no doubt that it was the mind of the man, and not his character, that was worthy of admiration and was, 'to this day, and to the most dispassionate readers of his works, the object of the same sort of wonder with which it was regarded by his contemporaries'.[129] On the topic of scientific method, De Morgan and Brewster had more in common than might be expected and it was not a point of confrontation. Both emphasized that Newton's discoveries were made possible by the work of predecessors and contemporaries and that Newton was fallible. Neither credited Newton's success entirely to the 'patience and perseverance' available to all men, although De Morgan was less inclined to hint at inspiration. De Morgan used the word 'sagacity' rather than 'genius' to describe Newton's ability, to avoid a word that he believed mythologized the act of discovery:

> The world at large expects ... to hear of some marvellous riddles solved, and some visibly extraordinary feats of mind. The contents of some well-locked chest are to be guessed at by pure strength of imagination: and they are disappointed when they find that the wards of the lock were patiently tried, and a key fitted to them by (it may be newly imagined) processes of art.[130]

Like Priestley, De Morgan thought Newton partly to blame for this erroneous conception, both for the 'veil of obscurity' that covered his writings and his combination of talents, which meant that he was 'so happy in his conjectures, as to seem to know more than he could possibly have had any means of proving'. Newton, he believed, might 'serve to illustrate what a popular reader would hardly suppose, namely, that the wonder of great discoveries consists in there being found one who can accumulate and put together many different things, no one of which is, by itself, stupendous'.[131]

Although De Morgan concluded his 1846 biography with the assertion that Newton 'remains an object of unqualified wonder, and all but unqualified respect', his writings on Newton were always so weighted as to leave the criticisms uppermost in the reader's mind. When reviewing Brewster's *Life of Newton* in 1832, Malkin had told Napier that, because '[c]ensure necessarily consists of detail', the critique of Brewster would take up more space than his appreciation.[132] The same excuse was put forward by De Morgan, who claimed that his emphasis on the injustices committed against Leibniz and Flamsteed were required to correct Brewster's bias, for, if Newton was judged by his book, 'we could only infer that his moral character had suffered from no one instance of human infirmity, and that every action had been dictated by feelings of benevolence and the love of truth'.[133] It was, therefore, the fault of other biographers that Newton's unjust behaviour had to be emphasized, for, 'As long as Newton is held up to be the perfection of a moral character, so long must we insist upon the exceptional cases which prove him to have been liable to some of the failings of humanity'.[134] De Morgan's subsequent research on specific controversial issues in Newton's life and work meant that the weight of criticism grew and increasingly little attention was paid to his admirable qualities and achievements.

Morality and 'Impartial' History

I have used the word 'impartial' to describe the history that De Morgan and others aimed to write, for this is the word they used themselves. As Lorraine Daston has shown, the concept of objectivity was in the process of formation and the word only achieved its modern meaning in the later nineteenth century. Daston and Peter Galison have described the increasing desire to minimize errors and individualized experiences of experimenters, and the simultaneous 'moralization of objectivity'.[135] These moral values are visible when Herschel wrote of the need for the man of science to

> strengthen himself, by something of an effort and a resolve, for the unprejudiced admission of any conclusion which shall appear to be supported by careful observation and logical argument, even should it prove of a nature adverse to notions he may have previously formed for himself ... Such an effort is, in fact, a commencement of

that intellectual discipline which forms one of the most important ends of all science. It is the first movement of approach towards that state of mental purity which alone can fit us for a full and steady perception of moral beauty as well as physical adaptation.[136]

A similar desire to produce 'objective' history and an appreciation of the moral implications of the attempt can be seen in those who published texts with as little interpretation and editorial intervention as possible. As in the 'mechanical objectivity' that increasingly entered experimental science, 'Imagination and judgement were suspect'.[137] As shown above, Baily and Rigaud played on the similarities between their historical and scientific work, bringing the authority of the latter to the former. This was especially important when contributing historical 'facts' to a controversial debate. The attempt at the 'impartial' presentation of historical data and the removal of the author's presence from the text was not the less genuine because authorial intent can be clearly discerned in either choice of material or footnote commentary.

De Morgan also based his writings on primary sources, and in his bibliographies, for example, made much of the fact that he only included books that he had actually seen, allowing him to verify all details. However, he did not, like Baily, Rigaud, Edleston and others, simply reprint texts, believing that historians must also produce critical analyses. For those writing critical or discursive essays of this type, the range of evidence considered and the way in which it was presented were all important. It might be tackled in the manner of either the judge or the advocate, and these two words frequently appeared in commentary on historical writings. Earlier in the century Macaulay had described Henry Hallam's work as 'eminently judicial. Its whole spirit is that of the bench, not that of the bar.'[138] The judge surveyed the evidence presented by both sides in a dispute and attempted to come to an impartial decision, while the advocate, always morally suspect, presented the evidence that could best back his position. The historian-judge listens to evidence presented by history's advocates, and must therefore judge their bias as well as the case before him. De Morgan's writings are full of such judgments of the value of evidence from historical and contemporary figures. The self-disciplined maintenance of impartiality was paramount, and the historian had to surrender himself to the historical facts and, like Herschel's astronomer, accept his findings even if they were contrary to his theory or his hopes.

George Levine has considered this ethos in relation to both science and literature in the nineteenth century. On the one had there is what he calls 'dying to know', in which the self-restraint and even self-annihilation of the seeker of knowledge is required, and on the other is a necessity of being prepared to accept the unpleasant consequences of that knowledge. Quoting Mill's phrase, 'One must seek the truth even when it threatens to produce unpleasant results', Levine suggests that for some Victorians, especially those influenced by the positivism and altruism of Comte, 'Facing the amorality of the world entailed ... a higher morality than that of traditional religion'.[139] In science this entailed accepting experiments or observations that removed man from the centre of creation. In history it meant accepting evidence that demonstrated, for example, that immorality had prospered while morality suffered. In literature, Levine connects this stance to the realist novel, and a similar dichotomy can be found between 'romantic' and 'realist' history. The difference between Brewster's and De Morgan's positioning in this respect is clear from their very different judgment of J. S. Bailly's history of science. De Morgan called Bailly's work a 'romance under the name of a history'. For Brewster, however, it was 'one of the most interesting books that has ever been written upon a scientific subject', full of 'ingenious speculations'. He particularly admired the 'copious brilliancy of his descriptions', the 'eloquence with which he pleads his cause, and paints the sufferings of neglected genius', his 'glowing imagery' and the 'lively fancy ... [which] embellishes the general narrative' and throws 'enchantment round the most common details'.[140]

Claims and calls for impartiality were hardly new, and the simile of the historian as judge also had a long history. However, what was changing was the type and amount of evidence that was required to make an 'impartial' judgment, and the role of the historian's own beliefs. This approach was not thought possible or sufficient by all. In discussing the liberal Anglican historical approach, which was influential in the writing of Whewell, Duncan Forbes has suggested that

> Impartiality, for the Liberal Anglicans, did not mean the absence of a standpoint (the objectivity of the scientist was ultimately impossible in the study of history), it meant having the best standpoint, on which, though itself beyond the domain of the processes of purely political history, was yet in history, and this Christianity alone could provide.[141]

As well as providing space for religion, a fundamental difference was the role of the imagination, which for the liberal Anglicans was essential to reaching historical truth, but was deeply distrusted in the rationalist-utilitarian tradition. Whewell's *History* had defended both the active role of the human mind in scientific discovery, in supplying what he termed Fundamental Ideas, and the usefulness of broad interpretive schemes in historical analysis.[142] However, for De Morgan, Whewell's work lacked the detail necessary to investigate 'what actually happened', and had therefore departed from the humble search for the truth.

Conclusion

History of science could be the vehicle for a variety of moral positions. In the British tradition it was, of course, very commonly linked to the support of natural theological arguments. For Whewell the history of man's attainment of knowledge through the combination of experience and Fundamental Ideas taught that man's mind was God-given.[143] In popular histories the assumption that intellectual ability and moral worth went hand-in-hand made tales of great men morally improving. However, as this chapter has shown, the moral position of the author might be subsumed into the chosen historical format. Some chose to publish texts with as little editorial commentary as possible so that the personality of the historian was removed from the presentation of 'facts'. Although such works could still present a particular viewpoint, if the editor had done his job with care the work would have a natural authority. De Morgan was equally convinced of the need to base historical writings on original texts, but wrote within a different tradition. His style was, essentially, that of either the judge, when presenting his own research, or the critic, when commenting on the works of others. His review of Brewster's second biography of Newton, discussed in the following chapter, demonstrates that De Morgan saw it as the duty of the reviewer to correct bias. Indeed, he saw all of his research and writing on Newton in this manner, believing that he must balance the pro-Newton propaganda of the previous century and allow the reader to join him in making a judge-like decision.

The impetus given to Newtonian studies in particular and history of science in general in the nineteenth century came as much from writers operating within these two genres as from those writing more traditional biographies, like Brewster, general histories, like Powell, or philosophically-driven histories, like Whewell. Baily had used archival sources to challenge the received tradition and thus encouraged writers like Edleston and Rigaud to look to equally authoritative forms of writing and archival evidence as a defence. However, in their focus on such sources, they were typical of historical studies of the period, when such texts were being printed or otherwise made accessible through private individuals, societies and government initiatives. The lesson that histories must be based on archives had been taught by German historians, but some felt caution was needed. An English historian, in a generally laudatory review, felt Niebuhr's writings had 'taught scholars to appreciate more justly the traditional records of all ancient nations' but perhaps 'creating in some, no doubt, excessive scepticism'.[144] General histories of science were also responding to this requirement, as Weld's *History of the Royal Society* and Grant's *History of Physical Astronomy* testify. However, despite their differences it was the detailed, one-subject studies, of Rigaud and De Morgan in particular, that generated a sophisticated canon on which future historians could build.

5 DAVID BREWSTER'S *MEMOIRS OF SIR ISAAC NEWTON* (1855): THE 'REGRETFUL WITNESS'

> BREWSTER with delight is glowing, laurels won from NEWTON showing
> 'The House of Fame' (1853)[1]

After the publication of Baily's *Account of Flamsteed*, Rigaud, Edleston and De Morgan further extended Newtonian scholarship through the investigation of defined topics and publication of manuscript collections. Brewster's second biography of Newton, the *Memoirs of the Life, Writings and Discoveries of Sir Isaac Newton*, had to tackle the perceived charges against Newton's character and incorporate the new information made available in these texts. The reviews of his 1831 *Life of Newton* had criticized him for relying solely on secondary sources so, because of this and the fact that he was tackling a series of works that were based on large amounts of primary material, it is unsurprising that Brewster began his research among the Portsmouth Papers. His new reliance on such sources can therefore be seen as dictated by his need to find an authoritative means of defence. The contents of his book were also reactive to the work of other writers, tackling the problematic issues identified in the previous three decades. This counteractive and defensive element of the *Memoirs of Newton* is identified in this chapter through Brewster's research and writing process and through his treatment of the controversial themes in comparison to the approach of his critics, especially De Morgan. The topics that caused Brewster most anguish were Newton's quarrels with Flamsteed and Leibniz, his Antitrinitarianism and his interest in alchemy. Brewster published new evidence on all these points which, although it significantly altered his portrayal of Newton, was presented as positively as possible. Brewster also considered occasional suppression justifiable. However, despite Brewster's continued insistence that Newton's 'social character' was 'modest, candid, and affable, and without any of the eccentricities of genius', his evidence projected a very different image.[2] What Brewster left implicit was to be presented explicitly by many of his reviewers. Brewster's defensiveness appeared in one further area: his treatment of the history of optics. Although a considerable feature of the *Life of Newton*, this later work reveals

Brewster's strategy at a time when he was one of the last advocates of the corpuscular theory of light.

The Gestation of Brewster's *Memoirs*

While the 1831 *Life of Newton* was still in proof, Brewster had written to his publisher to suggest the publication of a longer biography of Newton that was 'more scientific & containing his correspondence' in the Oxford and Cambridge archives.[3] Brewster felt that this would sell but Murray probably considered it too soon for such a project. It was, therefore, not until Brewster was spurred by the need to respond to Baily's *Account* that he recommenced his research. In May 1837, Brewster undertook 'a weeks intense labour among Sir Isaac Newtons MSS.' at Hurtsbourne Park, the seat of Lord Portsmouth. By December of the following year, the biography was envisaged as a work of 'at least two large 8vo volumes' and Brewster had written to Murray regarding a project he thought 'will excite universal Interest'.[4] In 1841 he told J. O. Halliwell, who provided him with material relating to Newton, that he hoped to begin printing soon.[5] In 1843, the *Edinburgh Review* reported that the book 'has, for some time, been ready for the Press'.[6] It was, however, to be delayed for a further decade. Work began again in 1853, the first volume was printed by July 1854, and the second, delayed again as Brewster awaited the arrival of various manuscripts in Scotland, was printed and revised by March 1855.[7]

The delay may in part be due to that fact that in 1838 Brewster became a College Principal in the University of St Andrews. He was quickly involved with attempts for the reform of the university, putting him on bad terms with many and leading to the creation of the Royal Commission on St Andrews in 1840.[8] In addition, in 1843, Brewster played a prominent role in the Disruption, the schism in the Church of Scotland leading to the formation of the Free Church.[9] However, he did not forgo either scientific research – Maria Gordon lists 107 scientific papers published between 1838 and 1854 – or popular writing.[10] His *Martyrs of Science* was published in 1841 and Brewster not only continued to contribute to literary reviews but founded the Free-Church *North British Review* in 1844. He also published an article on Newton in the 1842 *Encyclopaedia Britannica*.[11] A significant problem seems to have been differences with Murray as to the terms of publication. Murray wished to see the whole manuscript up front and later offered to publish 1,000 copies, taking on the risk and cost and halving the profits.[12] Brewster evidently wished for the same arrangement as his earlier books, when Murray, taking all the risk and profit, purchased the copyright of a projected work.[13] Ultimately Brewster chose to publish with Thomas Constable in Edinburgh.

It is clear that Brewster always envisaged his work as an extended biography, which would bring further manuscript sources into print. He reported to Macvey Napier in February 1837, a month after Baily had published his *Supplement to the Account*, that he had heard from Newton and Henry Fellowes that he was 'to have the unreserved use of all the *Portsmouth Papers*, namely *Newton's MSS & Correspondence*. I shall therefore gird myself for the labour of producing an elaborate & Scientific Account of his Life & discoveries.'[14] However, because Brewster emphasized his exclusive use of the manuscripts in *The Times* and the *Britannica* article, and perhaps also because it was understood that he wished to counter the manuscript-based *Account*, several contemporaries appeared to think Brewster would publish an edited selection. For example, Whewell referred to 'Brewster's edition of Newton's letters from the Portsmouth papers' in 1849. He expressed concern that the publication of Trinity's manuscripts (Edleston's *Newton and Cotes*) might be thought by Brewster to interfere with his projected work, as did Rigaud with regard to his publication of the Macclesfield manuscripts.[15] Brewster, however, assured Rigaud that, 'so far from thinking that your work w^d interfere with mine', he would rather 'have withheld my volume in order to get the benefit of yours'. He later told Edleston that he was glad his book had been delayed, 'as I have had access to the valuable information contained in your admirable "Synopsis of Newton's Life"', in Dawson Turner's publication and elsewhere.[16] Although Brewster shared the prevailing sense of the importance of manuscript sources, the trend for publishing undigested documents did not serve his purpose.

In his *Britannica* article, the only material from the Portsmouth Papers that Brewster referred to was the 'new and valuable information ... relative to the early life of Sir Isaac, which had been collected by his nephew-in-law Mr Conduit [sic]'.[17] The Preface to the *Memoirs* described the search of the Portsmouth Papers as 'particularly directed to such letters and papers as were calculated to throw light upon [Newton's] early and academical life' (vol. 1, p. vii). These were found in Conduitt's collection, and this was the material that Brewster highlighted in his correspondence. In several letters Brewster wrote excitedly about these materials, 'which were not supposed to exist' and contained the 'most interesting facts and anecdotes'.[18] We cannot know exactly what he saw while at Hurtsbourne Park but he recorded that Henry Fellowes assisted him 'night and day in Copying from the MSS' and that they 'went over all the Papers *twice*'.[19] His interest in personal details was again highlighted in his report of a 'love letter', apparently from Newton to Lady Norris, while his sense of identification with Newton was boosted by the discovery of Newton's proposed reform of the Royal Society, 'placing it on the footing for which I have been contending for the last 15 years'.[20] Brewster's letters indicate that he saw a wide range of material but he chose to privilege certain discoveries, particularly when advertising the work or asking for assistance.

Brewster's findings meant that he had to begin revising his ideas. For example, when he came across a 'most interest^g & kind letter of *Hooke's* with Sir *Isaacs* answer in the same right[?] spirit', he was *'constrained* to take a more favourable view of Hooke's conduct than has been usually done'.[21] Brewster also told Rigaud that Newton's 'Chemical, particularly the Alchymical Papers are extremely numerous, as well as the Theological ones'. Following this statement, the letter contains five obliterated lines which, from the context, probably contained further details concerning Newton's religious beliefs. However, the letter ends reassuringly: 'Every thing that I have seen among the Portsmouth MSS has contributed to exalt Newtons Character in my estimation, high as my impression of it had previously been'.[22] Brewster was also able to tell Brougham that he had 'looked anxiously for letters or Papers connected with the year 1692' and found 'that persons had seen him or corresponded with him with out suspecting that his mind was in any way disturbed' and that there was no evidence that Newton had interrupted his studies.[23] However, this 'anxious' search tells its own story about Brewster's approach to the available material.

Brewster had a second chance to see some of the Portsmouth Papers in 1854. He wrote to Brougham to say that he had finished the first volume of the *Memoirs* but 'cannot proceed a step further till I get the Manuscripts', or at least, if they did not arrive in St Andrews soon, he would 'be obliged to proceed without them'.[24] In the event they arrived safely on 8 August. The selection included the Conduitt material and 'numerous' documents relating to the calculus dispute. The letter gives us the sense that he was overwhelmed by this material; it was not until November that he reported the discoveries that led him to change his mind on this topic.[25] In considering mathematics, Brewster was stretched to his limit. He exclaimed to Brougham 'I wish our Friend Prof. De Morgan had these Papers in his hands. He is so thoroughly Master of the subject, that he would bring out their substance far better than I can do.'[26] Already in 1842 Brewster had written to De Morgan, saying 'I should be very glad also to know your views on the subject of the controversy as I am anxious to give a fair account of it'. De Morgan's papers contain a long draft on the subject of the controversy that was written in reply to Brewster's request. It is remarkable that Brewster should have desired the assistance of an individual who believed that during this dispute Newton had 'acted in a manner not becoming to a gentleman' and believed 'Newton saw Leibnitz's *system of notation* before he had one of his own'.[27] It is remarkable, too, that Brewster should have recommended De Morgan as his reviewer in the *North British Review*, when from both published writings and private correspondence he was aware of the extent to which they disagreed.

Brewster was on easier ground with other individuals from whom he requested assistance. He benefited from Rigaud's publications and used him as a source of miscellaneous information, such as details of Lady Norris, the recipient of the 'love letter'.[28] In the 1850s, Brewster was in correspondence with Edleston, requesting his opinion on controversies and his assistance in minor points. In his Preface the highest words of praise were for Edleston, both for his 'important contribution to the History of Mathematical and Physical Science' and 'his judicious criticisms and useful suggestions' (vol. 1, p. xiii). Others from whom he sought assistance included Brougham and Halliwell. In part, this help was needed because Brewster lived many miles from his main sources. He did not, for example, even have access to a first edition of the *Principia* and had to ask Edleston to check one on his behalf.[29] However, it is clear that Brewster also asked for help because he found the composition of the work difficult as a result of the vastly increased range of sources, his own prejudices and, in the case of the calculus controversy, his lack of expert knowledge. His daughter remembered that the composition of the book 'involved severe labour', although she insisted that it 'was most congenial work'.[30]

It is easy to find fault with Brewster's *Memoirs* and nearly every commentator has mentioned his partiality, although those who have discussed it more recently have tried to balance such criticism. Brewster has been commended for the amount of work undertaken and evidence introduced, especially that which pointed to flaws in Newton's character or evidence of his interest in alchemy or unorthodoxy.[31] John Christie has pointed to the influence of the *Memoirs* on subsequent works, believing that 'Brewsterian categories and dichotomies still infect the very framing of the questions we pose'.[32] Westfall also sees the book as having 'defined most of the problems of Newton's personal life in the terms with which succeeding scholarship has continued to discuss them' and Brewster as, for example, 'the first who indicated Newton's secret role in the campaign against Leibniz'. Such statements are, however, misleading. While it is true that Brewster tackled a number of difficult issues in an authoritative and lengthy work, most of his new sections responded directly to the work of Biot, Baily and De Morgan, or to long-established traditions, rather than striking a new path. We have already seen that Brewster acknowledged De Morgan as the pioneer regarding the calculus controversy. It was these writers and traditions rather than Brewster who 'defined ... the problems of Newton's personal life'.[33]

In fact, in comparing the 1831 *Life* and the 1855 *Memoirs*, the reader is struck by their similarity. Wherever possible, Brewster left phrases, paragraphs and sections untouched. This led to further contradictions in Brewster's assessment of Newton's character and to occasional errors. For example, in the first of two chapters dealing with the invention of and controversy over the fluxions, Brewster wrote that it would be 'inconsistent with the nature of this work to enter into a detailed history of the dispute between Newton and Leibnitz' but that he would give a 'brief and general account' (vol. 2, p. 23). This sentence was taken from the *Life*, with only the removal of the word 'popular' before 'nature', and does not in the least describe Brewster's subsequent lengthy and involved discussion.[34] The main alterations to the 1855 book were made first to give a more extended account of events or discoveries, second to include details published by Rigaud, Edleston, De Morgan or others, and third to add information found in the Portsmouth Papers or elsewhere. By and large this new material addressed issues that had been raised by others, frequently casting Newton in an unflattering light, and that Brewster felt 'a sacred duty to investigate' (vol. 2, p. 227). Thus the basic framework, conceptual and literal, was unchanged and, at times, conflicted with the new evidence.

De Morgan believed that Brewster's 1855 biography was moulded by exterior forces, including his own work:

> we live, not merely in sceptical days ... but in discriminating days, which insist on the distinction between intellect and morals. Our generation, with no lack of idols of its own, has rudely invaded the temples in which science worships its founders: and we have before us a biographer who feels that he must abandon the demigod, and admit the impugners of the man to argument without one cry of blasphemy. To do him justice, he is more under the influence of his time, than under its fear: but very great is the difference between the writer of the present volumes and that of the shorter life ...[35]

However, he pointed to the differences between the two books, predicting that in the future 'it will be said of our day that the time was not come when both sides of the social character of Newton would be trusted to his follower in experimental science'. They were in an intermediary stage, where, though 'biography be no longer an act of worship, it is not yet a solemn and impartial judgment'. He accused Brewster, just as reviewers had in 1831, of producing an '*ex parte* statement'.[36] This was, however, how Brewster conceived the task of the biographer, and his surprise at De Morgan's critical review suggests that he did not grasp the nature of these objections.

The *Memoirs* and the History of Science

To Brewster's annoyance his reviewers chose to pay little attention to the more scientific parts of the *Memoirs*, focusing instead on biographical details and controversies.[37] The reviewers understood Brewster's interest to be in the personal history of Newton, therefore his discussion of science, which included little new material, was largely dismissed. Powell feared his article on Newton for the *Edinburgh Review* might tread on Brewster's toes but he reassured himself that, unlike Brewster's, his 'narration is less a *personal* one, than of the times

& the history of discovery'.[38] De Morgan noted that Brewster's account of the history of science was a 'cursory glance', much of which was similar to that in the *Life*. While 'the book is very readable' and useful for the 'general reader', he felt that Brewster should have moved beyond the elementary to address an informed audience.[39] Both De Morgan and Biot suggested that Brewster could, and should, have taken the chance to add something new and valuable to the available knowledge of Newton's intellectual development. De Morgan felt the Portsmouth Papers might hold the key to understanding this and regretted that Brewster published few papers of purely scientific interest. Biot made the same complaint, ridiculing the fact that Brewster instead paid attention to minute and mundane details.[40] Brewster, however, was not attempting to produce the kind of sophisticated investigation constructed by Rigaud. Tellingly, he suggested that the only reason to establish the dates of Newton's discoveries was to counter the Leibnizians' charges of plagiarism and Biot's claim that the 1692–3 breakdown permanently affected his intellect. 'The historian of science has not a more painful duty', he wrote, 'than that of fixing the date of discoveries, but it is a duty which he is never called upon to perform unless there are conflicting claims submitted to his judgement' (vol. 2, p. 367). Brewster in fact frequently allowed his chronology to break down and he did not respond to Biot's earlier criticism that by treating Newton's discoveries out of order he failed to reveal the development of these ideas. In 1831 the *Life* had generally been praised for its history of science, but the field had changed significantly and the expectations of his reviewers had increased.

In one area, apart from the calculus controversy, Brewster did produce a detailed study. As De Morgan noted, 'Brewster gives Newton's career in optics at great length'.[41] The longest extract from Newton's scientific work that Brewster printed was his 1675 'Hypothesis' on light and colours (vol. 1, pp. 390–419). This previously published material was reprinted because it had 'led some writers to suppose that [Newton] had abandoned the corpuscular or emission theory' and Brewster wished his readers to reinterpret it. He showed that in 1673 Newton 'summarily' rejected Hooke's hypothesis of light (vol. 1, p. 135) and, in judging the 1675 Hypothesis, 'it is necessary to keep this in view, as it appears to be quite clear that this hypothesis is not what he believes, but what he found it necessary to draw up for the information of many of his friends' (vol. 1, p. 136). Newton, he boldly claimed, used an undulatory hypothesis merely for illustration and made suggestions about the ether 'as if he were amusing himself with the extravagance of his speculations' (vol. 1, p. 137).

One of those who had interpreted this paper wrongly, in Brewster's eyes, was Thomas Young. In his 1801 paper – which Brougham had vehemently criticized – Young had referred to this Hypothesis in order to demonstrate that his wave theory was closer to Newton's thinking than generally understood, showing 'that

Newton considered the operation of an ethereal medium as absolutely necessary to the production of the most remarkable effects of light'. Brewster contradicted this interpretation and pointed to instances in which Newton maintained that light was material (vol. 1, p. 147). He referred to the second edition of *Opticks* (1717) as 'the mature and the latest judgement of Newton on the subject of light' (vol. 1, p. 149). Ignoring much of Newton's speculation about the ether, Brewster quoted Query 28, which asked, 'Are not all hypotheses erroneous in which light is supposed to consist in pression or motion propagated through a fluid medium?' (vol. 1, p. 148). As with Biot and Brougham, Brewster's stance seems to prove the truth of Whewell's claim that reverence for Newton slowed the acceptance of the wave theory.[42] However, this is too simplistic: Brewster had to work hard to demonstrate Newton's unwavering corpuscularianism and, by this late date, Newton's authority was not sufficient to refute the well-established wave theory.[43]

Brewster's main correspondent regarding this section of the *Memoirs* was Brougham, to whom he wrote with pride that he thought he had 'been able to throw some new light on the subject of Newton's opinion respecting the Emission and Undulatory hypotheses', showing that the undulationists 'have entirely mistaken the parts of Newton's writings in which he speaks of an *Ether*'. Brewster also asked for an account of Brougham's recent experiments in order to 'have an accurate account of what has been done[?] by Newton's succession'.[44] These experiments had been undertaken as part of a deliberate strategy designed to reignite the debate between the rival theories of light, which had died down as the wave theory became generally accepted.[45] In 1847 Brewster persuaded Brougham to 'take up the subject of the Emission versus Undulatory theory of light', hoping to cause 'the downfall of that presumptuous theory' by exploiting a perceived 'division of sentiment' among the undulationists and recent experimental results that were inexplicable by the wave theory.[46] In line with his own strategy at BAAS meetings (Figure 7), the aging Brewster sought facts that would raise questions the wave theory could not easily answer. He explained that, 'as one of the important class of *Rienistes*, I place greatest value on the results of observation, and on experimental laws'.[47] Brewster, and subsequently Brougham, had borrowed the term *rieniste* from Biot as a term of tactical convenience. Brougham's 1850 paper claimed to 'purposely avoid all arguments and suggestions upon the two rival theories – the Newtonian or Atomic, and the Undulatory' and to provide conclusions 'wholly independent ... of that controversy'. Despite this, the paper was 'heavily contaminated by his emission framework'.[48] The undulationists were forced to consider Brougham's experiments, but were not unduly troubled by its implications for the wave theory. To Brewster this was typical of the undulationists, who did not give sufficient prominence to experimental results, especially if they were inconvenient. In 1849 he told Brougham, 'the undulationists of this country are such fanatics that they have no faith in physical truths beyond their pale'.[49]

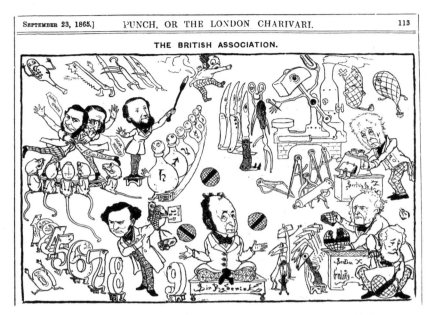

Figure 7. 'The British Association', *Punch*, 49 (23 September 1865), p. 113. The elderly Brewster demonstrates his stereoscope on the right. Permission Special Collections, Senate House Library, University of London.

Brewster's use of empirical results raised questions about the wave theory – and, in 1855, he was still hopeful that a 'critical mass' of evidence would be reached – but also helped him to 're-establish the importance of experiment as an autonomous discourse'.[50] This greater emphasis on empiricism prompted some further changes to the *Memoirs* that only enhanced the contradictions in Brewster's depiction of Newton's scientific approach. In criticizing Hooke, and his 'present day disciples', the undulationists, he wrote, 'It would have been well for the progress of science and the tranquillity of its friends, if experiment and observation had been, more than they have, our guides in philosophical inquiry'. It was only a 'small remnant in the Temple of Science', presumably including himself, 'who, while they give to theory its due honours and its proper place, are desirous, as experimental philosophers, to follow in the steps of their great Master' (vol. 1, p. 92). Newton now appears as a solid empiricist, but Brewster still wished to make this a heroic enterprise. Differences between a passage in the 1831 and 1855 texts also demonstrate this desire:

> The variety in the <objects and> phenomena of nature call forth<, summons to research> a variety of intellectual gifts: <Observation collects her materials, and patiently plies her humble avocation: Experiment, with her quick eye and ready hand, develops new facts:> The <lofty> powers of analysis and combination are applied to

~~the humbler labours of observation and experiment~~ <generalize insulated results, and establish physical laws;> and in the ordeal of ~~rival inquiry~~ <contending schools, and rival inquirers,> truth is finally purified from error. (vol. 1, p. 251)[51]

However, as shown in the discussion of Brewster's treatment of the Flamsteed controversy below, Brewster ultimately admired theoretical speculation over practical labour. His fundamental beliefs were not sufficiently altered to require him to change the 1831 text on either Baconian methodology or the character of Newton's genius.

Controversies: The Second Volume of the *Memoirs*

De Morgan's review noted that it was the second volume of the *Memoirs* 'on which we more especially differ from Sir D. Brewster'.[52] This volume, even more than the first, lost its intended chronological structure because it was here that Brewster marshalled his evidence in his attempt to challenge the writings of Biot, Baily and De Morgan. In Christie's opinion, Brewster was reasonably successful in refuting both Biot's claim that Newton's mind was permanently weakened and Flamsteed's multifarious complaints about Newton's conduct. It was on the topics of alchemy and Newton's religion that insurmountable problems were revealed. That these were created by the inclusion of new manuscript material is viewed by Christie, as it was by Brewster's contemporary reviewers, as evidence of Brewster's essential honesty. On both topics, Christie believes that Brewster 'bit the bullet and produced a brilliant, if impossibly tendentious reading of the evidence'.[53] It is at times difficult to decide if Brewster arranged his evidence in a calculated attempt to convince his readers of Newton's innocence or whether there was only one version of events he could admit to himself. De Morgan was probably right to see 'a pledge of earnest sincerity in the wildness with which the barbed arrow is fired at Leibniz or at Flamsteed' and which made Brewster's partiality transparent to the reader.[54]

Flamsteed

Brewster's idea of writing a second biography of Newton was forced into a firm intention by the publication of Baily's work on Flamsteed. As Brewster's daughter explained,

> That [Newton] should be attacked more than a hundred years after his death, was to Brewster's mind a personal grief and an English scandal. He therefore for twenty years made it one of his objects in life to search out every proof and evidence by which he could defend Newton from the charges against his sanity, his probity, and his justice, which were circulated when the hand and the tongue of the accused and his contemporaries were safely mouldering in the grave.[55]

Brewster told Brougham he agreed that 'Mr Baillie [sic] and the Admiralty acted a most unjustifiable part in publishing the Flamsteed papers' but believed he would 'be able to counteract any injurious affects which they may have produced upon the public mind'.[56] Baily, like Newton 'safely mouldering in the grave', was his chief target. De Morgan wrote little about this in his review, but it was a topic to which he later returned with some anger.[57] Again, we note the strange fact that Brewster was 'silly enough to recommend De Morgan' as his reviewer. He was aware that 'De Morgan is the only friend of Baily's likely to defend him' and should have anticipated that this review would give him the opportunity 'of accumulating and exaggerating all his previous calumnies against Newton'.[58]

However, Brewster undoubtedly believed he was as generous to Flamsteed as possible. He acknowledged the importance of his data to Newton, his initial readiness to share this information, and suggested that he could 'find some apology for his conduct in the infirmities of his health and of his temper' (vol. 1, p. 313). Brewster was likewise prepared to acknowledge that, on the question of whether the two comets of 1680 were the same object, Flamsteed was correct and Newton wrong (vol. 1, p. 303). Regarding an incident in the subsequent dispute, Brewster admitted to Edleston, 'I find I must modify my attack upon Flamsteed. I had misapprehended a remark of his'.[59] The word 'attack' is, of course, very revealing, but this comment also demonstrates that Brewster believed he should be faithful to his sources. However, he continued to think that 'Flamsteed did not sufficiently appreciate the importance of Newton's labours'. This judgment accords with his view of the relative worth of practical and theoretical astronomy, which is made clear in his comparison of two astronomers:

> Recumbent on his easy-chair, the practical astronomer has but to look through the cleft in his revolving cupola, in order to trace the pilgrim star in its course; or by the application of magnifying power, to expand its tiny disc, and thus transfer it from among its sidereal companions to the planetary domains. The physical astronomer, on the contrary, has no such auxiliaries: he calculates at noon, when the stars disappear under a meridian sun: he computes at midnight, when clouds and darkness shroud the heavens; and from within the cerebral dome, which has no opening heavenward, and no instrument but the Eye of Reason, he sees in the disturbing agencies of an unseen planet, upon a planet by him equally unseen, the existence of the disturbing agent, and from the nature and amount of its action, he computes its magnitude and indicates its place. If man has ever been permitted to see otherwise than by the eye, it is when the clairvoyance of reason, piercing through screens of epidermis and walls of bone, grasps amid the abstractions of number and quantity, those sublime realities which have eluded the keenest touch, and evaded the sharpest eye. (vol. 1, pp. 369–70)

Thus, despite the increased emphasis on experimentation in optics noted earlier, Brewster preferred in this context to privilege the theoretical.

However, Brewster's main concern was not the theory/practice debate but the perceived injury to Newton's moral character. His chief tactic was to undermine the *Account* by indirectly accusing Baily of suppression and Flamsteed of altering evidence. In his preface, Baily claimed that he looked for evidence 'that might tend either to extenuate or explain the conduct of Newton and Halley' but had found nothing relevant among the Portsmouth Papers.[60] Brewster felt this inexplicable, as Flamsteed's letters to Newton 'had been carefully preserved'. He had 'now before me nearly *forty*' of these letters, while Baily 'was able to publish only *eleven* of Flamsteed's letters to Newton, and these not correct copies of the originals' (vol. 2, p. 161). On several subsequent occasions Brewster noted where the text in the *Account* differed from the Portsmouth manuscripts, and on one occasion wrote

> This letter of Flamsteed's, as published by Mr. Baily, differs entirely from the letter actually sent to Newton, and must have been a scroll [i.e. draft], which he greatly altered and enlarged. *We cannot, therefore, place confidence in the abstracts of his letters to Newton, as printed by Mr. Baily.* (vol. 2, p. 172)

Brewster did not explain, as Baily had, that the *Account* included drafts written by Flamsteed at the bottom of Newton's letters and he hinted that there was something sinister in the discrepancies between the sources. At times he appeared to accuse Flamsteed of intentionally leaving a false version of the correspondence, at others the finger of blame pointed at Baily, alternately for mis-transcription or for suppression. Flamsteed and Baily, 'two English astronomers, the one a contemporary and the other a disciple', were jointly accused of a cross-century conspiracy 'to misrepresent and calumniate their illustrious countryman' (vol. 1, p. xi).

In suggesting that Baily deliberately ignored letters at Hurtsbourne Park, Brewster was conveniently forgetting something he admitted in his Preface. Here he described how, in 1837, he and Fellowes had 'anxiously searched, but in vain', for Flamsteed's letters and other documents with which he could defend Newton. He explained that it was only later that he was relieved from this 'embarrassment ... by the receipt of all Flamsteed's letters and other important papers which Newton had carefully preserved, and which Mr. Fellowes had discovered and set aside for my use' (vol. 1, p. xii). It thus seems likely that in 1835 Baily had also been unable to find the crucial manuscripts. In this light, Brewster's dark hints of 'causes which I cannot explain' behind Baily's failure to print the full correspondence seem disingenuous (vol. 2, p. 161). Similar suspicions were implied during Brewster's discussion of the agreement made between Flamsteed and the Royal Society for the printing of his catalogue of stars. Baily had printed the version of the agreement found among Flamsteed's papers, but Brewster found an alternative draft among Newton's. As De Morgan commented in his review, 'wonderful to relate, the unsigned draughts actually differ; Flamsteed's draughts bind him less, Newton's draughts bind Flamsteed more'.[61] Brewster, however, claimed that Flamsteed left no signed copy of the articles 'because he had wilfully violated them' and remonstrated with Baily for having failed to note that the document he published was only a draft. Brewster asserted, without evidence, that the drafts found among Newton's papers 'cannot be very different from those really signed' (vol. 2, p. 223).[62] This section, for De Morgan, highlighted the errors in Brewster's approach: 'When Sir D. Brewster not merely *opines*, but *narrates*, that Flamsteed left no copy because he had wilfully violated [the articles], he is our very good friend, and lightens our task very much'.[63]

Fluxions

In tackling the evidence on the controversy over the invention of the calculus, Brewster solicited an outline of events from De Morgan. With his mathematician's hat on, De Morgan told him that, rather than Leibniz borrowing from Newton, the latter might have benefited from knowledge of the former's notation:

> The matter of Newton's mind and writings bears out this view. *He was not a master of expression*: in all his writings he is obscure and clumsy in his handling of his tools ... I should as soon have expected Leibnitz to have worked out gravitation as Newton to have originated any thing worthy of himself in notation.

As a bibliographer, De Morgan had focused on the publication of the *Commercium Epistolicum* and the partiality of the Royal Society Committee that produced this report. De Morgan told Brewster, 'I think that Newton himself acted in a manner not becoming to a gentleman in several particulars', including in his claim that the anagram he sent Leibniz, in which he hid his method, was a comprehensible declaration. In addition, he thought 'Newton shewed an habitual disregard of accuracy, or strange want of memory (if the latter, to an extent which renders his evidence weak on all points)'.[64] In an article of 1852, De Morgan declared he had 'never come fresh to this controversy of Newton and Leibniz without finding new evidence of the atrocious unfairness of the contemporary partisans of Newton'.[65] Although lacking positive proof, he was of the opinion that Newton had taken an active, though back-seat, role in the controversy. The proof was provided by copious papers by Newton relating to the dispute and, as De Morgan told Herschel, this was presented by Brewster, the 'unwilling, or rather regretful witness'.[66]

As with his account of Flamsteed, we can detect Brewster's attempt to demonstrate balance in his judgment of Leibniz by hinting at the blame that could be laid at Newton's door. In a passage that would not look out of place in an essay by De Morgan, Brewster warned: 'We are too apt to regard great men, of the order of Newton and Leibniz, as exempt from the common infirmities of our nature, and to worship them as demigods more than to admire them as sages' (vol. 2, p. 3). Brewster's duty to 'historical truth' led him to admit that 'Newton was virtually responsible for' the contents of the *Commercium Epistolicum* (vol. 2, p. 75). He acknowledged that at times Newton's behaviour was hardly candid and that there was some excuse for Leibniz in his 'excited feelings' and 'in the insinuations which were occasionally thrown out against the originality of his discovery' (vol. 2, p. 83). However, blame was comprehensively placed on Leibniz's shoulders. Brewster believed that it was he who took the 'first false step', in 1684, when he wrote of his calculus but failed to mention Newton's discovery directly (vol. 2, p. 28). This Brewster took as 'suppression', but De Morgan asked whether Leibniz should have 'promulgated what Newton was doing everything in the power to conceal?'. Rather, De Morgan indicated that the other side was guilty of suppression, pointing to a paper of 1686 'which Newton did not cite ... which the Newtonians are very shy of citing, and of which, apparently, Sir David knows nothing'. This paper not only referred to Newton, but expressed a hope that he would publish his method.[67]

Rather than taking De Morgan's writings as a guide, Brewster told Edleston that his account was 'taken almost wholly from a carefully drawn up MSS, not in [Newton's] own hand, but abbreviated from his own manuscripts'.[68] He wrote,

> He who dares to accuse a man like Newton, or indeed any man holding a fair character in society, of the odious crime of plagiarism, places himself without the pale of the ordinary courtesies of life, and deserves to have the same charge thrown back upon himself. (vol. 2, p. 43)

He dwelt on the occasions when Leibniz and his supporters appeared in a poor light and, with the 'anonymous shafts of the slanderer, denied what they had written, and were publicly exposed through the very rents which they had left in their masks'. Newton's role was described in much more muted language and his defenders appeared as 'men of station and character, who gave their names and staked their reputation in the contest' (vol. 2, p. 81). Even though, as De Morgan acknowledged, Brewster's admission of Newton's involvement was important, he did not give 'chapter and verse' to balance the previous bias against Leibniz.[69] However, for Brewster, Leibniz's worst crime was that he subsequently attacked Newton's philosophy for irreligion. He was outraged that Leibniz had:

> dared to calumniate that great and good man in his correspondence with the Princess of Wales, by whom Newton was respected and loved, – when he ventured to denounce his philosophy as physically false and dangerous to religion, – and when he founded these accusations on passages in the *Principia* and *Optics*, glowing with all the fervour of genuine piety, he cast a blot upon his name which all his talents as a philosopher will never be able to efface. (vol. 2, p. 83)

Brewster's belief that Leibniz opposed Newton's philosophy out of malice meant that he did not pause to consider the merits of his arguments, or try to understand why Newton's ideas could be considered dangerous. In this, however, he was not alone. De Morgan, while a great admirer of Leibniz's mathematical innovations, described his metaphysics as 'extraordinary fancies'.[70]

Religion

The question mark over Newton's religious beliefs had a long history, fuelled by rumour and his published religious writings. As De Morgan wrote,

> There was a strong and universal impression that Horsley had recommended the concealment of some of the Portsmouth papers, as heterodox: and here and there was to be found, in every generation, a person who had been allowed to see them, and who called them dubious, at least.[71]

Brewster's statement in 1831 that Newton believed in the Trinity was a response to sects who had claimed Newton as a spiritual brother. Whiston, an Arian or Eusebian, and Hopton Haynes, a Humanitarian, were contemporaries who each claimed that Newton had shared his beliefs. Their testimonies inspired subsequent generations and, in the early nineteenth century, there was alarm in the High-Church community about the use of Newton's name by Unitarians in support of their creed.[72] Remarkably, in 1831 Brewster was accused by Thomas Burgess, the Bishop of Salisbury, of having '*done the same injustice to the memory of Sir Isaac* by his restatement and revival of the general contents' of Newton's unorthodox essay *Two Notable Corruptions*, without emphasizing that it had been suppressed by Newton (vol. 2, pp. 523–4).[73]

In 1846, De Morgan 'brought together all the evidence for Sir I. N. being a Unitarian – of a deeper cast than Arianism' and was the first to examine these claims in a non-sectarian publication.[74] He first stressed, as Brewster did not, that at this period Antitrinitarians were barred from official positions or even imprisoned. He argued that this should be borne in mind when judging Newton's apparent anger when Whiston called him an Arian and the fact that he made no clear statement against the doctrine of the Trinity. Newton's vauge, 'formula' expressions were carefully chosen to be non-incriminating.[75] However, in a private letter to Brougham, De Morgan suggested that in *Two Notable Corruptions* 'Newton absolutely sneers at the Trinity'.[76] Although more cautious in public, De Morgan stated that Newton's 'infirmity' – his hatred of opposition and dislike of publication – proved the depth of his feeling on this topic.[77] Perhaps denying his own feelings, De Morgan asserted his impartiality by claiming to have tackled the question not because of 'any particular interest' but because he had come across 'a curious matter of evidence, and an instructive view of party methods of discussion'.[78] This related to Whiston's statements on the matter, which De Morgan felt had been misinterpreted. His support for Whiston, whom he described as 'all honesty and no discretion', was undoubtedly influenced by his admiration for Whiston's refusal to betray his beliefs and his sympathy for one who was expelled from Cambridge because of unorthodoxy.[79] His view was very different from that of Whewell, who had represented Whiston's judgment as worthless.[80]

As with Newton's involvement in the writing of the *Commercium Epistolicum*, De Morgan presented no positive proof but believed the circumstantial evidence enough to convince the unbiased reader of Newton's Unitarianism. It was therefore De Morgan's work, as well as the evidence held in the collections of Lord Portsmouth and Jeffrey Ekins, that meant Brewster felt he could 'hardly avoid' discussion of Newton's religious opinions.[81] Brewster was more explicit in his Preface than he proved to be in the main body of the book, perhaps because, as he stated, he wished to submit the evidence 'to the judgement of the reader'. Here alone, he wrote that Newton's beliefs were 'adverse to my own, and I believe to the opinions of those to whom his memory is dearest'. He claimed he did not feel justified in 'conceal[ing] from the public that which they have long suspected, and must have sooner or later known' (vol. 1, p. xv), but this inevitability was one factor in persuading him to publish. Thus Brewster, in a section bare of commentary, printed or gave an account of Newton's *Paradoxical Questions concerning the Morals and Actions of Athenasius and his Followers* (vol. 2, pp. 342–6), *Irenicum, or Ecclesiastical Polity tending to Peace* (vol. 2, p. 347; Appendix, pp. 29, 526–31), *A Short Scheme of True Religion* (vol. 2, pp. 347–8) and *On our Religion to God, to Christ, and the Church* (vol. 2, pp. 349–50). Brewster at least found these materials comforting in that they demonstrated Newton's faith and their shared preference for the 'simplicity of apostolic times'.[82]

Although his daughter wrote that Brewster 'seemed to cling to what he considered a fact, that there was no distinct declaration of Newton's rejection of the doctrine of the Trinity', he undoubtedly realized that the new material would speak for itself.[83] In a letter to Brougham he was perfectly candid: 'I have the most ample proof, in the form of numerous Extracts from Sir Isaac's Theological MSS, that he was an Unitarian. No person that has seen his MSS. can entertain the slightest doubt upon this subject'.[84] However, in print he continued to assert that Newton's publications 'warrant us only to *suspect* his orthodoxy' and thus managed to back his own conclusion of 1831 (vol. 2, p. 337). Brewster continued to claim Newton's faith as a triumph for religion and the closing pages of his chapter on theology were identical to those of the *Life*. As Christie has pointed out, the revealed unorthodoxy cannot but affect the religious imagery that Brewster used here and elsewhere: 'Retrospectively, then, the whole rhetorical drive of the biography, the sacred image of Newton as High Priest, is nullified'.[85]

It was an issue over which Brewster struggled between 1837 and 1855. His discomfort is revealed in a letter to a friend, on the topic of the 'publication of unsound opinions on Religion' in the letters of David Hume. He did not wish, he said, to vote at the Royal Society of Edinburgh over the question of their publication because 'there would be a chance of my voting with the Majority', who wished publication to go ahead:

> I considered Religious truths to be too firmly established to be shaken by the publication of infidel speculations. In reference to MSS of Sir Isaac Newton which I found ... & which are very heterodox, I had occasion to consult some leading members of the English Church, and also of our own, and I found that they agreed with me in thinking that the opinions of great men are not to be suppressed, however erroneous, with the view of protecting Great Truths which need not fear the assaults of man.[86]

Although hoping to express the certainty of his beliefs, Brewster's anxiety is evident in his need to consult several divines and in his desire not to have to cast his vote on this occasion. In his writings, he continued to assert the authority of Newton as favourable to religion and 'leaves it to be implied that he does not any longer dispute the heterodoxy of Newton's creed'.[87]

In his review, De Morgan described the issue of Newton's Antitrinitarianism as 'a vexed question no more'. Unlike Brewster, he provided an interpretation of the new evidence and concluded that it 'would be difficult ... to bring [Newton] so near to orthodoxy as Arianism'.[88] An admirer of the outspoken honesty of Whiston, he could at least say of Newton: 'we have no doubt, that in his theological opinions, Newton was as uncompromising and as honest as in his philosophical ones'.[89] As with the policy at UCL, De Morgan believed in liberty of conscience and a separation of religion from public life.[90] He kept his own beliefs so private that his wife was unsure where his sympathies lay. The clearest evidence is contained in two letters to his Evangelical mother, in which he defended his right to reach his own conclusions by the application of reason, and to avoid further discussion of the 'painful' subject on which their views so differed. Sophia De Morgan wrote that, although he never joined a particular sect, she thought he respected the Unitarians most, 'as being most honest in their expression of opinion, and having most critical learning'. In a footnote she added that, when discussing the proposal to include a declaration of religious denomination in the census, De Morgan had told her that he would describe them as 'Christians unattached'.[91]

Brewster, however, was an Evangelical whose religious opinions bore 'decided "orthodoxy"'. The experiences of the Disruption – in which he sided with the Free Churchmen who emphasized doctrine and discipline – and, according to his daughter, the death of his wife in 1850 had served to increase Brewster's religious concerns.[92] According to Paul Baxter, another important epoch was heralded by the anonymously published *Vestiges of the Natural History of Creation* in 1844. Hitherto, Brewster and other Scottish Evangelicals had been confident enough in their natural theological outlook to accept new scientific theories, without fearing that they might lead to a conflict with scripture. Brewster had initially been prepared to accept the nebular hypothesis for the formation of stars, but abandoned it when it became 'the basis of mischievous speculation' in the *Vestiges*.[93] By the 1860s, he had serious doubts, and told his daughter:

> It is difficult, if not impossible, to reconcile certain statements in Scripture with what is accepted by many persons as science, and imperfect and unsuccessful attempts to do this are more injurious than beneficial to religion. The only mode of dealing with this matter is to show that the science which is opposed to Scripture is not truth …

An acquaintance remembered that, 'jealous' for the interests of science, Brewster's main defence was to 'put into my hands a list of scientific men of high standing who had avowed their faith in Scripture. [Brewster] referred to this as a token that there was no natural tendency in science to shake the faith of men in the Word of God.'[94] The authority of talented individuals had become even more important to him, and Brewster continued to highlight Newton's genuine belief, despite his rejection of the 'grand old orthodox truths'.[95]

Alchemy

Like Newton's religious unorthodoxy, his interest in alchemy had long been a subject of speculation. Brewster had been forced to tackle the subject in 1831 because William Law, a follower of the mystic Jacob Boehme, claimed that there were copious extracts from Boehme's writings among Newton's papers.[96] In addition, correspondence with Locke, published by Lord King in 1829, demonstrated the interest of Locke, Newton and Boyle in alchemical practices.[97] However, in 1831 Brewster still maintained there was 'no reason to suppose that Sir Isaac Newton was a believer in the doctrines of alchemy'. The topic had again arisen when Brewster treated the life of Tycho Brahe in *Martyrs of Science*. He clearly found the subject distasteful but explained the fact that Tycho was 'misled by its delusions' by reference to the context of the times. He pointed to the financial benefit – thus blaming the neglect of science by Tycho's contemporaries – and suggested that, in credulous times, the wonders of chemical reactions seemed to suggest something magical. However, Brewster wondered 'how far a belief in alchemy, and a practice of its arts, have a foundation in the weakness of human nature; and to what extent they are compatible with the piety and elevated moral feeling by which our author was distinguished'.[98]

Brewster faced incontestable evidence of Newton's interest in alchemy among the Portsmouth Papers. In the *Memoirs*, as in his life of Tycho, Brewster referred to the unenlightened 'taste of the century' and was insistent that it was 'ambition neither of wealth nor of praise' that tempted Newton, Boyle and Locke to alchemy (vol. 2, p. 375), but 'a love of truth alone, a desire to make new discoveries in chemistry' and a laudable 'wish to test the extraordinary pretensions of their predecessors and their contemporaries' (vol. 2, p. 374). However, although he felt he could excuse Newton's research into the transmutation of metals and even the 'universal tincture', he ultimately admitted defeat: 'we cannot understand how a mind of such power, and so nobly occupied with the abstractions of geometry, and the study of the material world could stoop to be even the copyist of the most contemptible alchemical poetry ...' (vol. 2, p. 375).[99]

De Morgan seems to have been uninterested in the issue of alchemy. He did not mention it in his review of Brewster nor in his other articles on Newton. The evidence for Newton's alchemical interests was, however, collated in Powell's review of Brewster, Edleston and Brougham. Here, for the first time, it was suggested that while Newton was at Cambridge alchemy was 'the absorbing passion of his life'. The image Powell summoned up was striking:

> Engaged in such an engrossing pursuit he threw aside fluxions, optics, and gravitation; and, with the glowing vision of the philosopher's stone before his eyes, was blind to all prospects of sublunary fame or distinction, and desired nothing in life but the peaceful seclusion of his laboratory and the uninterrupted enjoyment of the pursuit of the grand arcanum.[100]

Although J. M. Keynes is usually credited with revealing Newton as the 'last of the magicians', and popular writers and broadcasters still suggest it is surprising to find that Newton was an adept, this is a portrait that the modern reader would recognize.[101]

Suppression and Bias in the Memoirs

All of Brewster's reviewers informed their readers that this was a biased biography. Most of them would have agreed with De Morgan's statement that the *Memoirs* were 'very much superior' to the 1831 *Life*, but they again dwelt on Brewster's treatment of controversy. He was hurt by these judgments and, in his letters to Brougham, he fumed about De Morgan, accused the *Edinburgh* reviewer of being 'incapable either of seeing [the book's] defects or intimating what is new and important in it', claimed that the *Times* reviewer 'could not possibly have read my Preface' and said he would not read Biot's review at all.[102] He told Brougham:

> it is an undoubted fact that not one of [Newton's] biographers has found so much fault with him as I have. I have taken the part of Hooke, & Flamsteed in one important case, and in several points both scientific and social I have animadverted[?] upon his opinion and conduct.[103]

He urged Brougham to find another reviewer, for he believed the *Memoirs* had 'not yet been honestly reviewed, that is, by a person who had read it, and was capable of appreciating the value of the many new facts, letters &c regarding Newton'. De Morgan's review, he said, 'has given great offence to some of the leading men at Cambridge, and I hope that some of Newton's admirers will take up the subject in a serious tone'. He even wrote to Peacock with 'hope that some member of Trinity w^d feel it his duty to reply to the Review'.[104] But no defence was forthcoming.

Because of the bias in his account, we might question Brewster's disappointment. He both presented and, on occasion, suppressed his evidence so as to deceive the reader. As shown above, De Morgan accused him of ignoring certain pieces of evidence with regard to the calculus controversy and he was not entirely open about Baily's sources. A similar lack of candour appears elsewhere, for example when Brewster wrote of the Royal Society dispute between Hans Sloane and John Woodward.[105] A letter to Newton in relation to this argument referred to his having called Sloane '*a tricking fellow*; nay *a villain* and *rascal*'. Although this reference was probably correct, since the letter was written to persuade Newton to act against Sloane, Brewster commented in a note: 'Without better evidence than that of a partisan, we cannot believe that these words were in Newton's vocabulary'. To back this point he claimed that when Newton was provoked by Flamsteed, 'he could not command a harsher term than that of *Puppy*' (vol. 2, p. 246). His earlier account of this incident ignored the fact that Flamsteed claimed 'Puppy' to have been the *least* of the insults used by Newton, and chose instead to see this mild insult as evidence of Newton's 'simple-mindedness' (vol. 2, p. 239).[106] Brewster also failed to mention one of the principal sources, previously cited by De Morgan, for the rumours surrounding Newton's niece, claiming instead that the story was 'unknown to any contemporary writer' (vol. 2, p. 281). Likewise, in his account of Newton's breakdown, Brewster at times discounted de la Pryme's story of a fire as referring to events of an earlier period, and at other times continued to use it to back his claim that Newton did not 'run mad' in 1692 (vol. 2, pp. 137–41).[107] He also chose not to include his impression that the various drafts of the Scholium in the *Principia* were 'so expressed that it w^d have greatly offended Leibnitz'.[108] Since it is difficult to be sure exactly what Brewster saw on his visit to Hurtsbourne Park, we can only speculate about other similar decisions. For example, Brewster saw the notebook in which Newton, in shorthand, had confessed his sins in 1662 (vol. 1, pp. 31–4), but we cannot know if he could read that shorthand and, if he could, whether he decided the contents were too shocking or too private to be printed.[109]

Defence of the biographical subject was commonly seen as a virtue in the nineteenth century; one of Brewster's reviewers approved of the fact that the Portsmouth Papers had been used 'in defending Newton against a system of calumny and misrepresentation unexampled in the history of science'. Suppression could be a legitimate part of this defence, and Brewster was prepared to accept 'credit for having withheld some of the letters in my possession'.[110] His choices about how much information to include were subject to the pressure of his own desires and beliefs, but also those of society and, perhaps, those of the custodians of Newton's manuscripts. Henry Fellowes's attitude is indicated in a letter from Ada Lovelace, written after Brewster's 1837 visit to Hurtsbourne, which commented that Fellowes surprised me much by saying that 'of theological papers, only such will be published as are sufficient to *prove that Newton believed strictly* in the Trinity, so as completely to answer the *Unitarians and Deists*, who had hitherto gloried in his authority, & appropriated him to themselves' ...[111]

However, Brewster had more than one standard by which to judge suppression. He approved of Newton's decision to remove his acknowledgment of Leibniz's discovery of the calculus from the third edition of the *Principia*, but accused Leibniz of 'suppression' in not referring to Newton's unpublished method. Similarly, he seemed to accuse Baily and Flamsteed of hiding evidence, but also wished that the Admiralty had suppressed the *Account*. Undoubtedly he considered some acts of suppression morally valid, particularly in instances where Newton had refrained from publishing, while others were deceitful. De Morgan's review implied that those who approved of the publication of the *Account* had a different moral outlook to those demonstrating this kind of double standard. Should the government pay for the publication of Newton's papers, 'Those who were scandalised at the idea of the nation paying for the printing of an attack upon Newton would take it as reparation: while those who entirely approved of the proceeding would as heartily approve of the new measure'.[112]

Newton's Personality in the *Memoirs* and its Reviews

While Brewster continued to claim that Newton's 'noble and generous mind' was characterized by 'equanimity' and a lack of 'personal invective' and 'vulgar jealousy' (vol. 1, p. 86), he also finally included Locke's evidence that he was 'a nice [i.e. tricky] man to deal with' (vol. 2, p. 409). This had been published in 1829 but Brewster had avoided quoting it in the *Life*.[113] It was probably too well known by 1855 to allow a second omission. Brewster wrote that when Locke described Newton as 'a little too apt to raise himself in suspicions where there is no ground',

> he referred to an imperfection of character which we have not scrupled to notice, whether in his controversies with Hooke or with Flamsteed. It would be a sacrifice of truth, and an empty compliment to the memory of so great a man, to speak of him as exempt from the infirmities of our common nature ... It is far from the duty of a biographer, who has been permitted to inspect the private and sacred relics of the dead, to sit in judgement on the failings they may disclose. It is enough that he deals honestly with what is known, and makes no apology for what is socially or morally wrong. (vol. 2, p. 409)

This paragraph pays lip-service to De Morgan, who had 'pointed out more conspicuously than other biographers the failings to which we have referred', but Brewster went on to cite Alexander Pope as an authority who did 'not doubt that [Newton's] life and manners would make as great a discovery of virtue and goodness and rectitude of heart, as his works have done of penetration and the utmost stretch of human knowledge' (vol. 2, p. 410). Brewster's desire to both see the best in Newton and to respond to developments in Newtonian biography – in the use of primary sources and the acknowledgment of Newton's flaws – led to irresolvable contradictions within his text.

For most of Brewster's reviewers, with the partial exception of C. R. Weld, it would appear that the lasting impression of Brewster's volumes was not the eulogized Newton but the novelties, which were beginning to build an altered picture of Newton the man. Brewster wrote of Newton's 'extreme sensitiveness' to criticism and, although this was still referred to as his 'dread of controversy' and 'feeble ... love of wealth and fame' (vol. 2, pp. 1–2), other writers could begin to reinterpret aspects of Newton's life in this light. Brewster wrote to Brougham regarding the breakdown, saying,

> Newton's temper was peculiar, and he was often so thoroughly abstracted in study that he did things that scarcely any man in his senses would have done. It is quite certain that he often neglected to take food, believing that he had taken it, and if we suppose that this happened when his general health was affected, and when nervousness and sleeplessness were combined, it is easy to understand how he wrote his letter to Locke.[114]

While there was nothing this plain in the *Memoirs*, the elements that might create a similar picture in the reader's mind were present. In particular, the breakdown began to be viewed not as an 'aberration' but as another element in Newton's personality.

Since Newton's letters to Pepys and Locke in the autumn of 1693 had been published, it was no longer possible to deny that Newton suffered some sort of illness, although much of the rest of the account originally supplied by Biot was now considered 'apocryphal'.[115] Although writers pointed to proximate causes – insomnia, physical illness, the difficulty of the lunar theory on which Newton was engaged – there was also a tendency to see continuity in Newton's behaviour and similarity in the type, if not the intensity, of accusations levelled at Locke to those against Flamsteed, Hooke and others. De Morgan told the reader to imagine Newton's morbidity as 'the constant attendant of the whole life', so that his 'known exhibitions of it' would not amount to much beside the 'strong self-control' required to 'suppress its effects'.[116] Powell, although he felt Biot had gone too far in insisting that all of Newton's post-1693 works 'betray an enfeebled intellect', felt he should 'admit that the morbid sensitiveness which was a prominent feature in Newton's *original constitution* may have been acted upon to so injurious an extent by bodily ill health and mental labour, as to leave him liable to nervous irritability of mind under peculiar exciting circumstances ...'.[117] From the mid-1830s, in a revival and reinterpretation of the seventeenth-century descriptions of scholarly melancholy, Newton's basic character was commonly considered sombre and abstracted from daily life.

This picture of Newton made one of the novelties introduced by Brewster, the putative love letter from Newton to Lady Norris, difficult to credit. This letter, which appeared to propose marriage to the widow Norris, had been copied by Conduitt and annotated by another hand, 'A Letter from Sir I. N. to —' (vol. 2, pp. 211–13). Brewster was inclined to believe that the 'remarkable epistle' was written by Newton to press his own suit, although he told Brougham that it was 'so unlike [the] production of any other mind that it might be held as a proof of mental aberration'.[118] Brewster's reviewers were not convinced by his interpretation. De Morgan found the letter 'amusing' but considered that there was 'no authority for it coming from Newton'.[119] The reviewer in *The Times* was also sceptical: 'Sir Isaac in love! – it is incredible, it is impossible'.[120] Weld considered the evidence 'so feeble that we are compelled to give the reader fair notice that our great philosopher will not be found "sighing like his furnace" or even playing the part of the lukewarm lover'.[121] Brewster's belief that Newton had written this letter is perhaps indicative of his wish to place Newton in the sociable London world.[122] The reviewers quoted above could not accept his claim because of the lack of proof, but also because the letter could not be squared with their view of Newton's personality. The idea of Newton in love, at least after his youthful affections for Catherine Storer, was a comic creation rather than a historical likelihood (Figure 8). De Morgan saw Newton as 'a man of feeling, right or wrong', who could never have addressed someone he wished to marry in the detached manner of this epistle.[123] The author of the article in *The Times*, however, took the opposite view, writing of 'the negations of Newton's animal and emotional nature'.[124] Powell, although he did not refer to the Norris incident, hinted at a similar image when he suggested that Newton's 'temper was of that negative kind which arose from intense absorption within himself and insensibility to things around him'.[125] Even Weld, in a review that took Brewster's side in all the controversies, could not picture Newton as an elderly lover.[126]

As suggested by the quote above, the negative view of Newton presented by anonymous reviewer in *The Times* was quite different from that of De Morgan, who envisaged a man with a fluctuating temper.[127] Here Newton was described as 'a stoic without the merit of a stoic, for he had no feelings to contend with' yet, inconsistently, this 'vacuum' of a man was jealous of his rivals and a coward, who kept his discoveries for 'his own private satisfaction', feared opposition and 'shunned mankind'. These remarkable judgments seemed to encompass the effects of science itself:

> The only qualities in Newton that were positively unamiable were his suspicious temper and his impatience of contradiction. All else was negative. His goodness even was negative, with the exception of his piety and his veracity. He was good, because he was passionless; and he was not loveable, because he was void of emotion.

For this author, biographies of men of science held little interest because 'the more completely a man devotes himself to science he becomes the less a social being; the

Figure 8. G. Cruikshank, 'Sir Isaac Newton's Courtship', in R. Bentley (ed.), *Bentley's Miscellany*, vol. 4 (London, 1838), between pp. 166–7. Permission British Library (W16/0513 DSC).

less, therefore, a man, and the more a philosophical instrument'. The lives of men of science had 'passed into an algebraical formula' and, with the possible exception of the literary politician Benjamin Franklin, made for inherently dull biographies.[128] All Brewster's attempts to display the knighted, presidential and sociable Newton had little effect on this reader.

The apparently opposing depictions of the sociable and the sombre Newton could be read into contemporary portraits of him. What has subsequently become the most popular portrait of Newton, painted by Kneller in 1689, only became well known in the 1860s (Figure 9).[129] Samuel Crompton described it to the Manchester Literary and Philosophical Society, clearly bewitched by this image of 'the immortal Newton, as distinguished from Queen Anne's Newton', painted 'in the plenitude of his intellectual power'.[130] This was the figure described by Humphrey Newton, whom Crompton quoted at length, and it underlined once again the perception of Newton as the solitary, abstracted and brilliant mind. Joseph Edleston had used a similar image as a frontispiece to his book despite the fact that the letters between Newton and Roger Cotes that it contained referred to the second edition of the *Principia* (Figure 10). This had appeared in 1709 and therefore the later portrait by Kneller would have been more appropriate, but the earlier picture, 'representing [Newton] at a time of his life the least remote from those memorable eighteen months which it cost him to produce the great work that has immortalized his name', proved too tempting to reject.[131] Crompton considered that the 1702 Kneller portrait (Figure 11), used as a frontispiece by Brewster,

> by no means gives a desirable representation of Newton the philosopher. It was rather an affected representation of Newton the dandy, and of Newton the prosperous man of the world, with a carriage and horses, and with three male and three female servants. [Crompton] looked on these prints with pity; and could not, for one moment, allow that any one of them represented that Isaac Newton, the yeoman's son, while at work in the wells of truth, and wresting from nature secrets hidden from the foundation of the world.[132]

It was appropriate that Brewster should have used this image, for he wished to reveal to the reader the successful Sir Isaac. However, he too was taken with the image of the younger Newton. Writing to Edleston in 1855, he explained that he and his publishers were searching for a suitable portrait and he wondered if, 'Failing our getting another', he might acquire a copy of the portrait used by Edleston, 'the finest I have seen'.[133] Fine though it was, Brewster did not use it, and thus displayed 'Queen Anne's Newton' at the beginning of his work. Creating another chronological glitch, the image of the solitary scholar was relegated to the second volume, in an engraving of the Trinity College statue by Louis-François Roubiliac.

Figure 9. Isaac Newton by Godfrey Kneller, c. 1689. Permission The Portsmouth Estate. Photography by Jeremy Whitaker.

Figure 10. J. Edlestone, *Correspondence of Sir Isaac Newton and Professor Cotes* (Cambridge and London: John Deighton and J. W. Parker, 1850), frontispiece. From the personal collection of the author.

Figure 11. Isaac Newton by Godfrey Kneller, 1702. Permission National Portrait Gallery, London (NPG 2881).

Conclusion

In an article on the centennial and bicentennial celebrations of Antoine-Laurent Lavoisier, Bernadette Bensaude-Vincent has argued that both hagiographical commemorations and 'objective' historical accounts 'contributed equally to the fight against the myths attached to the memory of Lavoisier'.[134] A similar case can be made for Brewster's 'commemoration' of Newton. Brewster produced an authoritative, primary source-based account that remained the standard biography of Newton until Richard Westfall published *Never at Rest* in 1980. Although Brewster was criticized for bias, nearly all commentators have been prepared to accept his presentation of manuscript material, as if a factual historical record could break through a partisan narrative untainted. As demonstrated, this is not a safe assumption. Brewster not only attempted to present his material in the best possible light but was also selective in what he revealed. Suppression was sometimes chosen as the most worthy option, as shown by Brewster's correspondence, the *Memoirs* themselves and by subsequent knowledge of the contents of the Portsmouth Papers.

The differences between the 1831 *Life* and the 1855 *Memoirs* were largely dictated by the writers Brewster wished to refute. He approached the archives because he needed to find evidence that could counter the negative image of Newton that was beginning to develop. In addition, primary source-based texts like Baily's *Account* could best be confronted with similar material. The new evidence that Brewster presented was biased towards those areas of controversy that had been identified by Biot, Baily and De Morgan, or that had long been subjects of speculation in certain circles. The nature of these materials was painful for Brewster but he believed their inevitable revelation should be controlled by an advocate. As De Morgan commented, however, the fact that Brewster was so openly defensive in his approach made the impact of the new evidence all the greater.[135] It led reviewers to speculate freely about Newton's personality and to embroider Brewster's reluctant admissions. The more complex figure gleaned from the *Memoirs of Newton* appeared, therefore, to have a stamp of authenticity from the scientific community. However, as the next chapter suggests, this figure always coexisted with a simpler image of Newton the icon.

6 THE 'MYTHICAL' AND THE 'HISTORICAL' NEWTON

> Let a flaw be a flaw, because it is a flaw: Newton is not the less Newton
> ... Augustus De Morgan[1]

In 1858 and 1867 there were two events relating to Newton and his reputation that received thorough coverage in the daily and weekly press. They were of popular interest but were also to receive attention from two experts, Brewster and De Morgan. The first event was the erection of a statue of Newton in Grantham. The second was a literary *cause célèbre* that saw a challenge to Newton's position as the discoverer of the laws of gravitation from the publication of a number of forged documents. The very different attitudes of Brewster and De Morgan will be compared with the reactions of other men of science and a wider public in order to highlight the consistency of outlook from the two biographers of Newton and the relationship of their work to a non-expert perception of him. A close examination of the research relating to Newton with which De Morgan was engaged at the time of these events demonstrates that historians who have viewed it as having a morality and a reverence for Newton that would be more consistent with the work of Brewster have misunderstood his intentions.

De Morgan's bugbears in fact remained those explored in his 1855 review of Brewster's *Memoirs of Newton*. As time went on he became, if anything, more hard-line in his approach and his criticism of Brewster was increasingly severe. De Morgan's last and longest work relating to Newton was not published in his lifetime but the two events on which this chapter focuses gave him the opportunity to repeat his message publicly. Although the two men did not correspond after the publication of the 1855 review, 1867 saw them involved in a communication of sorts, carried out through the pages of a literary weekly and a national newspaper. The concerns of both men harked back to the debates of the 1830s, and, by the late 1860s, would appear to have been of limited appeal or importance to other men of science and the general public. The division between popular conceptions of Newton and those created through critical source-based history had widened, and the affair sparked by the forged manuscripts demonstrates the precariousness of knowledge produced by such histories.

Placing Newton on his Pedestal: The Grantham Statue (1858)

In May 1853, before the publication of Brewster's *Memoirs*, the Grantham town council met to consider granting a site for a monument to the most famous pupil of their grammar school. They resolved to allocate £100 for the preparation of the proposed site and formed a local committee to oversee the project. Their former Mayor, Thomas Winter, was sent to London to 'request the Royal Society to take or suggest such measure for the promotion of the proposed Memorial as they may think proper'.[2] There he was told merely that the President and Council would 'be happy to be further informed from time to time of the progress which has been made in the fulfilment of this design'.[3] This brush-off suggests that the Royal Society, as a body, considered this a purely local affair. National interest was, however, raised and subscriptions from outside Grantham totalled £1,000. The inauguration of the statue by William Theed in September 1858 was attended by eminent men from Oxbridge and the metropolis as well as the immediate neighbourhood of Grantham.[4] Local committees raised subscriptions in several counties and cities and, although the Royal Society did not officially support the project, they were represented on the day by Sir Benjamin Brodie.[5] Also attending was Walter White, assistant secretary to the Royal Society, who was in charge of the reflecting telescope lent by the Society for the occasion.[6] The guest of honour and keynote speaker at the inauguration was Lord Brougham, who had recently, with Edward Routh, published his *Analytical View of the Principia* (1855).[7] In the elaborate parade to the statue before the unveiling, Brougham was flanked by Whewell, representing Trinity College, and Richard Owen, President of the BAAS. As well as the telescope, boys of the Grammar School carried a copy of the *Principia* and Newton's prism. Once the parade reached the platform Brougham was seated on a chair believed to have belonged to Newton (see Figure 12).[8] Grantham gave honour to the distinguished guests and speakers as much as to Newton.

Possibly also attending was Joseph Edleston, as secretary to the Cambridge Committee.[9] Brewster was invited and told Brougham that he hoped to attend.[10] In the event he appears not to have been present, possibly because he was recovering from bronchitis.[11] Certainly one would expect him to have been there, and the *Illustrated London News* in fact included his name in their report.[12] He was glad that Brougham was to play such an important role at the inauguration, passed on suggestions and information to aid the composition of his address and expressed his regret that the Royal Society refused close involvement.[13] He also

Figure 12. 'Inauguration of the Statue of Sir Isaac Newton' *Illustrated London News*, 33 (1858), p. 299. Brougham is in mid-speech: note the prominence of the prism, *Principia* and telescope. Permission Senate House Library, University of London.

thought the occasion important enough to be recorded in the second edition of the *Memoirs* (1860).[14]

The speeches made by the main protagonists of the inauguration drama were printed, together with a biography of Newton, by Edmund King of Clare College, Cambridge, a native of Grantham.[15] These consisted of Brougham's open-air address and those made by Whewell, Brodie, Winter and the master of Grantham School at the lunch held afterwards at the town hall. Both Brougham and Whewell emphasized Newton's debt and connection to Cambridge over that to Grantham, the latter claiming that Trinity men 'feel as if Newton had passed from among us yesterday'.[16] Brodie was able to 'reinforce current campaigns aiming to improve the [Royal] Society's public image', perhaps particularly necessary after their decision not to become involved with the scheme.[17] Winter took the opportunity to thank those, from 'the school-boy to the senator', who had donated time and money.[18] Hoping for the widest possible response, his inclusiveness was also manifest in the initial call for subscriptions, where Newton was presented as a hero for all men:

> All civilized nations may ... claim an interest in Newton, and there is no class of men to whom the proposal to do honour to his name may not be fairly addressed. It recommends itself to the beneficent, for he was one of the greatest benefactors to our species – to the philosopher, for he was, and must ever be, the pride of philosophy – to the scientific, for he is the glory of science. His was no mere flash of genius, but the steady process of untiring industry which the laborious of every class may appreciate. His was not the mere contemplation of wisdom, for he was the most practical of philosophers. The appeal is therefore laid before the great, the wise, and the good, throughout the world.[19]

King's life of Newton, like the speeches, was informed by the publications detailed in the previous chapters. Given the occasion, it is unsurprising that Newton's weaknesses, as detailed over the past three decades, were glossed over. King largely avoided difficult discussion by pointing the reader to the works of Edleston, Brewster and De Morgan on issues such as the calculus dispute, Newton's breakdown and his religious opinions. His most critical comment reflected the reviews of Brewster's *Memoirs*, for he noted Newton's 'impatience of opposition' and disinclination to publicize his discoveries. This was, however, explained as being due partly to 'an intense dislike to disputes and contentions' and also 'a desire to push his researches still further in the same direction'.[20] The Newton both he and the speakers preferred to promote was largely a more traditional and popular formulation.

Naturally De Morgan did not attend the unveiling of the statue. His opinion of the matter is clear from a review he wrote of King's book and Brougham's speech.[21] This article, which queried the propriety of national funding for a local monument, paid little attention to the two works ostensibly under review, merely acknowledging that King's work was 'put together for the occasion, to which it is suited' and stating that the author had:

> nothing to criticize in Lord Brougham's speech. We cannot object to one or two points of biography, as occurring in a speech made for such an occasion. It is sufficient that they are backed by Sir David Brewster, whose book is justification enough for the purpose.

His main gripe was rather different. Much of the commentary on the proposed statue had been to the effect that it was to England's shame that no national monument to Newton had previously been erected. De Morgan was not, however,

> prepared to look upon [the] inauguration as reparation of neglect, or as the proof of an awaked sense of Newton's merits. Newton is one of those whose merits have never been slighted; they were acknowledged in a tone approaching to idolatry during the last thirty years of his life; and every country of the civilized world has recognized a peculiar greatness in his intellect from the day of his death to the present time.[22]

De Morgan pointed to developments in the understanding of Newton's character which he claimed had 'silenced the defenders of what we call the mythical Newton'. Again he rehearsed Newton's weaknesses, but placed emphasis on one that he had not previously addressed. He claimed that, despite Newton's 'real shrinking from publicity, there was a love of fame and desire for public employment' and that 'what in the young man was impatience of opposition' was 'in the old man love of power'. Once again he maintained that 'the scientific glory of Newton ... can stand very hard attacks upon personal character' and that the 'theory of gravitation would not be altered in value, or in opinion of value, though its author were proved to have committed murder and robbery'. As in the earlier review, he expressed the wish that the Portsmouth Papers be examined fully and published, later telling Brougham that the 'monument I should like to see is a publication of *all* the Portsmouth papers. Statues are very local things. If Abraham had lived in our day, he would not have stuck up a stone: he would have advertised in the Times.'[23]

Newton: His Friend: And His Niece (1853–1870): Misreadings and Reassessment

During the period between the first proposal and the inauguration of the statue, De Morgan's research on Newton centred on the relationship between his niece, Catherine Barton, and his friend and patron, Charles Montagu, later Earl of Halifax, through whom he became Warden of the Mint. Barton had lived with Newton in London and in 1717 married John Conduitt, who was to succeed Newton at the Mint. Halifax, who died in 1715, willed Barton a large amount of money, 'as a token of the sincere love, affection, and esteem I have long had for her person, and as a small recompense for the pleasure and happiness I have had in her conversation'.[24] This seemed to suggest that their relationship had been more than platonic, but it was unclear if Barton had been Halifax's wife or mistress, or whether Newton had encouraged a potentially profitable liaison. De Morgan sent two letters on this topic to *Notes and Queries* in 1853 and 1856, the second being provoked by new evidence and Brewster's treatment of the subject in the *Memoirs*.[25] This periodical, as *A Medium of Inter-Communication for Literary Men, Artists, Antiquaries, Genealogists, etc.*, De Morgan considered the 'most appropriate place of deposit for the provisional result of unfinished inquiries' to which others might contribute additional information. In 1857 he drew up a longer article for publication in the *Companion to the British Almanac*, to which he had contributed for over two decades. The article was rejected, but he kept and added to his manuscript throughout the 1860s, and it was published posthumously by his wife as *Newton: His Friend: And His Niece*.[26]

More recent historians have stressed the obscure nature of the question that so interested De Morgan and, because he concluded that the most likely of the various possibilities was that Barton and Halifax had been privately married, his work has been seen as an attempt to 'exonerate his hero' by denying Newton's complicity in an immoral relationship.[27] Rice writes of De Morgan's 'instinctive sense of propriety', Paul Theerman of his 'solitary moral stands' and Richards of his 'moralistic musings'.[28] Frank Manuel believed that De Morgan was 'unable to face the possibility of the chaste Newton's tolerating a sinful relationship between his young friend and patron' and that, in general, 'Victorians could not suffer the intrusion of illegitimate sex in the family of the hero'.[29] E. A. Osborne saw De Morgan's book as an attempt to vindicate Newton, 'whom he admired this side of adulatory', and to avoid the unthinkable conclusion that Newton had allowed his niece to become Halifax's mistress, perhaps to ensure his continuing patronage.[30] It is not easy, however, to square these judgments of De Morgan's motivation with his other work on Newton, his criticism of Brewster and his opinion that Newton's scientific glory would not be diminished if he was proven to be a murderer. If we acknowledge De Morgan's insistence on the 'distinction between intellect and morals' it is difficult to explain his motives in this manner.[31]

It appears that these writers have followed Sophia De Morgan's introduction to *Newton*, which reported that De Morgan considered the question important, as 'nearly concerning Newton's moral rectitude'.[32]

> His intense reverence for Newton would have given way before the clear conviction of a grave defect in moral character. He had been compelled to admit the culpable weakness, to call it by the mildest name, of Newton's conduct to Flamsteed; but a continued countenancing of immorality, by which he was supposed to have gained, would have been, in my husband's view, enough to darken even Newton's intellec-

tual brightness ... The question is not an unimportant one to those who hold moral excellence to be of more value than intellectual power, and who are sometimes apt to believe the possession of the one implies that of the other.[33]

Sophia may have been privy to De Morgan's private anguish over Newton's moral weakness, but this account is utterly at odds with all of De Morgan's written comments, including those in *Newton*. We should not, because of his wife's interpretation, ignore the forceful message that he was actually presenting in this book. At times it appears that previous commentators have not read it in its entirety, or have wilfully ignored much of its contents.

De Morgan: Lord Halifax: And Catherine Barton

If not for the purpose of rescuing Newton's moral character, why did De Morgan spend so much time on this obscure topic? The relationship between Barton and Halifax had been contemporary scandal, referred to in a number of eighteenth-century texts. The 'Life' of Halifax appended to *The Works and Life of ... Charles, Late Earl of Halifax* (1715) reported that, after the death of his wife, Halifax had 'cast his eye upon' Barton 'to be Super-intendent of his domestick Affairs' and that because she was 'young, beautiful, and gay, so those that were given to censure, pass'd a Judgement upon her which she no Ways merited'.[34] Delarivier Manley's *Memoirs of Europe Towards the Close of the Eighth Century* (1710–11), a scandal-mongering political satire, introduced the pseudonymous 'Bartica' as mistress of Julius Sergius, a caricature of Halifax.[35] Voltaire also referred to Barton and suggested that Newton owed his post at the Mint less to his scientific abilities than to the charms of a pretty niece.[36] Although the rumour appears to have remained extant, the issue was raised again in the nineteenth century by Baily's *Account of Flamsteed*. This contained a letter from Flamsteed commenting on the large sum of money Barton had received on Halifax's death, '*for her excellent conversation*', and, in a note, Baily quoted the opinion that Barton 'did not escape the censure of her contemporaries'.[37]

De Morgan first referred to the question in his review of C. R. Weld's *History of the Royal Society* (1848), which had cast scorn on Voltaire's 'flippant conclusion'. Insisting that the investigation of rumours was a 'serious duty' for the historian and that readers should be wary of 'anything on the subject of Newton which emanates from the Royal Society', De Morgan argued that Weld's instant dismissal of the rumour only raised suspicions, while a fuller account would have protected Newton better:

> supposing the worst of Mrs. Barton, nothing concerning Newton can be brought up to reasonable likelihood except that he, in an age when men as strict as himself could not move in public life without tolerating the most open contempt of decency,

received and countenanced a relative whose natural protector he was, and who, whatever she might before have been, was then in a respectable position.[38]

At this date, De Morgan assumed from the will that Barton had been Halifax's mistress, for, if this were not the case, she 'was very much to be pitied and Lord Halifax and his solicitor very much to be blamed'.[39] His interest in the question appears to have been rekindled when he acted as Baily's literary executor in 1852. Baily had made enquiries about Barton between 1836 and 1838, receiving responses from Rigaud, Dawson Turner and Joseph Hunter.[40] As De Morgan wrote in *Notes and Queries* (1853), Rigaud's letters to Baily showed, among other things, that Barton had erroneously been thought to be the widow of a Colonel Barton who was, in fact, her brother.[41] The realization that she was not a widow immediately raised questions about her status, given the recorded freedom of her discourse with contemporaries such as Jonathan Swift.

Previous biographers of Newton had, according to De Morgan, treated the topic 'with the utmost reserve', as if 'they were afraid that, by going fairly into the matter, they should find something they would rather not tell'.[42] De Morgan's intention was to sweep this hypocritical fear aside by presenting 'the only unreserved statement of the existing case which has ever been printed'. Others had, with trepidation, decided to 'Let well alone' and De Morgan admonished them for not having 'shown their faith by a most searching examination'.[43] It was satisfyingly ironic that his conclusion regarding a private marriage indicated that this faith might have been rewarded, although De Morgan was careful to state that 'we are by no means justified in throwing off at once, with disgust, the bare idea of the possibility of a distinguished philosopher consenting to an illicit intercourse between his friend and his niece'. Rather than assuming there could be no impropriety because Newton ought to be morally pure, he argued that because everything we have heard of Newton suggests he *was* pure, 'the *most probable*' conclusion was that no impropriety had taken place.[44] Realizing that this still sounded as if he could not countenance an immoral Newton, he added that, even if we ignore this argument,

> there is that in the weaker part of his character which is of itself almost conclusive. Right or wrong, Newton never faced opinion ... This morbid fear, which is often presented as modesty, would have made him, had he acted a part with regard to his niece which he could not avow, conduct it with the utmost reserve. The philosopher who would have let the theory of gravitation die in silence rather than encounter the opposition which a discovery almost always creates, would not have allowed his *name* to be connected with the annuity which was the price of his niece's honour, or which carried the appearance of it ...

The annuity referred to was described in Halifax's will as having been bought in Newton's name. De Morgan, for several reasons, assumed that it was purchased by Halifax and that he brought Newton's name into the will as a means of attesting his approval of the connection between Barton and Halifax.[45]

De Morgan returned to the subject in his review of Brewster and in the second *Notes and Queries* communication. On both occasions he reasserted his interpretation, which he felt Brewster had misstated by suggesting that his 'respect for the memory of Newton has led him to what he regards as the only conclusion which is compatible with the character of a man so great and pure'. In Brewster's opinion, Halifax's affection for Barton did not exceed 'the love and affection which married men, and men of all ages, ever feel in the presence of physical and intellectual beauty'.[46] Brewster sounded half in love with Barton himself and, after receiving a photograph of her miniature from Jeffrey Ekins, he declared 'Mrs Conduitt is beautiful, and the possession of her likeness adds greatly to the interest I feel in her history'.[47] He argued strongly against a secret marriage, quite fairly seeing no reason why it could not have been made public, and the possibility, which De Morgan considered a certainty, that Barton had lived in Halifax's lodgings. In the review De Morgan was required to maintain his anonymity, but he suggested that Brewster would have avoided the subject entirely if he had not pointed out in his *Notes and Queries* article that such issues are bound to come out in the end 'let biographers be as timid as they will'. He claimed that he devoted space to this minor topic 'because it will enable us to show to every reader the kind of reasoning which can be pressed into the service of biography, when biography herself has been tempted into the service of partisanship'.[48]

In the 1856 letter to *Notes and Queries* De Morgan gave a more personal response and acknowledged himself as the author of the 1855 review. Against Brewster's assertion that his conclusions resulted from his respect for Newton's memory, De Morgan asked:

> When did I ever show any respect for the memory of Newton, in any sense in which respect for the *memory* of the dead means something different from respect for *merit* in the living? Respect for memory, in the sense in which Sir D. Brewster appears to use the words, generally includes willingness to cast a veil over faults for the sake of excellences. Now, of all Englishmen living, I am the one who has most dwelt upon Newton's faults, and most strongly insisted that respect for his memory should not prevent the clearest and fullest exposition of them. I have always insisted that *greatness*, intellectual greatness, should be no cover whatever for delinquency of any kind. And I confidently appeal to those who have read any of my writings on the subject of Newton, whether they will not believe me when I make the assertion following. I say that if I had on close reflection seen reasonable to think Newton had connived at a dishonourable union between his friend and niece, I would no more have been deterred from giving that opinion to the world ... than I would have been deterred from giving evidence that a man had gone down into a coal-mine by my knowledge of his having at another time gone up to the top of St. Paul's.

He defended the reasoning of his 1853 article, pointing out that his conclusions were probabilities, not, as Brewster had implied, positive assertions.⁴⁹

Another reason that De Morgan revisited the subject was because he had come across what he considered 'a clincher'.⁵⁰ This was a letter, shown to him by Guglielmo Libri, from Newton to a Sir John of Lincolnshire, assumed to be Sir John Newton. Writing four days after Halifax's death, Newton excused himself from a visit to Lincolnshire because the 'concern I am in for the loss of my Lord Halifax, and the *circumstances in which I stand related to his family*, will not suffer me to go abroad till his funeral is over'.⁵¹ De Morgan argued that Newton had no reason to add the second part of the explanation – his relationship to the family – unless it was a particularly significant one, comprehensible to a relative such as Sir John. There was no other evidence of a connection between Newton and the Montagu family and, since De Morgan believed that he had already proven that Barton and Halifax had cohabited, this letter, added to the rest of his '*presumptive evidence*', was for him sufficient 'secondary evidence of a marriage for the Courts'.⁵² Although, as he told Herschel, he could not send the evidence to Brewster to gauge his opinion because he 'has never written to me since I reviewed his book', he claimed that 'Macaulay, who used to battle the point, and fought for the Platonics, now says he does not *entirely reject* my hypothesis' and 'Lord Brougham is brought up by it; says it is very curious, and he must think about it'.⁵³

The 1857 article intended for the *Companion to the Almanac* was a much longer consideration of the evidence, setting out 'matters relative to the parties separately, as well as to their connexion' in the hope that these might open new paths of enquiry.⁵⁴ Lengthy asides on the history of marriage law, genealogy, Swift and the early eighteenth-century meaning of the words 'conversation' and 'lodgings' helped add weight to the argument, as did new evidence of Newton's condemnation of his half-nephew, Benjamin Smith. Both Rigaud and Whewell, in the period in which they were attempting to combat the deleterious effects of Baily's *Account*, became interested in Flamsteed's apparent slur regarding Barton's '*excellent conversation*'. When Rigaud began research for an essay, Whewell 'exhorted [him] to poise a lance for the honour of Mrs. Barton', although Rigaud admitted that 'Newton's character is my real object'.⁵⁵ Rigaud approached this topic as Newton's advocate, but there are some interesting similarities to De Morgan's essay, for example in the accumulation of contemporary usages of the word 'conversation'. However, he began by asserting:

> We are not to look merely to the glory, which England derives from having given birth to the author of the Principia; his moral character is a beacon, to which the eyes of the whole Christian world are turned; & to clear off any extraneous obstacle to its brilliancy is to extend the influence of his great & bright example.

His attitude to historical gossip was also the reverse of De Morgan's: 'When tales to the injury of individuals are, at the time of their propagation, left uncontradicted, it is best in general to avoid adding to their circulation by unnecessary discussion'.[56] Writing two decades before De Morgan, Rigaud did not consider the hypothesis of a secret marriage and his essay was a direct confrontation of claims that Barton was the mistress of Halifax.

The two men differed significantly in their acceptance of the evidence of the 1715 biography of Halifax, published by Edmund Curll. While De Morgan called this 'a serious and eulogistic biography', especially since it was appended to the *Works* of Halifax, Rigaud considered Curll generally inaccurate, although in the draft of his essay he deleted the claim that he 'wallowed in scandal & falsehood'. Also, unlike De Morgan, Rigaud argued that the wording of Halifax's will showed there could have been no illicit relationship, for 'criminal love & affection can hardly be called "sincere", and are absolutely contradictory to esteem'. The Victorian prudishness attributed to De Morgan appears far more clearly in Rigaud's (pre-Victorian) essay. The possibility that Newton received the wardenship of the Mint because of his niece, dismissed also by De Morgan, struck Rigaud as involving 'circumstances … of inconceivable atrocity', for at this date (1696),

> the poor girl was only 16 years of age, & yet we are to be called upon to admit that he … whose own conduct & manners were eminently pure from any licentious indulgence, was to pervert the virtue he had himself instilled, & sacrifice to avarice (which he had never felt) the unripened innocence of the child he loved & of whom he had undertaken to be the guardian & instructor![57]

Rigaud declined to publish his essay until Brewster 'shall have published what he thinks fit on the subject'.[58] Writing in 1837, this appeared imminent but the lengthy delays to the *Memoirs* meant that he did not live to see its publication.

Combating Brewster

De Morgan returned to the manuscript of *Newton* between 1865 and 1868, perhaps because, as he began to retire from an active role in the scientific community, he had more leisure.[59] He was also motivated in 1867 by the dispute that flared once again between him and Brewster, detailed in the following section. De Morgan itemized Brewster's 'defective reasoning and want of precision', again showing him to be *counsel* for Newton, not judge of his case'.[60] To demonstrate the latter point, claiming it as an aid to readers of the *Memoirs*, De Morgan included six letters he had received from Brewster before its publication. This he may have decided to do only after Brewster's death in 1868 – and he may never have considered actually publishing them – but he believed the letters showed that Brewster 'knew he was under a self-imposed necessity of defending Newton; not of defending him right or wrong, Heaven forbid! but of finding out that he did nothing but what was right'. Brewster was therefore produced 'against himself', having written that,

> In the matter of Catherine Barton, whatever the view we take of it, Newton, I think, must be held blameless. I cannot adopt the marriage theory, because ... it was certainly the duty of Newton ... to have protected the memory of their relative from the suspicions which the concealment of the marriage has produced.

'That is', De Morgan wrote, 'no theory must be adopted except one under which Newton is blameless'.[61] Brewster's comment could be construed as little different to De Morgan's own argument from Newton's moral squeamishness, but in De Morgan's mind the differences were clear. They become so to the reader when Brewster is quoted as saying that, because Newton's character would have been ruined if his niece had been living in Halifax's house, *'Every means of defence ... becomes obligatory on me as his biographer'*.[62]

In *Newton* De Morgan describes the 'mythical Newton' as the creation of the 'respectable', 'useful', middle class of society rather than the man of science:

> Who does not know the smug individual of this species, as he sees him picking his way through the world? His highest model is aristocracy; his social life is silver-forkery; his main pursuit is money-grubbery; and his whole religion is Sunday-prayery ... This class is, in every case in which its members knew the name of Newton, the one in which you were safe to be reckoned as in the broad way if you imputed anything wrong to the man who bore that name at the Mint ...
>
> 'And, so you think that Newton told a lie;
Where do you hope to go when you die?'

De Morgan clearly describes the kind of attitude that his wife attributed to him in the introduction to *Newton* as hypocrisy:

> I am sure that if I had found what I held to be sufficient proof that Newton had been a Sir Pandarus of Troy, I should not have shrunk from unreserved exposure. I detest the fictitious association, as a matter of course, of moral with intellectual greatness; and I laugh at it into the bargain, as I should at the attempt to prove that all great minds have long noses. There is but one thing of the kind which is more mischievous, and that is the tendency to throw dirt at the morals, or to impute damnation into the destiny of those who differ from the thrower or the importer in religious belief.[63]

His association of the protection of a mythical Newton with religious intolerance suggests the strength of his feeling, especially as this statement was made about the time that he offered his second resignation to UCL.

We can read De Morgan's book as a defence of the need to separate public and private life, a policy which, as at UCL, included the separation of religious practice and instruction from intellectual endeavour.[64] Here, as in other writings, he argued that the biographer should accept 'the fragmented nature' of his subject and that private morals did not affect the intellectual life of an individual.[65] However, despite his insistence that moral and religious issues should be left at home, he chose to explore these very points, apparently driven by his greater desire to expose the hypocrisy of those who venerated an idealised image of Newton and, by extension, of a whole class who, concealing immorality behind a facade of propriety, feared the consequences of honest enquiry. This call for unhampered investigation also meant that the book defended a particular approach to historical research and writing. De Morgan chose here to promote the research of Baily, Peacock and others as the antithesis to that of Brewster. The book was perhaps partly intended as a virtuoso performance in the search for, assessment and presentation of historical evidence.

Newton concludes with the observation that the 'name of Newton, with reference to his discoveries, has been brought before the public in my day in two very remarkable but very different ways'. The first of these was the discovery by John Couch Adams and Urbain Le Verrier of the planet Neptune in 1846. This came at a time when astronomers had 'rather given over expecting anything very great in the future' but heralded a new vigour in the subject. The second brought De Morgan to this manuscript for the last time to add his most damning words regarding Brewster. In a passage written after Brewster's death, De Morgan noted that he 'had a mind in which the prevailing impression acted very strongly upon the facts before him'.[66] He was, in other words, a wholly unreliable and inevitably biased narrator who was unable to judge historical evidence. This book remained unpublished during the lifetimes of both disputants, but the affair of the Pascal forgeries was to provide an opportunity for their conflict over Newton's reputation and the history of science to be aired in public one last time.

'*Newton dépossédé!*': The Affair of the Pascal Forgeries (1867–1870)

'*Newton dépossédé!*' ('Newton dispossessed') was a headline in a Brussels newspaper of August 1867.[67] This article described how Michel Chasles had presented to the Académie des Sciences a number of manuscripts which he claimed gave priority for the discovery of the law of universal gravitation to Blaise Pascal, thus dispossessing Newton of his former glory. Chasles, Professor of Higher Geometry at the Sorbonne, was a respected member of the international scientific community both for his original researches in pure geometry and in the history of mathematics.[68] He had been a corresponding member of the Académie since 1839, becoming a full member in 1851, and was a foreign associate of the Royal Society (1854) and the London Mathematical Society (1867), the former awarding him its Copley Medal in 1865. In the same year that Chasles revealed his manuscripts, he began preparing a report on the history and progress of geometry for the minister of public education.[69] As was said at the time, Chasles's reputation was a strong argument in favour of the authenticity of the remarkable documents, which consisted of notes and letters addressed to Robert Boyle and signed 'Pascal'. The texts were inserted into the Académie's *Comptes rendus* and provoked intense criticism at the following meeting. Chasles's defence was based on the numerous other manuscripts that he claimed to own, including letters from Pascal to Newton. The dispute over their authenticity was to drag on for more than two years and involve scholars from all over Europe. There was general agreement that the manuscripts were forgeries, but little consensus about how this could be proved. Those who joined the debate did so for a variety of reasons, which reflected personal, scientific or national loyalties and differing ideas about the purpose of the history of science.

The Affaire Chasles

Between 1867 and 1869 a large number of Chasles's manuscripts were printed in the *Comptes rendues*, including letters bearing the signatures of Newton, Galileo, Huygens, James II and numerous other historical figures. Each time objections were raised, Chasles arrived at the following meeting of the Académie with a document that appeared to answer his critics. When it was asserted that Pascal could not have achieved the results he was apparently conveying to Boyle without knowledge of the fluxional calculus, Chasles introduced a letter in which Pascal acknowledged his receipt of 'a treatise on the calculus of the infinite' from an eleven-year-old Newton. When it was pointed out that Newton had deduced his laws from Galileo's observations, Chasles produced letters demonstrating that Galileo had communicated with Pascal. Galileo was known to have been blind by the period in question, but at the next session Chasles supplied a letter in which Galileo claimed to have feigned blindness in order to receive gentler treatment at the hands of the Inquisition.[70] It is hard to understand how any of the academicians could have been fooled. Apart from the fact that each objection was answered with another letter, they were all written in French and on French paper. The forger had artificially aged his paper and used archaic forms of language and handwriting, but wrote in what was essentially modern French (see Figure 13).[71]

Some academicians thought the letters must be genuine because of the amount of scientific detail they contained. It was difficult to accept that a common forger would have knowledge of the names, let alone the achievements, of so many natural philosophers. Even when it became apparent that sections of the letters had been copied from textbooks, many still believed that there must have been more than one forger. Support for Chasles was even forthcoming from the *Revue des deux mondes*, the historian Louis Adolph Thiers and Elie

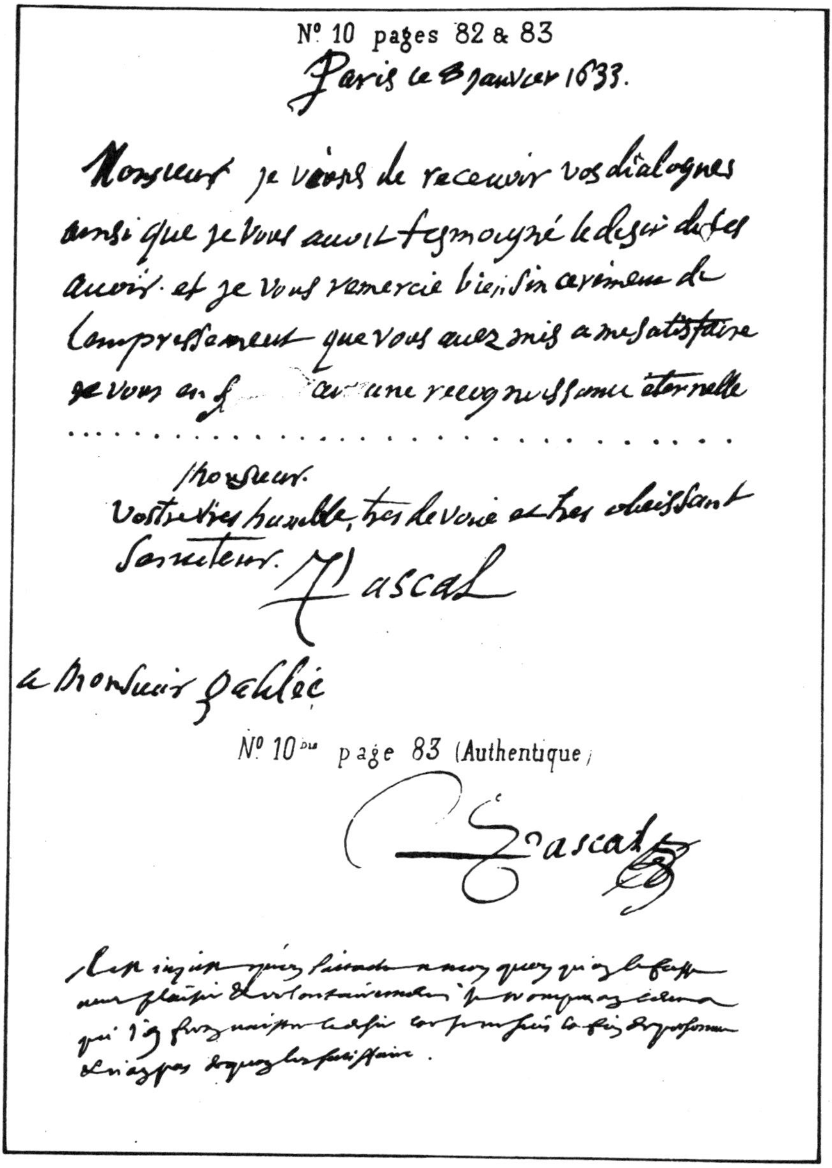
Nº 10 pages 82 & 83

Paris le 8 Janvier 1633.

Monsieur je viens de recevoir vos dialogues
ainsi que je vous avois tesmoigné le desir de les
avoir. et je vous remercie bien sincerement de
l'empressement que vous avez mis a me satisfaire
je vous en [illegible] une recognoissance éternelle

. . . .

Monsieur.
Vostre tres humble, tres devoué et tres obeissant
Serviteur. Pascal

a Monsieur Galilée

Nº 10 bis page 83 (Authentique)

Pascal

Figure 13. Forged (top) and genuine (bottom) examples of Pascal's handwriting. Facsimiles of a forged letter from Pascal to Galileo – 'Sir, I have just received your dialogues ...' – and a genuine fragment from Pascal's *Pensées*, in H. L. Bordier and E. Mabille, *Une fabrique de faux autographes, ou récit de l'affaire V. Lucas* (Paris, 1870), Plate 6. Permission British Library (11900.k.17).

de Beaumont, permanent secretary to the Académie. The last asserted that the letters attributed to Louis XIV were of 'a noble simplicity' and free from the kind of 'inconsistencies that could not have failed to emerge from forgers'.[72] Others, convinced that the documents were forged, sought satisfactory proof: handwriting was scrutinized, content analysed and ink tested. Only the last test could be said to have supported Chasles's position, for it suggested that the ink could indeed be seventeenth century, but Chasles demanded that his accusers prove that the events, meetings and communications he reported could not have occurred. Chasles introduced more and more supporting evidence in the form of hundreds of additional letters from a bewildering variety of individuals, making the construction of a clear proof difficult and adding to concerns about the source of the papers. Commentary in the *Athenaeum* – probably by De Morgan – declared that Chasles was 'seriously compromised by this suspicious appearance, from time to time, of what looks like stop-gaps and trumps shuffled in after the deal'. *The Times* pointed out that this reticence 'would alone have been fatal to their value as evidence in a court of law'.[73]

Chasles's chief critics within the Académie were Prosper Faugère, an expert on Pascal, and Urbain Le Verrier.[74] They pointed to the problems of style and handwriting, scientific data and lack of external evidence. Although he refused to name the dealer, Chasles eventually bent to pressure and passed on the tale that he had been spun regarding the vendor. He was said to be an old man, descended from the Comte de Boisjourdain, with a vast collection of books and manuscripts assembled during the French Revolution – a period during which there had been ample opportunities for collectors of all kinds. The truth was only discovered when, two years after first reading the letters to the Académie, Chasles had the man who sold him the documents arrested. Incredibly, his charge was non-receipt of items purchased, but it was fraud for which Vrain-Denis Lucas was tried in February 1870. It came out in the trial that Lucas spent his days in the Bibliothèque Imperiale, where he researched the individuals and stole the paper and words that formed his forgeries. The judge who sentenced him to two years in prison declared, 'You have abused in the most brazen manner the passion of an old man, of a scholar, his passion as a collector and his love for his country, in order to deceive him shamefully'.[75]

The Pascal forgeries are symbolic of the enormous rise of interest in the collection of autographs and other literary curiosities that had occurred throughout Europe.[76] A new market existed and the unscrupulous could take advantage of the willingness of such collectors to be duped by their desire to possess. The situation was made more tempting to thieves and forgers by the fact that libraries and other collections were frequently poorly catalogued and supervised. Chasles's blindness to the facts was undoubtedly caused by his ardent wish to believe what the manuscripts claimed. It is unclear when he acquired the Pascal letters, but they were an inspired production by the forger, for Chasles was soon preparing a monograph on Pascal's discovery of the laws of gravity. Henri Bordier and Emile Mabille, who took part in the trial and published an account of the affair in 1870, somewhat charitably suggested that Chasles was 'naturally imbued with the desire to prove a thesis, [and] saw only that which agreed with his argument'.[77] Lucas, a provincial autodidact who had been disappointed in his attempts to earn a legitimate living in the libraries and publishing houses of Paris, was an expert in flattering the vanities of his clients. His initial foray into forgery had been the provision of false genealogies to the nouveaux riches.[78]

The British Response to the Pascal Forgeries

Several British periodicals reported the Chasles saga, and its progress can be followed in The Times and the Athenaeum, which, between mid-August and the end of November 1867, carried some thirty relevant letters, articles or commentaries.[79] Discussion thereafter died down until a brief revival of interest in September 1869 when Lucas was arrested. As might be expected, there are differences in the tone of the commentary in the two periodicals. The Athenaeum, which frequently printed miscellaneous scholarly information, including reports of disputes and forgeries, chiefly discussed the matter in its 'Weekly Gossip' column. Its treatment of the affair was in line with its tradition of satire and exposure of literary error.[80] The Times was more straightforward in its reportage and printed readers' letters without commentary, although it tended towards a nationalistic tone, suggesting that the allegation against Newton 'touches our national pride' and should be 'repelled by his countrymen'.[81] Although the commentary in both publications appeared anonymously, some can be safely attributed and letters to the editor were signed.

One of the correspondents was Brewster, who, as Newton's biographer, felt 'called upon' to respond to the French reports. He outlined his objections in letters to the Académie, published in the *Comptes rendus* on 12 August and 30 September, and in *The Times* and the *Athenaeum*. In addition to taking up his pen, Brewster attempted to involve the BAAS and the Royal Society. His last comments on the affair were published on 30 November 1867, only months before his death.[82] On 17 August De Morgan dismissed the forgeries in an article entitled 'Newton Ousted!' in the *Athenaeum*.[83] De Morgan was perhaps best known to the public at this time for his regular *Athenaeum* column on the history of mathematics, the 'Budget of Paradoxes', and most of the commentary on the affair that appeared in this periodical was his. He also contributed some longer articles on the subject and a review of Faugère's *Défence de B. Pascal*. Through the partial anonymity of the *Athenaeum* De Morgan continued to criticize Brewster both for defending Newton and for poor historical research. These comments were probably also among the last words De Morgan published in relation to Newton, for he died in March 1871.[84] Two other British men of science became publicly linked with the exposure of Chasles: Robert Grant, Professor of Astronomy and director of the observatory at Glasgow University, and Thomas Archer Hirst, De Morgan's successor as Professor of Mathematics at UCL.[85] Both men had undertaken much of their scientific education abroad and had attended scientific lectures in Paris, and the latter had close links with Chasles. Thus, although much of the dispute over the forgeries was characterized by chauvinism, the affair also highlights international scientific networks.

It is clear that the issue would not remain a local one, involving as it did the reputations of various national heroes. Just as the British wrote in defence of Newton, so Italian scholars defended Galileo and the Dutch defended Christiaan Huygens. Chasles's status demanded first that he be heard and second that he be granted a considered response. De Morgan was well placed to judge Chasles as 'an historian of science distinguished by research and acuteness'.[86] No one publicly suggested that Chasles himself was the forger and there was, rather, a sadness that he had been so obviously and completely fooled. In addition, by revealing the letters to the world through the Académie, Chasles had tied them to the reputation of that body. Many Frenchmen feared the Académie would be ridiculed and Brewster, De Morgan, Grant and Hirst were all concerned for its reputation. Grant, for example, regretted that 'the Academy had lent the high sanction of its authority to the publication of papers so injurious to the memory of Newton', while De Morgan and Hirst pointed out that few academicians believed Chasles's claims.[87] However, these men each took a different approach to repudiating the forgeries, reflecting their understanding of what values were under attack.

Brewster: 'champion of the illustrious dead'

'As the biographer of Newton, and the only living person to have examined his literary remains, it was natural and appropriate that Sir David Brewster should again come forward as the champion of the illustrious dead'.[88] Such, at least, was the statement of Brewster's daughter, and it is true that his opinion was given as an undisputed expert on Newton's life, although the 'intense earnestness' of his response to Chasles reflects his personal investment in the issue.[89] He continued to view himself as the defender of Newton's reputation and his defence on this occasion was a continuation of that against Leibniz, Hooke, Flamsteed, Biot, Baily and De Morgan. His daughter's explanation of his involvement, written after Brewster's death and while Lucas was in prison awaiting trial, rings true. If the forged evidence was admitted it 'would have tarnished the fame of his noble and beloved master, and broken the shrine of seventy years' fervent admiration'.[90] The letters had to be proved false because they accused Newton of fraud and plagiarism and attacked his character as well as his position as scientific hero. Brewster attributed the forgeries, and the high-profile discussion surrounding them, to envy of Newton's reputation. He suggested that the notion that Pascal had priority over the Englishman was 'highly gratifying to the French nation'.[91]

Brewster's concern for Newton's reputation continued beyond the point at which most British commentators were content to leave the French to deal with their own embarrassments. He could not remain idle so long as any Frenchmen continued to see Chasles as 'the determined champion of France' and Newton as 'a rascally robber'.[92] As well as sending a series of letters to *The Times* and the *Athenaeum* and speaking on the subject at the BAAS, Brewster enabled comparison of the forgeries with genuine documents and made a plea to the President of the Royal Society, Edward Sabine, to create a committee that would 'protect the character of Newton' against these charges.[93] His stated objections to the forgeries demonstrated his familiarity with Newton's life and manuscripts and can thus be seen as the product of a biographer. They successfully lead the reader to conclude that the letters were forged but, interestingly, do not comment on their scientific content. He noted first that the handwriting was clearly not Newton's, and second that there was no mention of Pascal in Newton's correspondence. Third, far from being likely to have carried out a correspondence on infinitesimals and gravitation at the age of eleven, Newton – whose intellectual birthplace was, for Brewster, Cambridge – had been something of an idler with an interest in mechanical models. Fourth, Newton never wrote in French, but communicated with the international scientific community in Latin. Fifth, the letters name his mother Anne Ascough Newton when, after her remarriage, she was called Hannah Smith. Sixth, there were phrases in the letters that were unlikely to have been used by an Englishman. Finally, the letters refer to an individual in Paris being in communication with Newton, for which there was no evidence. Brewster's opinion of the handwriting was formed from memory and so he enlisted the authority of the British Museum's Sir Frederick Madden and the two principal private holders of Newton's manuscripts, the Earl of Portsmouth and Lord Macclesfield, by requesting and quoting their opinion of the letters. All agreed that, in Madden's words, the letters were 'a palpable fake'.[94]

Brewster rather strangely came to believe that the forgeries were the work of Pierre Desmaizeaux (*c.* 1672–1745), who, according to evidence produced by Chasles, had at one time possessed them. Brewster's reasoning seems to have been based on his belief in a Continental conspiracy against Newton's reputation and his poor opinion of Desmaizeaux's character. It was, however, a suggestion that ignored errors in fact and language that Brewster had already highlighted, especially since Desmaizeaux had lived in England for fifty-three years.[95] This was the opinion that Brewster maintained in his last letter on the subject, although he might have revised his opinion if death had not intervened. Given his passionate interest in Newton's reputation and his outspokenness in countless other disputes, it is likely that he would have maintained his interest in this controversy.

De Morgan: A 'budget of deceptions'

De Morgan's response to Chasles's revelations differed in motivation and resonance to that of Brewster. That his commentary on the subject appeared anonymously in the *Athenaeum*, for which he wrote his humorous though sometimes barbed 'Budget of Paradoxes', suggests that he wished to downplay the significance of the forgeries. This was made explicit when he noted that they included 'a bit of mathematical reasoning, supposedly by Newton, not surpassed by any of the paradoxers in the Budget', and predicted 'some amusing end to this budget of deceptions'.[96] De Morgan did not feel personally attacked by the accusations against Newton, as Brewster clearly did, nor did he feel that Newton required a defence. Indeed, he felt that Brewster's approach was dangerous, suggesting that

> he may end by doing what M. Chasles certainly will not do, that is, by giving a moderate number of half-informed persons in England, against all common sense, a notion that the Hannah-Smith-signing-herself-Miss-Anne-Ascough-Newton papers are really things to be seriously argued against.[97]

De Morgan insisted that 'the French *savants*' (as opposed to the credulous, who he classed as '*sous-savants*', or the sensationalist press, dubbed '*sous-journaux*') 'make very light of the Pascal papers' and that all scientific Frenchmen should not be condemned because one was making a fool of himself. Equally, however, he was concerned about Chasles, blaming his poor judgment on illness and telling Herschel, 'He and I are friends of some standing: but it is not the time to mince matters, and plain speaking will be kindness'.[98] De Morgan had little patience with those who stood on the fence or gave the documents more dignity than they deserved. His involvement became even more personal when his friend Libri was accused of being the forger, the suggested motive being ridicule of the Académie, 'to show how easily the French are deceived'.[99]

This was an ideal opportunity De Morgan as well as for Libri to ridicule the assumptions and errors made by some academicians, and to show off their own knowledge. In a letter to Brougham, another Libri advocate, De Morgan pointed out the difference between Libri and the other academicians, calling him 'a *scholar* among the *technologists* of the Institute'.[100] He suggested that one reason why the influential François Arago had become Libri's adversary was that, during their disputes, Libri had been able to demonstrate his wide knowledge, while 'Arago was, in history, of very little depth and very little judgement'.[101] Interestingly, the one person in the Académie that De Morgan felt might have understood and appreciated Libri's talents was Chasles:

> The only *scholar* among them, Chasles, the only man who had any chance with Libri (*not much*), was kept out of the Institute for many years – and in fact until Libri was removed. When I say scholars, I mean having a scholar like knowledge of math[cs], its history, & literature.[102]

Thus for De Morgan the debate over the forgeries was central to the status of the history of science and demonstrated the importance of such knowledge. In private he also admitted his concern for the trustworthiness of Chasles's historical studies:

> What a miserable mess has been made by Chasles, Lucas, and Co! I am obliged to give up Chasles until he clears himself, which I have small hope of his doing. The different accounts he has given at different times are such as must be reconciled, or otherwise explained. If there be no explanation except sub-human credulity, then arises the question which is so important in lunacy inquiries, *When* did this defect begin? For Chasles has a lifetime of memoirs full of references to MSS., many of them unseen as yet except by himself. It will be unsafe to quote him – at least to a *better-not* extent.[103]

In his article, De Morgan firstly declared that both he and the majority of *savants* 'smile at the newspaper idea of Newton's being *dispossessed*'. Attacks on Newton's status occurred because of ignorance and would continue to occur 'so long as his claim is stated in the incorrect way that prevails both at home and abroad'. As he says, 'When the newspaper writer dispossesses Newton by the help of two sentences of Pascal, it is because those two sentences contain *all he knows about Newton*'. Historians, however, understood that, while Newton owed much to Kepler, Pascal, Fermat, Huygens and others, he had demonstrated what others had guessed and developed these ideas to a different level of sophistication.[104] De Morgan thus offered his readers a brief lesson in the history of science and discovery: 'These newspaper mares'-nests are useful', he said, 'they tend to call public attention to exact specification of discoveries, of which there is, we believe, not one which was not preceded by some hint, guess, conjecture, surmise, *aperçu*, or conceit'.[105] His remaining points are the product not of the mathematician or biographer of Newton, but the scholar and bibliophile. He pointed out, for example, that the style of writing was unlike that of Pascal, that there is no suggestion of communication with Pascal in the five-volume edition of Boyle's works, and that Boyle's known references to Pascal are distant. The final point, to which De Morgan devoted most space, is the kind of obscure historical detail with which he loved to entertain himself and his readers. One of the letters referred to coffee, in a manner suggesting familiarity with the substance, but was dated 1652, the year when coffee was first said to have been introduced into England and was still a rarity. De Morgan noted with amusement that a 1670s coffee-house had been established by an Armenian named Pascal, and that the merchant who first introduced coffee to England 'brought with him a Greek *named Pasqua*, who knew how to roast it and boil it'.[106]

It is consistent with the light tone of his comments in the *Athenaeum* that De Morgan did not search for further evidence of forgery, either in the form or content of the letters. As time went on he continued to report on and ridicule the saga but was adamant that no evidence in favour of the papers could make up for the initial errors. He felt that the forgeries would be dignified by further analysis, for they could be adequately swept aside with simple points such as the 'Anne Ascough Newton' error.[107] By repeating a few, obviously erroneous, details he attempted to steer a course through Chasles's obfuscations and the needless complications introduced by Brewster. It was the latter who subsequently provoked De Morgan to his most hostile attack.

Grant and Hirst: The Men of Science

That Brewster and De Morgan each had a very particular view of the affair becomes evident when the other objections to the forgeries are analysed. An examination of the scientific and mathematical content of the letters was not produced by a British writer until Grant wrote to *The Times*. He was motivated to write because the 'direct tendency of the numerous statements contained in these Pascal documents is to degrade Newton from the high position he has heretofore occupied as a natural philosopher and a mathematician'.[108] Grant's authority came from the recognition he had achieved for his *History of Physical Astronomy*, which placed Newton in his historical context and saw his work as a product of his time.[109] However, he was particularly annoyed by the fact that Newton 'is presented to be constantly sitting at the feet of Descartes ... concealing, however, his obligations to the French Philosopher'.[110] Grant, like Brewster, wrote in defence of Newton, but focused on his scientific position rather than his moral status. He owed much of his scientific development to the Académie, and worried that its reputation was suffering as a result of the protracted debacle. In addition, while in Paris he had attended lectures by Le Verrier, who was now one of Chasles's most vociferous opponents. The immediate cause of Grant's intervention was the conclusion, reached by the committee appointed by the Académie to investigate the forgeries, that the available evidence was not 'capable' of producing a definite opinion regarding their veracity. Grant believed that few concurred with this and he pointed to the work of Le Verrier, Brewster, De Morgan and Faugère as providing 'the most conclusive evidence that they are not genuine'.[111]

Perhaps despairing of what proof the Académie would accept, Grant offered something concrete, calculated to appeal to these 'technologists', and based on 'certain numerical results contained in these documents which do not appear to have been yet subjected to examination, "although they are capable" of furnishing a decisive test in the investigation of the question of the authenticity of the documents'. He showed that the details provided in the Pascal letters regarding the densities of the Sun, Earth, Jupiter and Saturn were 'pure forgeries of the corresponding numbers contained in the third edition of the *Principia*'. Thus Newton, in the previous two editions of 1687 and 1713, was supposed to have provided less accurate data than Pascal had achieved in the mid-seventeenth century, while the 1726 edition had provided exactly the same figures. In a detailed and carefully argued piece, Grant explained that Newton's results had depended on the observations of first Flamsteed and later the more accurate data of Giovanni Cassini, James Pound and James Bradley. The most accurate observations available in Pascal's time were made by Huygens and, using these, Grant calculated the densities that Pascal ought to have been able to achieve. These very different results pointed to the conclusion that 'the numbers communicated by M. Chasles to the Academy of Sciences must be pure forgeries'.[112] Commentators in *The Times* and *The Athenaeum* welcomed the clarity of this letter, and the latter declared that it 'contains an absolute proof that the whole of the documents in question are forged'.[113]

Like Grant, Hirst had closer connections with European science than most of his contemporaries. He had stayed in Paris during the winter of 1857–8 and attended the lectures of Gabriel Lamé and Chasles. His links with Chasles remained strong; in 1865 he had campaigned to ensure that he received the Royal Society's Copley Medal and he acted on the Frenchman's behalf at the BAAS. Hirst also visited Chasles in Paris and watched his performance under attack at the Académie.[114] However, although Hirst felt loyalty towards his former teacher, he could not support him over the forged documents and, as De Morgan told Herschel, he was 'trying to get Chasles to give up his authority'.[115] At the meeting of Section A of the BAAS, Hirst said that, if true, what the documents claimed was 'simply astounding' and, 'to say the least of it, very difficult to believe'.[116] However, he did allow them the dignity of careful investigation and underlined Chasles's good faith by showing that he was 'desirous of submitting his newly-acquired papers to every possible test'. His letter to *The Times*, which appeared on 1 October, describes how he went to the Royal Society where they examined the Newton manuscripts. He described the results of this investigation as 'perfectly conclusive' but, in addition, they came across the 'annihilating fact' that passages of the letters had been copied from a work by Desmaizeaux.[117] However, even after this Hirst requested a Royal Society grant to cover the cost of photographing the letters, with which Chasles had entrusted him, in order to compare them with genuine manuscripts.[118]

In his journal Hirst recorded the interest shown in his investigations and the help he was given by a number of individuals. He was aided by Lady Lubbock, who copied genuine letters in the British Museum for comparison, and had an interview with Edmund Bond, keeper of manuscripts at the British Museum, with whom he compared genuine and fake letters of Leibniz. The following week he was in 'the City making enquiries' on the subject.[119] The forgeries were of general interest to London society that season and Hirst found himself having to report the latest details to the members of the Philosophical Club and later showed the photographs he had commissioned to the guests at a dinner party. He was even almost disturbed whilst taking a Turkish bath by the architect Owen Jones, who was 'much interested in the Pascal-Newton question' and desired to talk with Hirst. Jones had to be turned away by the proprietor of the baths.[120] However, after Christmas 1867, interest seemed to wane and Hirst did not return to the matter until he once more visited Chasles, in the summer of 1869. He noted then, with sadness, that Chasles was 'still combating in the Academy about his documents. He has now however a terribly severe opponent Le Verrier and meets with no mercy.'[121]

Grant and Hirst had reasons to be involved in the debate over the forgeries that bore little relation to the reputation of Newton. Hirst wished to find a speedy resolution to an issue that caused him pain and embarrassment on behalf of his former teacher. Grant was interested in the perception of the Académie, and his links were to one of Chasles's principal adversaries. He certainly admired Newton but his historical work emphasized his position in relation to forebears, contemporaries – including observers like Flamsteed – and followers: a scheme that the forgeries threatened to upset. Both Grant and Hirst produced clear evidence against the forgeries in a single letter. Although Chasles brought forward documents that attempted to answer their points, they did not feel the need to return to the subject as Brewster did. Nor, in this context, were they, like De Morgan, concerned with the status of research in the history of science. Like Grant, De Morgan wanted people to understand Newton's real significance, but he was far more concerned by the fact that the documents were treated as genuine by some whom the public considered expert in the field. Brewster had failed miserably in this test of historical capability and thus had to be exposed.

'Brewster and the Athenaeum'

The debate over the forgeries gave rises to a second dispute, in which De Morgan and Brewster continued the disagreement that went back at least to the former's review of the *Memoirs of Newton*. By November 1867 De Morgan felt that 'the time is past for serious opposition' to the forgeries themselves. Instead he moved on to criticize the two individuals, apart from Chasles himself, who were keeping the issue alive – the Abbé Moigno, editor of the *Revue des deux mondes*, and Brewster. The former was dismissed as supporting Chasles for patriotic reasons and with very little skill. He wished rather to counter Brewster's letters, for he was particularly irritated by Brewster's 'weak' suggestion that Desmaizeaux was the forger.[122] Brewster responded by writing a letter to *The Times* that complained about the 'anonymous but well-known writer in the Athenaeum' and his 'very peculiar tone'. In attempting to clarify his argument, Brewster revealed himself to be almost as muddled as Chasles, accepting evidence from some of the letters if they supported his theory against others. His letter ended with the hope that the dispute would not be carried on 'with jokes, gibes, and sneers flashed in meteoric showers in the faces both of friends and foes'.[123] Naturally De Morgan was quite content for things to proceed thus and claimed that the whole scientific world, with the exception of Chasles, Moigno and Brewster, shared the joke.

In an article entitled 'Sir D. Brewster and the *Athenaeum*', De Morgan noted Brewster's sincerity and enthusiasm throughout forty years of work on Newton's biography, but claimed this was accompanied by 'a hasty temperament' rather than 'judgment and accuracy'. He noted the 'peculiar rashness' of Brewster's mind, and called him 'a reckless and inaccurate writer who cannot learn his own faults' and 'who bungles everything'. Brewster was charged with every crime against sound historical method, including minor inaccuracies, making sweeping claims from dubious evidence and inability to make useful comparisons between sources: 'Sir David now maintains that the forgery is more than a century old. The rest of the world seems to be tolerably unanimous in believing that the forgery is now in the course of manufacture from week to week.'[124] De Morgan continued the theme in private, telling Herschel 'Brewster is the king of slapdashery'. While he had mitigated his condemnation of Chasles on account of his old age and illness, he was unforgiving of the even more elderly Brewster: 'He is now very old, and writes without any thinking: he never wrote with much'. With some hyperbole he claimed that in his lives of Newton Brewster had 'made the two greatest biographical mistakes of our day'. The sins of misattributing the author of a letter and miscalculating dates were perhaps serious enough for De Morgan, but worse was the fact that Brewster was 'not very candid' in dealing with matters such as the relationship between Barton and Halifax.[125] For De Morgan, Brewster went about the celebration of Newton's life and achievements in an entirely misguided manner. He felt that there was virtue in honest investigation and had long hoped that these were now 'discriminating days, which insist on the distinction between intellect and morals'.[126]

Conclusion

Debates about Newton's status as a scientific hero continued into the 1860s and De Morgan in particular became increasingly critical of the continuing tendency to mythologize. In 1855, he had suggested that times were changing and that ties between intellectual brilliance, morality and orthodox religion were no longer assumed. In many ways he was right, for even Brewster had been forced to accept many details regarding Newton's life that he found difficult.[127] As minute researches, like De Morgan's on Newton's niece, were prosecuted and an increasing amount of manuscript material was traced and published, a more nuanced picture of the past and its heroes was developed. For De Morgan such research could be an end in itself, containing its own morality tales and exemplars. The occasion of the unveiling of Newton's statue, however, demonstrated that, in terms of popular presentation, little had changed. The image of Newton, separated from a historically sophisticated account of his life and character, remained idealized. Newton's statue at Grantham was intended to be both an offering of thanks from and an inspiration to 'the great, the wise, and the good', but also the boys of Grantham School and the ordinary, 'respectable' people of Lincolnshire.[128]

Although, by the 1860s, a great deal was known about Newton, his life, his works and his character, this was to remain the preserve of historians of science. Manuel suggested that Baily's and De Morgan's '[d]isconcerting revelations about Newton's personality ... were, after an initial shudder, forgotten by most of their contemporaries'.[129] If their work was forgotten it was largely because it was not of general interest and was irrelevant to the understanding of Newton that was, for non-historians, worth perpetuating. By 1861 Whewell thought the editor of the *Quarterly Review* unwise 'to load his pages with long discussions of such stale subjects as Newton and Shelley. Readers are weary of them.'[130] He was referring to a tardy review of Brewster's *Memoirs* by George Hemming, who, noting the difference between the common opinion of Newton and the more nuanced figure that had been created by historians, enquired into the 'laws which govern the award of fame'. He suggested that common opinion regarding Newton was influenced by the fact that his discoveries were made within a sphere of awe-inspiring grandeur, by national prejudice and by the 'all-pervading power of theological sentiment' which 'works with facts, or in spite of facts'.[131] This was contrasted with the 'sceptical' view generated by 'Paradoxical thinkers', presumably De Morgan, who had attempted to 'explain away the great pre-eminence of Newton' and elevate the claims of his contemporaries and predecessors.[132]

The process of the secularization and professionalization of science meant that Newton's importance was increasingly limited to his scientific role and that concerns about his social, religious and moral standards were of less obvious relevance. It is emblematic that Hirst was one of the founders of the X-Club, the members of which promoted the separation of science from natural theology. Likewise, in his famous Belfast Address of 1874, John Tyndall, also of the X-Club and a close friend of Hirst's, dismissed Newton's theological work and opposed those 'theologians' who 'found comfort and assurance in the thought that Newton dealt with the question of revelation'.[133] Despite this, he retained Newton as an iconic figure, available for rhetorical purposes. In 1870, he told the BAAS of 'Newton's passage from a falling apple to a falling moon' as 'a leap of the prepared imagination'.[134] It did not matter that Brewster and De Morgan had cast doubt on the story; Tyndall wished to illustrate his understanding of scientific discovery.[135] He perpetuated a Romanticized image of Newton, which was more or less unconnected to the picture that had been revealed over the previous decades, as part of a secularized treatment of the progress of science that was indicative of the change in debates about science and religion from the 1860s.[136]

The affair of the Pascal forgeries encouraged a challenge to and defence of Newton's scientific reputation in the British press. The issue raised public interest but, although the audacity of the forgery captured the imagination, it was of only passing significance. The letters were easily seen as forgeries and were effectively dismissed by the letters of Grant and Hirst. That the issue was kept alive in Britain even until 1868 was due largely to Brewster and De Morgan. For the former the attack was, as before, perceived as one on Newton's morals as well as his scientific reputation. The octogenarian Brewster was revisiting concerns of the 1830s and once again felt it his duty to shore up the defences. De Morgan, provoked by Brewster, revisited an attack on his historical writing that had lasted over a decade. His 1846 biography had opened by declaring that in matters of 'opinion' he differed from Brewster, 'as well as from those (no small number) whose well-founded veneration for the greatest of philosophical inquirers has led them to regard him as an exhibition of goodness all but perfect, and judgement unimpeachable'.[137] He believed that this number had diminished but did not consider the battle won.

CONCLUSION

The debates that surrounded writings on Newton from the 1820s to the 1860s were the result of the interest of an identifiable number of individuals and the availability of relevant manuscripts. They also show that interest in his moral character was the product of a period in which the structures of science were increasingly specialized, secularized and even professionalized. The preceding chapters illustrate Newton's recruitment in defence of a variety of positions, particularly when these were threatened by change. As supporters of the corpuscular theory of light, both Biot and Brewster were in an increasingly isolated position when they highlighted Newton's use of that hypothesis. More positively, Brewster also used Newton's biography in campaigns for government-funded science and against over-large claims for 'Baconian' methodology. However, because Newton had come to represent both Anglican and theoretical science, Baily and his supporters found an alternative hero in Flamsteed, who could be made to characterize their vision of the modern scientific labourer. The analysis of writings on Newton from this period has therefore proved a fruitful means of examining individual strategies and positioning within the scientific community at a key period in the development of modern science.

Newton's scientific and symbolic importance sparked research into his life, but it was the results of this research that generated a period of sustained debate. This was largely concluded by the publication of Brewster's 1855 biography, which, despite many similarities to his 1831 *Life*, provided evidence to confirm the truth of many of the suspicions regarding Newton's character. At least some details of his quarrelsome, secretive and suspicious nature, his interest in alchemy and his Antitrinitarianism were available to the interested reader. It should be remembered, however, that many other texts written at this time continued to propagate a less complex and more idealized image of Newton. As suggested in the final chapter, the depictions that were informed by original research or by knowledge of innovative scholarship coexisted with more traditional accounts of Newton's life. The texts examined in *Recreating Newton* represent the aims and struggles of authors who either chose or were forced to confront the sources relating to Newton's life. They recognized each other as co-creators of a new narrative of that life and, in footnote references and private correspondence, as fellow contributors to a developing field of the history of science.

The extent to which this new knowledge of Newton's life was appreciated by a wider public is a question that this book has not attempted to answer. While some knowledge can be gained regarding readership of a number of these texts, this reveals little of the impact and general acceptance of their novelties. It seems clear, however, that more traditional depictions of Newton, not least in Brewster's 1831 biography, were always more familiar to the general public. The modern interest in the darker side of Newton, suggested by media coverage of recent books and television programmes, demonstrates that details of Newton's unorthodoxy, interest in alchemy and personal peculiarities failed to become common knowledge either in the nineteenth century or the twentieth.[1] Therefore, although some of the writers discussed aimed to influence popular opinion, their works tell us more about the earliest origins of a field of history of science than about how the image of Newton was changed in the mind of the average Victorian reader. Beyond this, these texts, by and large, reflected rather than influenced changing perceptions of the man of science.

The selected group of writings has revealed the increasing interest in publishing and comparing archives relating to Newton, forming an important contribution to our understanding of the history of biography and the historiography of science. The publications of Baily, Rigaud and Edleston were heavily based on manuscript sources, Brewster changed his biographical approach with regard to such material between 1831 and 1855, and De Morgan emphasized its critical analysis. However, the interest in Newton's personality took this development down two paths. While for Baily and De Morgan using original sources was a means of critiquing traditional authority, Rigaud, Edleston and, especially, Brewster were looking for a convincing means of defending Newton once these original accounts had been found lacking. In both cases writers derived authority from an empiricist approach to the historical remains. While Brewster's biographies aimed to impart a clear moral message, those who sought to undermine traditional narratives of this kind also appealed to a conception of morality. This was invested in the historian and reader rather than in the subject, being attached to claims about their 'impartiality' and willingness to accept whatever the records revealed.

Most clearly in the case of Baily and Rigaud, this 'impartial' approach echoed their scientific methodology. An emphasis was placed in the writings of both on the processes of collecting and recording data, suggesting the reliability and even replicable nature of their findings. This approach, and the link to their meticulous style of work in practical astronomy, was made explicit not only by the authors but also by sympathetic commentators. This point is clearly worth further attention. It indicates first that the scientific method promoted by these men had become successful in creating trust within a wider community than the select world of astronomy. Second, this case also highlights a conscious attempt to produce 'scientific history' at an earlier date than historians have generally recognized the existence of such a trend.[2] It seems likely that scientist-historians would have been particularly alert to the possibility of producing history derived inductively from archival 'facts', although it seems less likely that they should subsequently have had a direct influence on general historians.

By highlighting the use of historical techniques in nineteenth-century biography, this book emphasizes an alternative means of understanding this form of writing to that of literary critics. De Morgan's work goes furthest to demonstrate that in this period biography was not invariably hagiographical, a viewpoint echoed in the commentary of writers such as Galloway, Malkin, Powell and Baily. Regarding both sides in the debates over Newton's reputation, the appropriation of historical techniques to life-writing was practised to a greater extent than has been hitherto appreciated. These methods were used to produce an authoritative, 'virtuous' and apparently impartial text, although they could support very varied positions. Because they believed in the importance of establishing their case, these men were driven to produce high-quality historical work that formed the basis for all future examinations of Newton's life. While it has been suggested that biographical approaches to the past tend to privilege the role of the individual in scientific progress and reveal concerns about personal morality, this book has shown that the writings of Rigaud and Edleston, which were more clearly 'historical' than 'biographical', involved the defence of the privileged position granted to Newton. Conversely, while approving this kind of detailed, source-based history and claiming that Brewster was 'too much of a biographer, and too little of an historian', De Morgan did not eschew the biographical genre.[3]

In considering the views of those individuals who contributed to these debates, the diversity of those who supported the reform of scientific and educational institutions in this period has become apparent. Activity in this area was evidently not a criterion by which an individual's opinion of Newton's character can be judged; nearly all of those I have examined can be identified with attempts to reform the Royal Society or to introduce new subjects into university curricula. Even the party political differences between those who defended or criticized Newton can be hard to locate. It seems unsurprising that the Tory Rigaud should have been alarmed by attacks on Newton and we can also point to the responses of the conservatives Whewell and Edleston. However, we find the reformist Whig Brewster as a strong supporter of the idealized image of Newton. Similar complications can be found when considering educational background. While the Flamsteed debates indicate differences between those with sophisticated mathematical understanding, usually Cambridge educated, and the RAS practitioners, De Morgan is an obvious example in breaking this pattern, as is W.

H. Smyth, a 'Scientific Serviceman' who regretted the damage done to Newton's reputation. Thus it appears that, while they could be important in dictating positions within these discussions, these were not the crucial factors.

It is because Newton had most frequently come to symbolize the link between Anglicanism and science, through the British tradition of natural theology, that I have suggested that the most reliable determinant of attitudes towards the historical figure of Newton was the individual reaction to this connection. Because natural theology existed to support the status quo as much as science, there was typically a political dimension to this position. Thus staunch establishmentarians like Whewell and Rigaud feared the consequences for Church and State if Newton were toppled from his pedestal. Edleston, whose book supported Newton and Cambridge, seems to have experienced doubts about the doctrine of the Trinity when he was young but the fact that his politics were conservative and that he remained at Cambridge, took orders and subsequently accepted a living from Trinity College suggests that he put these behind him. Brewster, of course, belonged to a different tradition and experienced a secession from the established Church, but he belonged to the Evangelical Free Church of Scotland, which saw itself as upholding the importance of doctrine and revelation. For Brewster, critical issues included the doctrine of the Trinity, making his ultimate acceptance of Newton's position particularly painful. For many other commentators the revelations about Newton were regretted because they threatened the assumption that intellectual merit and personal morality were connected.

On the other side of the dispute, De Morgan, if he adhered to any group, was a Unitarian, and the metropolitan circles in which he moved, especially at UCL, the SDUK and the RAS, included many who, if not actually Nonconformist, were advocates for the removal of obstacles for Dissenters. Baily can clearly be linked to this position, and was, or had been, close to a number of important Unitarians, including Priestley and James Martineau. The discussion of those who received his *Account* favourably suggests similar patterns, strongly linked to a middle-class and reformist constituency. Revealingly, Baden Powell's journey from High Church to religious tolerance and a rejection of natural theology was accompanied by a radical change of attitude towards the historical figure of Newton.[4] Brougham's position in the debate is somewhat difficult to gauge. He evidently idolized the scientific hero, and presented a eulogy on the occasion of the unveiling of the Grantham statue, but his review of Lord King's *Life and Letters of Locke* and his correspondence with De Morgan demonstrate his willingness to discuss the most problematic issues regarding Newton. His religious opinions are likewise attached clearly to neither camp. Although a promoter of natural theology, he was a religious rationalist who advocated tolerance and non-sectarian education.

The reception of Baily's *Account of Flamsteed* suggests that some radicals and Nonconformists questioned Newton's status and were willing to replace him with an alternative hero who was, ironically, an orthodox minister. John Britton suggested that there should be an acknowledgment of the democratic distribution of talent and that Newton should not be allowed all the laurels, while Richard Phillips objected to the dominance of the Newtonian framework in general and welcomed any weakening of its power. Unsurprisingly, in a study primarily concerned with the views circulating within scientific communities, complete rejection of the concept of scientific heroes barely arises, although this certainly existed in some circles, if less strongly than at the beginning of the century. De Morgan queried how such heroes were celebrated but his aim was to make Newton available to different interests. His own celebration of Newton's science and his attempts to divorce this from questions about religion and morality suggest that, although Theerman saw him as the guardian of the 'scientist as private man', he was in fact ensuring that Newton could be a scientific hero for the reforming Nonconformist.[5] As with the foundation of UCL, the attempt to break the Anglican monopoly of a resource, whether education or the historical figure of Newton, resulted in its secularization.

The debate among informed individuals was partially resolved once the contents of the Newtonian archive were known, but interest in Newton's life and character also lessened as concerns about the place of science in society changed. During the 1830s to 1850s, the question of the morality of Newton and other scientific figures was of genuine importance. However, one of the reasons that the issues remained under discussion even until the 1860s was the longevity of the principal combatants, Brewster and De Morgan, as well as other contributors to the debates such as Brougham and Whewell. In the wake of the publication of Charles Darwin's *Origin of Species* (1859), and with the stance taken by leading scientific figures such as the members of the X-Club, the contentious issues surrounding science moved away from the questions regarding an individual's religious belief and personal morality to the possibility that science was inherently dangerous to religious interests. In 1865 De Morgan noted that the 'interest of the day centres in Essayists or Colensos', and the *Athenaeum* battle that he fought immediately before the revelation of the Pascal forgeries was against the 'Scientists' Declaration', signed by the old enemy, Brewster.[6]

Steven Shapin suggests that the assumption, highlighted by Robert K. Merton, that the morals of scientists are no different to those of any other group of individuals only became commonplace in the middle of the twentieth century, but that it was stated with increasing frequency from the late nineteenth century. Central to this change in attitude was the recognition that science was institutionalized, professionalized and employing increasing numbers: it was a living rather than a calling.[7] A similar point is made by Yeo, who says that, if 'during the early Victorian period the person, as much as the scientific process, was a source of authority', by the end of the century the dominant notion was that 'truth in science was achieved by the application of codifiable procedures that circumvented the emotions, interests, and biases of the individual scientist'.[8] Biographies of men of science did not cease to celebrate the moral qualities of their protagonists but increasingly these qualities were less those of the Christian gentleman than those more closely related to the 'disinterestedness' highlighted by Merton. This is a parallel move to De Morgan's claims for 'impartiality' in his history and it is notable that Shapin points to his review of Brewster's *Memoirs* as 'one of the earliest and most reflective assertions' of what he terms the 'moral equivalence of the man of science'.[9] These points likewise support the thesis put forward by Daston and Galison that a growing emphasis on objectivity became the key epistemological feature of later nineteenth-century science.[10] They argue that practitioners of science increasingly presented their work within a framework of neutrality and impartiality, emphasizing the de-personalized rigour of numbers, graphs and experiments, and also sought to enhance the reputation of science as an objective exercise. This study shows that this emphasis on objectivity can also be found in the practice of scientific biographical writing.

The debates surrounding the character of Isaac Newton can be viewed as one element of the secularization and even de-mystification of science and thus as fundamental to the creation of a critical tradition within the history of science. The importance of Newton, his usefulness as a resource to a wide variety of positions and the force of the disputes this provoked, meant that Newtonian research reached a level of sophistication beyond that with which other topics within the history of science were treated in this period. Although requiring confirmation from further research into the historiography of science in the nineteenth century, it appears too that this scholarship provided a stimulus for interest in the lives and milieus of Newton's predecessors, contemporaries and successors. For Baily, such work was done to highlight the importance of non-theoretical contributions to the advance of science, for Rigaud it established the scientific tradition of Oxford and for De Morgan it contextualized and explained the achievement of Newton. The emphasis on the critical reading of the primary sources created a specialism that was a long way removed from popular biographical accounts such as Brewster's 1831 *Life of Newton* or those produced from the 1860s by Samuel Smiles. Within Newtonian studies a defining moment was when their call to publish the Portsmouth Papers, which went back at least to Baily and Galloway in the 1830s, was partially answered in 1872.[11] The 'scientific' part of the collection was offered to Cambridge University and, although the rest of the papers were to be returned, the Earl of Portsmouth allowed the whole collection to be removed to Cambridge for cataloguing.[12] The process of cataloguing and the enforced division of the papers meant that the archive was moulded by nineteenth-century classifications.[13] However, the availability of these papers at Cambridge allowed the development of expertise on Newton's science. Begun by W. W. Rouse Ball, this style of enquiry was continued in the following century by the newly-professionalized historians of science that formed the 'Newton industry'.[14]

During the twentieth century efforts have been made to understand Newton's achievement within histories of scientific theory and method or within social contexts but biographical approaches have never disappeared. Since the full extent of Newton's alchemical interests was appreciated, when the Portsmouth Papers were recatalogued for sale at Sotheby's in 1936 and J. M. Keynes presented Newton as the 'Last of the magicians' rather than 'the first and greatest of the modern age of scientists', there have been attempts to reconcile these apparently conflicting aspects of his personality.[15] Although more recently historians have argued that Newton's various interests should be considered as separate enterprises, conceptually distinct and perhaps irreconcilable, or, like this book, have considered his reputation as a creation of his disciples, it seems impossible to get away from consideration of his life as a unit for study. In part, no doubt, this is due to a human instinct to understand other lives but it is also dictated by the development of the discipline. Works such as those discussed here, and the archival collections made accessible in this period, have formed the frames of reference for subsequent generations.

This book was not conceived as a contribution to the modern field of Newton studies. While it does unpick the original appearance of various stories and sources that have become standard in any consideration of his life, it is not concerned with understanding Newton or his thought. The primary concern was to use writings of Newton as a means of revealing nineteenth-century attitudes regarding the role of science and its practitioners within contemporary society. Developing from this research, however, was an appreciation of the innovations within both scientific biography and historical practice that these writings represented. While this presents an interesting case within the history of these fields, it has also highlighted the extent to which Newton was, from the earliest dawn of expert writing in the history of science, a special case. Although comparative research is lacking, it appears that, alone among British scientific heroes of the mid-nineteenth century, it was Newton's manuscript remains that attracted the kind of sophisticated attention described in the preceding chapters. The principal reasons why Newton was, at this date, a special case – the continuing relevance of his scientific legacy and his status as a figurehead of Anglican science – do not survive today. Thus, on the one hand, this book highlights the variety of concerns that could find expression in writings about Newton and more widely the history of science and, on the other, alerts us to the extent that research in these areas appears to have fed off itself. The research relating to Newton in the nineteenth century, which was generated by particular debates, provided rich materials for the use of subsequent generations. Understanding this and the environment that produced these materials should provoke reflection on its meaning to both modern scholarship and among general readers.

Likewise this study has revealed important clues about the way in which the biography of an important figure is moulded through the interaction of a number of individuals. Sometimes as a result of consensus and sometimes of disagreement, various aspects of Newton's life received particular emphasis while others were relatively ignored. The shape of the life can by this means be significantly altered from the version propagated at an earlier date, or even from the version adhered to by each individual writer. Revealed in detail here, this is something that must always be considered by both writers and readers of such biographies. This point should not, however, detract from the importance of biography, widely understood, within historical writing. In this study it has been necessary to delve into the biography of many of the protagonists in order to understand their writings, motivations and positioning in relation to other writers. Full biographies are likewise essential for any sort of understanding of individual experience, whether apparently typical or extraordinary, within otherwise amorphous categories of society. Scholars are now better prepared than ever before for the dangers and difficulties of biographical writing, a situation which I hope will improve rather than diminish contributions to the genre.

NOTES

The following abbreviations are used in the notes:

BL   British Library, London.
BLO Bodleian Library, University of Oxford.
CUL Cambridge University Library.
RAS Royal Astronomical Society.
RGO Royal Greenwich Observatory.
RSL Royal Society of London.
TCL Trinity College Library, Cambridge University.
UCL University College London.

Introduction

1. G. C. Lewis, *An Historical Survey of the Astronomy of the Ancients* (London: Parker and Bourn, 1862), p. 2.
2. I discuss the biography by Frenchman Jean-Baptiste Biot in Chapter 1, but am principally concerned with the effect this had on the British debate.
3. D. P. Miller '"Puffing Jamie": The Commercial and Ideological Importance of Being a "Philosopher" in the Case of the Reputation of James Watt (1736–1819)', *History of Science*, 38 (2000), pp. 1–24, on p. 2. Other examples are D. Outram, 'The Language of Natural Power: The "Eloges" of Georges Cuvier and the Public Language of Nineteenth Century Science', *History of Science*, 16 (1978), pp. 153–78; J. Gascoigne, 'The Scientist as Patron and Patriotic Symbol: The Changing Reputation of Sir Joseph Banks', and G. Cantor, 'The Scientist as Hero: Public Images of Michael Faraday', both in M. Shortland and R. Yeo (eds), *Telling Lives in Science: Essays on Scientific Biography* (Cambridge: Cambridge University Press, 1996), pp. 243–65, 171–93 respectively. Galileo's trial is treated similarly in J. H. Brooke and G. Cantor, *Reconstructing Nature: The Engagement of Science and Religion* (Edinburgh: T. & T. Clark, 1998), ch. 4.
4. S. Ross, '"Scientist": The Story of a Word', *Annals of Science*, 18 (1962), pp. 65–85. It should be noted, however, that William Whewell intended the word to underline the common aims and methodology of the disparate disciplines, and it was emphatically not intended to reflect professionalization of the field. See also R. Yeo, *Defining Science: William Whewell, Natural Knowledge and Public Debate in Early Victorian Britain* (Cambridge: Cambridge University Press, 1993), pp. 109–11.
5. J. Morrell and A. Thackray, *Gentlemen of Science: Early Years of the British Association for the Advancement of Science* (Oxford: Clarendon Press, 1981); R. Porter, 'Gentlemen and Geology: The Emergence of a Scientific Career, 1660–1920', *Historical Journal*, 21

(1978), pp. 809–36. See J. B. Morrell, 'Professionalisation', in R. C. Olby, G. N. Cantor, J. R. R. Christie and M. J. S. Hodge (eds), *Companion to the History of Modern Science* (London: Routledge, 1996), pp. 980–9; and A. J. Desmond, 'Redefining the X Axis: "Professionals", "Amateurs" and the Making of mid-Victorian Biology – a Progress Report', *Journal of the History of Biology*, 34 (2001), pp. 3–50 for overviews of the literature on professionalization and the sciences.

6. R. Barton, '"Men of science": Language, Identity and Professionalization in the mid-Victorian Scientific Community', *History of Science*, 41 (2003), pp. 73–119.
7. R. Laudan, 'Histories of the Sciences and their Uses: A Review to 1913', *History of Science*, 31 (1993), pp. 1–34, on p. 7. See E. Hobsbawm and T. Ranger, *The Invention of Tradition* (Cambridge: Cambridge University Press, 2000).
8. A. Cunningham and P. Williams, 'De-Centring the "Big Picture": *The Origins of Modern Science* and the Modern Origins of Science', *British Journal for the History of Science*, 26 (1993), pp. 407–32, on p. 424. See O. Chadwick, *The Secularisation of the European Mind in the Nineteenth Century* (Cambridge: Cambridge University Press, 1975).
9. S. F. Cannon, *Science in Culture: The Early Victorian Period* (Folkestone and New York: Dawson and Science History, 1978), p. 2.
10. F. M. Turner, *Contesting Cultural Authority: Essays in Victorian Intellectual Life* (Cambridge: Cambridge University Press, 1993), pp. 76–80, 101–10; Brooke and Cantor, *Reconstructing Nature*, pp. 148–53.
11. D. Lowenthal, *The Past is a Foreign Country* (Cambridge: Cambridge University Press, 1985), p. xvi. See also A. D. Culler, *The Victorian Mirror of History* (New Haven, CT, and London: Yale University Press, 1985); and S. Bann, *The Clothing of Clio: A Study of the Representation of History in Nineteenth-Century Britain and France* (Cambridge: Cambridge University Press, 1984).
12. J. S. Mill, *The Spirit of the Age*, ed. F. A. von Hayek (Chicago, IL: University of Chicago Press, 1942), p. 2. See P. J. Bowler, *The Invention of Progress: The Victorians and the Past* (Oxford: Basil Blackwell, 1989).
13. D. S. Goldstein, 'History at Oxford and Cambridge: Professionalisation and the Influence of Ranke', in G. G. Iggers and J. M. Powell (eds), *Leopold von Ranke and the Shaping of the Historical Discipline* (Syracuse, NY: Syracuse University Press, 1990), pp. 141–53, esp. pp. 142–5.
14. Ibid., p. 142, P. Burke, 'Ranke the reactionary', in Iggers and Powell (eds), *Leopold von Ranke*, pp. 36–44, on pp. 36–7, Bann, *The Clothing of Clio*, pp. 2, 8–14.
15. The argument was over the role and creativity of the author. Two opposing approaches within biography are exposed in reviews by Robert Southey and Francis Jeffrey. The former wrote of the heroic, 'sagacious and impartial biographer', who should display 'manly sincerity' in his narrative. The latter approved of biography that consisted almost entirely of letters and journals, and commended the 'manly clearness and directness' of the 'Editor', who kept his input to a minimum. [R. Southey], 'Hayley's *Life and Posthumous Writings of William Cowper*', *Annual Review*, 2 (1804), pp. 457–62, on p. 457; [F. Jeffrey], 'Memoirs of Sir James Mackintosh', *Edinburgh Review*, 62 (1835), pp. 205–55, on p. 255. On the use of primary sources in nineteenth-century biography see R. D. Altick, *Lives and Letters: A History of Literary Biography in England and America* (New York: Alfred A. Knopf, 1966), pp. 191–8; in historical writing, see P. Levine, *The Amateur and the Professional: Antiquarians, Historians and Archaeologists in Victorian England, 1838–1886* (Cambridge: Cambridge University Press, 1986), pp. 70–100, and R. Jann, *The Art and Science of Victorian History* (Columbus, OH: Ohio State University Press, 1985), pp. xvii–xxv.

16. D. A. Stauffer, *The Art of Biography in Eighteenth-Century England* (Princeton, NJ: Princeton University Press, 1941); J. W. Reed, *English Biography in the Early Nineteenth Century, 1801–1838* (New Haven, CT, and London: Yale University Press, 1966); Altick, *Lives and Letters*; A. O. J. Cockshut, *Truth to Life: The Art of Biography in the Nineteenth Century* (London: Collins, 1974); P. J. Korshin, 'The Development of Intellectual Biography in the Eighteenth Century', *Journal of England and German Philology*, 73 (1974), pp. 513–23, M. Pachter (ed.), *Telling Lives: The Biographer's Art* (Philadelphia, PA: University of Pennsylvania Press, 1981).
17. [J. Morell], 'Biography', in D. Brewster (ed.), *The Edinburgh Encyclopaedia*, 18 vols (Edinburgh: William Blackwood, 1830), vol. 3, pp. 506–12, on p. 512. This author objected to the lengthy treatment of 'inferior subjects'.
18. R. M. Young, 'Biography: The Basic Discipline for Human Sciences', *Free Association*, 11 (1988), pp. 108–30; R. S. Westfall, 'Newton and his Biographer', in S. H. Baron and C. Pletsch (eds), *Introspection in Biography: The Biographer's Quest for Self-Awareness* (Hillsdale, NJ: Analytic Press, 1985), pp. 175–89; S. Sheets-Pyenson, 'New Directions for Scientific Biography: The Case of Sir William Dawson', *History of Science*, 28 (1990), pp. 399–410; T. Söderqvist, 'Existential Projects and Existential Choice in Science: Science Biography as an Edifying Genre', in Shortland and Yeo (eds), *Telling Lives in Science*, pp. 45–84. This field gained impetus from those wishing to counter the genre's low esteem after the establishment of academic history of science, T. L. Hankins, 'In Defence of Biography: The Use of Biography in the History of Science', *History of Science*, 17 (1979), pp. 1–16.
19. History of science in the early twentieth century has received some attention in, for example, the special issue of the *British Journal for the History of Science*, 30 (1997), see especially J. A. Bennett, 'Museums and the Establishment of the History of Science at Oxford and Cambridge', pp. 29–46; A. K. Mayer, 'Moralizing Science: The Uses of Science's Past in National Education in the 1920s', pp. 51–70; and W. A. Smeaton, 'History of Science at University College London', pp. 25–8.
20. Yeo, *Defining Science*, p. 117.
21. R. Yeo, 'Genius, Method and Morality: Images of Newton in Britain 1760–1860', *Science in Context*, 2 (1988), pp. 257–84.
22. P. Fara, *Newton: The Making of Genius* (London: Macmillan, 2002). Yeo, *Defining Science*, provides a more detailed analysis of the contribution of William Whewell, pp. 129–44.
23. Yeo, 'Genius, Method and Morality', p. 265.
24. E. Young, *Conjectures on Original Composition* (1759; Leeds: Scholar Press, 1966), p. 12.
25. A. Gerard, *An Essay on Genius* (London, 1774), reprinted and ed. B. Fabian (Munich and Amsterdam: Wilhelm Fink Verlag, 1966).
26. Ibid., pp. 96–7; this demonstrates the importance of this story beyond mere anecdote, see Fara, *Newton*, pp. 213–19.
27. See S. Schaffer, 'Scientific Discoveries and the End of Natural Philosophy', *Social Studies of Science*, 16 (1986), pp. 387–420, on p. 407; S. Schaffer, 'Priestley and the Politics of Spirit', in R. G. W. Anderson and C. Lawrence (eds), *Science, Medicine and Dissent: Joseph Priestley (1733–1804)* (London: Wellcome Trust, 1987), pp. 39–53; Yeo, 'Genius, Method and Morality', pp. 264–7.
28. J. Priestley, *The History and Present State of Electricity, with Original Experiments* (London, 1767), p. 575.

29. See W. Grisenthwaite, *On Genius: In which it is Attempted to be Proved, that there is no Mental Distinction among Mankind* (London, 1830).
30. On William Blake's combined fascination for and antipathy towards Newtonianism, see D. D. Ault, *Visionary Physics: Blake's Response to Newton* (Chicago, IL, and London: University of Chicago Press, 1974). Keats wrote of science 'unweaving the rainbow', a sentiment that Wordsworth initially shared, but came to reject, W. K. Thomas and W. U. Ober, *A Mind For Ever Voyaging: Wordsworth at Work Portraying Newton and Science* (Edmonton: University of Alberta Press, 1989).
31. See N. Kessel, 'Genius and Mental Disorder: A History of Ideas concerning their Conjunction', in P. Murray (ed.), *Genius: The History of an Idea* (Oxford and New York: Basil Blackwell, 1989), pp. 196–212; S. Shapin, 'The Philosopher and the Chicken: On the Dietetics of Disembodied Knowledge', in C. Lawrence and S. Shapin (eds), *Science Incarnate: Historical Embodiments of Natural Knowledge* (Chicago, IL, and London: University of Chicago Press, 1998), pp. 21–50; Fara, *Newton*, pp. 155–81.
32. J. M. Gully, *Lectures on the Moral and Physical Attributes of Men of Genius and Talent* (London, 1836).
33. B. le B. de Fontenelle, *The Life of Sir Isaac Newton, with an Account of his Writings ...* (London: J. Roberts, 1728), reprinted in R. Iliffe, M. Keynes and R. Higgitt (eds), *Early Biographies of Isaac Newton, 1660–1885*, 2 vols (London: Pickering & Chatto, 2006), vol. 1: R. Iliffe (ed.), *Eighteenth-Century Biography of Isaac Newton: The Unpublished Manuscripts and Early Texts*, pp. 109–21. See C. B. Paul, *Science and Immortality: The Éloges of the Paris Academy of Sciences (1699–1791)* (Berkeley, CA, and London: University of California Press, 1980), pp. 30–2.
34. Conduitt's 'Memoir' and various notes and correspondence relating to a biography of Newton are held at Kings College, Cambridge, Keynes MS 129–36 and transcribed in Iliffe et al. (eds), *Early Biographies*, vol. 1, pp. 57–107, 127–140, 155–235.
35. T. Birch, 'Sir Isaac Newton', in P. Bayle, *A General Dictionary, Historical and Critical*, 10 vols (London, 1734–41), vol. 7 (1738), pp. 776–802, reprinted in A. R. Hall (ed.), *Isaac Newton: Eighteenth-Century Perspectives* (Oxford: Oxford University Press, 1999), pp. 83–95.
36. Conduitt's 'Memoir', with additions by Catherine Conduitt (Keynes MS 129 A), in Iliffe et al. (eds), *Early Biographies*, vol. 1, pp. 98–106, quotations on pp. 98–9.
37. Fontenelle quoted a letter of Newton's that was published in the *Commercium Epistolicum* (1712), in which he claimed that he decided not to publish his *Opticks* because 'I should repent my imprudence in losing something so real as tranquillity, in order to chase a shadow', B. le B. de Fontenelle, 'Eloge de Neuton', in B. B. de Fontenelle, *Oeuvres* (Paris: Salmon, 1825), trans. A. R. Hall, in Hall (ed.), *Newton*, pp. 59–74, on p. 73; H. W. Turnbull, J. F. Scott, A. R. Hall and L. Tilling (eds), *The Correspondence of Isaac Newton*, 7 vols (Cambridge: Cambridge University Press, 1959–77), vol. 2, p. 133.
38. Conduitt's 'Memoir', Iliffe et al. (eds), *Early Biographies*, vol. 1, p. 102.
39. Ibid., p. 101.
40. Ibid., pp. 103–4 (punctuation added).
41. Ibid., p. 106.
42. E. Turnor, *Collections for the History of the Town and Soke of Grantham* (London: W. Miller, 1806), pp. 160–71. The Portsmouth Papers had passed to the Portsmouth family after the marriage of Newton's great-niece, Kitty Conduitt, to Viscount Lymington. For the history of this collection, and details of those who viewed it before Turnor, see R. Iliffe, 'A "connected system"? The Snare of a Beautiful Hand and the Unity of Newton's Archive', in M.

Hunter (ed.), *Archives of the Scientific Revolution: The Formation and Exchange of Ideas in Seventeenth-Century Europe* (Woodbridge: Boydell Press, 1998), pp. 137–57.

43. Keynes MS 131, Iliffe et al. (eds), *Early Biographies*, vol. 1, p. 138.
44. Conduitt circulated a letter requesting information about Newton, and received replies from Stukeley, John Craig, Thomas Mason, William Derham, Humphrey Newton and Nicholas Wickins, Keynes MS 131-7, Iliffe et al. (eds), *Early Biographies*, vol. 1, pp. 131–40. The extracts from the Royal Society included information about Newton's dispute with Leibniz, Turnor, *Collections*, pp. 181–6, 186.
45. Ibid., pp. 174–80, quotation on p. 179. Stukeley wrote a longer, unpublished biography of Newton, W. Stukeley, *Memoirs of Sir Isaac Newton's Life: Bring Some Account of his Family and Chiefly the Junior Part of his Life*, ed. A. G. H. White (London: Taylor & Francis, 1936). The two manuscripts of this biography have been transcribed and compared in Iliffe et al. (eds), *Early Biographies*, vol. 1, pp. 243–308.
46. Turnor, *Collections*, p. 173.
47. Ibid., p. 172. In this conversation Newton discussed his notion of a circulation of matter and light between sun, planets and comets, and his belief in 'intelligent beings superior to us, who superintended' these celestial revolutions.
48. Ibid., p. vi.
49. Turnor's brother Charles also made a collection of 'Newtoniana' that was left to the Royal Society on his death in 1853, see *Gentleman's Magazine* (May 1853), p. 519.
50. Fara, *Newton*, p. xv.
51. B. J. T. Dobbs, 'Review of A. Rupert Hall, *Isaac Newton: Adventurer* and R. S. Westfall, *The Life of Isaac Newton*', *Isis*, 85 (1994), pp. 515–17, p. 516.
52. [J.-B. Biot], 'Newton (Isaac)', in L. G. Michaud (ed.), *Biographie universelle, ancienne et moderne*, 83 vols (Paris: Michaud Frères, 1811–53), vol. 31 (1822), pp. 127–94. J.-B. Biot, 'Life of Newton', trans. H. Elphinstone, in *Lives of Eminent Persons*, Library of Useful Knowledge (London: Baldwin & Cradock, 1833), reprinted in Iliffe et al. (eds), *Early Biographies*, vol. 2: R. Higgitt (ed.), *Nineteenth-Century Biography of Isaac Newton: Public Debate and Private Controversy*, pp. 1–63.
53. D. Brewster, *The Life of Sir Isaac Newton*, The Family Library, vol. 24 (London: John Murray, 1831).
54. F. Baily, *An Account of the Revd. John Flamsteed, the First Astronomer-Royal; Compiled from his own Manuscripts, and other Authentic Documents, Never Before Published, and Supplement to the Account of the Revd. John Flamsteed* (London, 1835, 1837; London: Dawsons of Pall Mall, 1966).
55. D. Brewster, *Memoirs of the Life, Writings and Discoveries of Sir Isaac Newton*, 2 vols (Edinburgh: Constable & Co., 1855), reprinted in The Sources of Science, vol. 14 (New York and London: Johnson Reprint Corporation, 1965).
56. This is akin to Herbert Butterfield's suggestion that the necessary stimulus for critical historiography was controversy, as disputants found it 'really necessary to know what happened in the past', and 'the very bitterness of controversy and the violence of partisanship served the cause of impartial history', as each side 'drove the other to profounder and more careful researches', H. Butterfield, 'Delays and Paradoxes in the Development of Historiography', in K. Bourne and D. C. Watt (eds), *Studies in International History: Essays Presented to W. Norton Medlicott* (London: Longmans, 1967), pp. 1–15, on pp. 6–7.

57. Further revelations were made when the Portsmouth Papers were catalogued in 1872–88, but these tended only to confirm, and strengthen, those of Brewster, *Memoirs of Newton*.
58. Many writers note the faith invested by nineteenth-century biographers in documents and facts, but do not tend to link this to archive-based histories. This is generally because, coming from literary studies, authors are searching for biographies that can be considered artistically satisfying. See also R. Higgitt, 'Discriminating Days? Partiality and Impartiality in Nineteenth-Century Biographies of Newton', in T. Söderqvist (ed.), *The Poetics of Biography in Science, Technology and Medicine* (Aldershot: Ashgate, 2007), pp. 155–72.
59. Levine, *The Amateur and the Professional*.

1 Jean-Baptiste Biot's 'Newton' and its English Translation

1. 'Nemo', *Mirror of Literature*, 4 (1824), p. 399.
2. [Biot], 'Newton (Isaac)'. Biot's article is usually erroneously given as 1821, but the title page of the relevant volume is dated 1822. Quotation from D. Gjertsen, *The Newton Handbook* (London and New York: Routledge & Kegan Paul, 1986), p. 83.
3. T. S. Kuhn, *The Structure of Scientific Revolutions* (Chicago, IL, and London: University of Chicago Press, 1962), p. 89.
4. E. Frankel, 'J. B. Biot and the Mathematization of Experimental Physics in Napoleonic France', *Historical Studies in the Physical Sciences*, 8 (1977), pp. 33–72; E. Frankel, 'Corpuscular Optics and the Wave Theory of Light: The Science and Politics of a Revolution in Physics', *Social Studies of Science*, 6 (1976), pp. 141–84, on p. 141.
5. See E. Frankel, 'Jean-Baptiste Biot: The Career of a Physicist in Nineteenth Century France' (unpublished PhD thesis, Princeton University, 1972), and 'Career-Making in Post-Revolutionary France: The Case of Jean-Baptiste Biot', *British Journal for the History of Science*, 11 (1978), pp. 36–48; M. Crosland, *The Society of Arcueil: A View of French Science at the Time of Napoleon I* (London: Heinemann, 1967), pp. 295–8.
6. R. Fox, 'The Rise and Fall of Laplacian Physics', *Historical Studies in the Physical Sciences*, 4 (1974), pp. 89–136. See also R. Fox, 'Laplacian Physics', in Olby et al. (eds), *Companion to the History of Modern Science*, pp. 278–94; and C. C. Gillispie, R. Fox and I. Grattan-Guinness, *Pierre-Simon Laplace 1749–1827: A Life in Exact Science* (Princeton, NJ, and Chichester: Princeton University Press, 1997), pp. 199–215, 243–9.
7. Fox, 'The Rise and Fall', p. 109.
8. Frankel, 'Corpuscular Optics', p. 144.
9. Laplace's *Traité de mécanique céleste* (1799–1825), quoted in Fox, 'The Rise and Fall', p. 89.
10. Quoted in Crosland, *The Society of Arcueil*, p. 90.
11. See A. E. Shapiro, *Fits, Passions and Paroxysms: Physics, Method, and Chemistry and Newton's Theories of Colored Bodies and Fits of Easy Reflection* (Cambridge: Cambridge University Press, 1993).
12. Quoted in Frankel, 'Biot and the Mathematization of Experimental Physics', pp. 71, 72.
13. G. N. Cantor, *Optics after Newton: Theories of Light in Britain and Ireland, 1704–1840* (Manchester: Manchester University Press, 1983), p. 1.
14. Crosland, *The Society of Arcueil*, pp. 90–1, describes the Arcueil Group as 'pre-positivist'.
15. Quoted in J. Worrall, 'Thomas Young and the "Refutation" of Newtonian Optics: A Case-Study in the Interaction of Philosophy of Science and History of Science', in C. Howson (ed.), *Method and Appraisal in the Physical Sciences: The Critical Background to Modern Science, 1800–1905* (Cambridge: Cambridge University Press, 1976), pp. 107–79, on p. 160.

Notes to pages 22–6 201

16 Biot, *Traité* (1816), quoted in Worrall, 'Thomas Young and the "Refutation" of Newtonian Optics', p. 151.
17. Supporters of the wave theory of light also saw these phenomena as the 'most untractable', W. Whewell, *History of the Inductive Sciences, from the Earliest to the Present Time*, 3rd edn, 3 vols (London, 1857), vol. 2, p. 336.
18. Frankel, 'Jean-Baptiste Biot', pp. 318–20.
19. Biot, *Précis expérimentale de physique* (1824), quoted in Frankel, 'Corpuscular Optics', p. 162.
20. Fox, 'The Rise and Fall', p. 127.
21. Frankel, 'Corpuscular Optics', pp. 171–3; Fox, 'The Rise and Fall', pp. 126–7.
22. Gillispie et al., *Pierre-Simon Laplace*, pp. 245, 248. Biot did not convert to the wave theory until 1846, but, according to Frankel, in print he 'moved to an agnostic position as early as 1830' and subsequently referred to himself as a '*Rieniste*', Frankel, 'Corpuscular Optics', pp. 173–4; see also X. Chen and P. Barker, 'Cognitive Appraisal and Power: David Brewster, Henry Brougham, and the Tactics of Emission-Undulatory Controversy during the Early 1850s', *Studies in the History and Philosophy of Science*, 23 (1992), pp. 75–101, on pp. 81, 86. In fact Biot's statements were already those of a positivistic *Rieniste* in 1822, although this was clearly a position adopted for convenience.
23. [Biot], 'Newton (Isaac)', pp. 169, 165. Page references will hereafter be given in the text. For translations of the French quotations from Biot's text given in this chapter, see the Appendix. All other translations are given in the notes.
24. L. T. More, noting that there 'has been assumed, in the minds of many physicists, a rather contemptuous attitude towards Newton's theory of "Fits"', thought it 'worth while to quote the opinion of so eminent an authority on optics as Biot', without realizing the significance of this hypothesis to Biot in his fight against the wave theory, L. T. More, *Isaac Newton: A Biography* (London and New York: C. Scribner's Sons, 1934), p. 116n.
25. Newton to Oldenburg, 7 December 1675, in I. B. Cohen and R. S. Westfall (eds), *Newton: Texts, Backgrounds, Commentaries* (New York and London: W. W. Norton, 1995), pp. 12–34. Although Newton here discusses Hooke's vibrational theory, he suggests that he prefers the 'Hypothesis of light's being a body', ibid., p. 13.
26. On the early diffusion and development of fluxions see N. Guicciardini, *The Development of the Newtonian Calculus in Britain 1700–1800* (Cambridge: Cambridge University Press, 1989), pp. 11–37, 55–67. Guicciardini cautions against too simplistic an interpretation of the differing British and Continental traditions, dating the break to the 1740s rather than the period of the calculus dispute. On the development of the Leibnizian calculus, see N. Guicciardini, *Reading the Principia: The Debate on Newton's Mathematical Methods for Natural Philosophy from 1687 to 1736* (Cambridge: Cambridge University Press, 1999), pp. 195–249.
27. 'Contemptible imputations ... with the most futile arguments and the most unconvincing hypotheses, the great and wise philosophy that Newton had introduced into the study of the phenomena of nature', [J.-B. Biot, F. P. G.[?] Duvau, P. Maine de Biran and P. A.[?] Stapfer], 'Leibnitz, Godfroi-Guillaume, baron de', in Michaud (ed.) *Biographie universelle*, vol. 23 (1819), pp. 594–642, on p. 639. Translation by Caroline Higgitt.
28. Huygen's original text reads: 'Die 19 maii 1694, narravit mihi D. Colin, Scotus, virum celeb. ac rarum geometram Is. Neutonum incidisse in phrenitin abhinc anno et sex mensibus. An ex nimiâ studii assiduiate, an dolore infortunii, quòd in incendio, laboratorium chemicum et scripta quædam amiserat? Cùm ad archiepiscopum Cantabrigiensem venisset, ea locutum quæ alienationem mentis indicarent; deindè ab amicis cura ejus suscepta,

domoque clausâ. remedia volenti nolenti adhibita, quibus jam sanitatem recuperavit, ut jam nunc librum suum Principiorum intelligere incipiat', Biot, 'Life of Newton', p. 42.
29. This story first appeared in Thomas Maude's *Wendsley-Dale; or Rural Contemplation* (1780) and was used by Charles Hutton in his *Mathematical and Philosophical Dictionary* (1795–6), Hall (ed.), *Newton*, p. 175. See Fara, *Newton*, pp. 207–8 on Diamond and his alter-ego Tray.
30. Biot's personal attachment to this story is demonstrated by the fact that when he visited Woolsthorpe, he 'gathered a few leaves [from the tree] to carry them religiously back home', Fara, *Newton*, p. 216; J.-B. Biot, 'Revue de *The Life of Isaac Newton*', *Journal des savants*, (1832), pp. 199–203, 263–74, 321–39, on p. 265.
31. There were many tales recounting Newton's absent-mindedness, see e.g. Hutton, quoted in Hall (ed.), *Newton*, p. 178. See also Shapin, 'The Philosopher and the Chicken'; and R. Iliffe, 'Isaac Newton: Lucatello Professor of Mathematics', in Lawrence and Shapin (eds), *Science Incarnate*, pp. 121–55.
32. This account was based on that in W. Whiston's *The Longitude Discovered by the Eclipses, Occultations, and Conjunctions of Jupiter's Planets* (London: J. Whiston, 1738).
33. 'Such is the frightful condition of man. Genius and madness may exist in his mind side by side and simultaneously. Pascal, having once suffered a great physical terror, from that time imagined that he beheld a gulf yawning beside him. His mind, disturbed and terrified, presented him with ascetic visions, the incoherent details of which he fixed in writing. He concealed these pious scraps in his garments, carried them about with him, and preserved them till his dying day; and in this state of mind wrote his profound *Thoughts* on God, on the world, and on man, showing an infinitely judicious and acute observation and appreciation of human societies, and of the artificial conditions by which they are united. And, what completes our astonishment, the expression of these *Thoughts* is admirable for the force, the grandeur, and concision of the style', Biot, 'Revue de *The Life of Isaac Newton*', p. 333. Translation from [T. Galloway], 'French and English Biographies of Newton', *Foreign Quarterly Review*, 12 (1833), pp. 1–27, reprinted in Iliffe et al. (eds), *Early Biographies*, vol. 2, pp. 65–92, on p. 81.
34. Pierre-Claude-François Daunou (1761–1840) was a historian and politician, at this time Royal Archivist and Professor of History and Ethics at the Collège de France.
35. Crosland, *The Society of Arcueil*, p. 301.
36. R. Hahn, 'Laplace and the Vanishing Role of God in the Physical Universe', in H. Woolf (ed.), *The Analytic Spirit: Essays in the History of Science in Honor of Henry Guerlac* (Ithaca, NY, and London: Cornell University Press, 1981), pp. 85–95; R. Hahn, 'Laplace's Religious Views', *Archives Internationales d'Histoire des Sciences*, 30 (1955), pp. 38–40; Crosland, *The Society of Arcueil*, pp. 91–3.
37. Hahn, 'Laplace and the Vanishing Role of God', p. 89.
38. RSL, Herschel Papers, JFWH 11.102, letter of Pierre Simon Laplace to John Herschel, 18 April 1823. The relevant section of this letter reads: 'j'ai prié M. Gauthier, que vous seres bien aise de connaître[,] de prendre des informations sur une circonstance de la vie de Newton, assés remarquable & qui jusqu['] ici a eté peu connue[.] je veux parler de l'etat d'alienation ou il parait avoir eté pendant plusieurs mois après la publication de son livre des principes & l'incendie de ses papiers. M. biot a publié sur cela dans sa notice biographique de Newton inserée dans la biographie universelle une lettre d'huygens a leibnitz. il est probable que des informations prises a Cambrige ou Newton était alors eclairiront ce fait & vous etes[,] plus que personne, a portée de vous les procurer. je desirerais savoir encore a quelle epoque Newton a commencé a s'occuper d'objets theologiques. Le Scolie

qui est a la fin de son livre des principes ne se trouve point dans la premiere édition de ce livre, & je ne vois point dans ses leçons d'optique les idées qu'il a repandues a cet égard, dans son traité d'optique. il parait que les idées théologiques ont eté la principale occupation de ses dernieres années, & l'on a imprimé quelque part que ses herietiers après sa mort, avoient brulé plusieurs milliers de feuillets ecrits de sa main sur ces matieres. toutes les circonstances, de la vie des hommes celebres interessent l'histoire de l'esprit humain. Car [?], suivant l'expression de M. bailli, c'est dans ces grandes têtes, que l'esprit humain a vecu' ('I have asked M. Gauthier, whom you would be glad to know, to find out something about a very curious event in Newton's life, little known of until now. I am talking about the strange state of mind ('alienation') in which he seems to have been for several months after the publication of his book of principles and the burning of his papers. In connection with this, M. Biot, in his biographical entry on Newton in the *Biographie Universelle*, included a letter from Huygens to Leibnitz. It is likely that information taken [i.e. available?] in Cambridge, where Newton was at that time, will clarify this fact and you more than anyone are in a position to get hold of it. I would also like to know when it was that Newton began to be interested in theological matters. The Scholium which is at the end of his book of principles does not appear in the first edition of this book and I do not find in his lessons on optics ideas that he has put forward [?] on this subject in his treatise on optics. It appears that theological ideas were the principal occupation of his last years and it has been written somewhere that after his death his heirs burned several thousands of sheets of paper written in his hand on these subjects. All circumstances about the lives of famous men are of interest to the history of the human mind. For, to quote M. Bailli, it is in the heads of these great men that the human mind has lived'; translation Caroline Higgitt).

39. This query rejected the notion of light as a vibration through an ether, arguing instead for a universe composed of particles and vacuum. It ends by suggesting that God 'in infinite Space, as it were in his Sensory, sees the things themselves intimately, and thoroughly perceives them, and comprehends them wholly by their immediate presence to himself', Cohen and Westfall (eds), *Newton: Texts*, p. 189.
40. Crosland suggests that 'Biot's attitude as a young man to religion seems to reflect that of Laplace', but notes that by 1816 he refrained from a forthright condemnation of the Catholic Church in his article on Galileo, and that in 1825 he sought an audience with Pope Leo XII, writing afterwards of the deep impression that this had on him, Crosland, *The Society of Arcueil*, p. 93, [J.-B. Biot], 'Galilée Galilei', in Michaud (ed), *Biographie universelle*, vol. 16 (1816), pp. 318–31. Michaud, the editor of the *Biographie universelle*, and many of his writers were identified, in a review of its second edition, as pro-Catholic and Royalist, [R. H. Christie], 'Biographical Dictionaries', *Quarterly Review*, 157 (1884), pp. 187–230, on p. 211, reprinted in I. B. Nadel (ed.), *Victorian Biography: A Collection of Essays from the Period* (New York and London: Garland, 1986), unpaginated.
41. Brewster, *Life of Newton*; Biot, 'Revue de *The Life of Newton*'; Baily, *Account of Flamsteed*; J.-B. Biot, 'Revue de *An Account of the Rev. John Flamsteed*', *Journal des savants* (1836), pp. 156–66, 205–23, 641–58; J. Collins, *Commercium Epistolicum ... ou Correspondance ... relative à l'analyse supérieure, ré-imprimée sur l'édition originale de 1712 avec l'indication des variantes de l'édition de 1722 ...*, ed. J.-B. Biot and F. Lefort (Paris, 1856).
42. Fara, *Newton*, p. 216.
43. This translation was first published in 1829 and reprinted in an 1833 collection of SDUK biographical tracts, entitled *Lives of Eminent Persons*.

44. [Sir B. H. Malkin], 'Brewster's *Life of Newton*', *Edinburgh Review*, 56 (1832), pp. 1–37, on p. 7.
45. The SDUK committee included six future Whig cabinet ministers – Brougham, Russell, Althorp, Auckland, Spring-Rice and Abercrombie. J. N. Hays, 'Science and Brougham's Society', *Annals of Science*, 20 (1964), pp. 227–41, on p. 227.
46. The prospectus of the SDUK announced that no treatise would 'contain any matter of Controversial Divinity, or interfere with the principles of revealed Religion', in H. Smith, *Society for the Diffusion of Useful Knowledge 1826–1846: A Social and Bibliographical Evaluation* (London: The Vine Press, 1974), p. 56.
47. See M. C. Grobel, 'The Society for the Diffusion of Useful Knowledge 1826–1846 and its Relation to Adult Education in the First Half of the XIXth Century' (unpublished MA thesis, University of London, 1933), pp. 2–7; J. R. Topham, 'Science and Popular Education in the 1830s: The Role of the *Bridgewater Treatises*', *British Journal for the History of Science*, 25 (1992), pp. 397–430. On Brougham see also E. A. Storella, '"O, What a World of Profit and Delight": The Society for the Diffusion of Useful Knowledge' (unpublished PhD dissertation, Brandeis University, 1986); H. P. Brougham, *The Life and Times of Henry Lord Brougham, Written by Himself*, 3 vols (Edinburgh and London: W. Blackwood & Sons, 1871); and E. Royle, 'Mechanics' Institutes and the Working Classes, 1840–1860', *Historical Journal*, 14 (1971), pp. 305–21.
48. J. G. Crowther, *Statesmen of Science* (London: Cresset Press, 1965); G. A. Foote, 'The Place of Science in the British Reform Movement 1830–50', *Isis*, 42 (1951), pp. 192–208; R. D. Altick, *The English Common Reader: A Social History of the Mass Reading Public 1800–1900* (Chicago, IL: Chicago University Press, 1957), pp. 130–2.
49. Charles Knight, quoted in Altick, *The English Common Reader*, p. 270; William Cobbett, quoted in A. J. Desmond, *The Politics of Evolution: Morphology, Medicine, and Reform in Radical London* (Chicago, IL, and London: University of Chicago Press, 1989), p. 30. See S. Shapin and B. Barnes, 'Science, Nature and Control: Interpreting Mechanics' Institutes', *Social Studies of Science*, 7 (1977), pp. 31–74.
50. See for instance Anon., *The Consequences of a Scientific Education to the Working Classes of this Country Pointed Out; and the Theories of Mr. Brougham on the Subject Confuted; in a Letter to the Marquess of Lansdown, by a Country Gentleman* (London: T. Cadell, 1826).
51. H. P. Brougham, *Discourse on the Objects, Advantages, and Pleasures of Science*, Library of Useful Knowledge (London: Baldwin & Cradock, 1827), p. 6. This tract had sold 33,100 copies by the end of 1829, J. A. Secord, 'Progress in Print', in M. Frasca-Spada and N. Jardine (eds), *Books and the Sciences in History* (Cambridge: Cambridge University Press, 2000), pp. 369–89, on p. 378.
52. Brougham, *Discourse on the Objects of Science*, p. 47.
53. The SDUK's readers were probably mostly middle class but, at least at this early date, the treatises were aimed, albeit with little success, at workers, Altick, *The English Common Reader*, p. 276.
54. See SDUK prospectus, Smith, *Society for the Diffusion of Useful Knowledge*, pp. 56–63.
55. *Eminent Persons* included 'Galileo' and 'Kepler' (J. E. Drinkwater), 'Mahomet' (J. A. Roebuck), 'Cardinal Wolsey' (K. Thomson), 'Sir Edward Coke' (E. P. Burke), 'Lord Somers' (D. Jardine), 'William Caxton' (W. Stevenson), 'Admiral Blake' (J. G. Gorton), 'Adam Smith' (W. Draper), 'Carsten Niebuhr' (S. Austin), 'Sir Christopher Wren' (C. H. B. Ker) and 'Michael Angelo Buonaroti' (T. Roscoe).

56. Grobel, 'The Society for the Diffusion of Useful Knowledge', pp. 331–2; UCL, SDUK Papers 5, Publication Committee, Minutes of Sub-Committees, 1827–9, 29 November 1827, p. 19.
57. A. T. Malkin (ed.), *Gallery of Portraits: with Memoirs*, 7 vols (London: Charles Knight, 1833–7); G. Long (ed.), *Penny Cyclopaedia*, 27 vols (London: Charles Knight, 1833–46). The Society's final project, the cost of which precipitated its demise, was an enormous biographical dictionary, G. Long (ed.), *The Biographical Dictionary of the Society for the Diffusion of Useful Knowledge*, 4 vols (London, 1842–4).
58. See Grobel, 'The Society for the Diffusion of Useful Knowledge', pp. 332–45, on discussions regarding biographies submitted to the SDUK. Grobel suggests that 'biographies and histories provided more opportunity for divergence of opinion than any other subject undertaken by the Society', p. 345.
59. Sarah Austin (1793–1867), wife of John Austin, Professor of Jurisprudence at UCL, translated many French and German writings, notably Ranke's *History of the Popes* (1840). Carsten Niebuhr (1733–1815), father of the historian Barthold Niebuhr, was a traveller and surveyor, who explored India and the Middle East.
60. [S. Austin], 'Life of Carsten Niebuhr', in *Lives of Eminent Persons*, Library of Useful Knowledge (London: Baldwin & Cradock, 1833), pp. 1, 31. The different usages of the word 'genius' in these quotations illustrate the recent evolution of the word. See S. Shapin and B. Barnes, 'Hand to Head: Rhetorical Resources in British Pedagogical Writing, 1770–1850', *Oxford Review of Education*, 2 (1976), pp. 231–54, on the type of knowledge thought possible, or suitable, for the lower orders.
61. The SDUK prospectus's list of 'Treatises preparing for publication' included biographies of 'Self-exalted Men', in Smith *Society for the Diffusion of Useful Knowledge*, p. 59.
62. [J. A. Roebuck], 'Life of Mahomet', in *Lives of Eminent Persons*, Library of Useful Knowledge (London: Baldwin & Cradock, 1833), pp. 12–13.
63. See Shapin and Barnes, 'Science, Nature and Control'; Altick, *The English Common Reader*, e.g. pp. 130–1. Shapin and Barnes suggest that the Broughamites hoped to gain the trust, and so control, of the working classes by presenting a highly objectified natural science. I argue that the SDUK also wished to present 'impartial', trustworthy history.
64. [J. E. Drinkwater], 'Life of Galileo: With Illustrations of the Advancement of Experimental Philosophy' and 'Life of Kepler', in *Lives of Eminent Persons*, Library of Useful Knowledge (London: Baldwin & Cradock, 1833).
65. UCL, SDUK Correspondence, letter from John Elliot Drinkwater to Thomas Cotes, 19 November 1834.
66. [Drinkwater], 'Kepler', p. 1. See also Yeo, 'Genius, Method and Morality', p. 267.
67. [Drinkwater], 'Galileo', pp. 101–2.
68. See e.g. F. E. Manuel, *A Portrait of Isaac Newton* (London: Muller, 1980), p. 213; Gjertsen, *The Newton Handbook*, p. 77; and Yeo, 'Genius, Method and Morality', p. 280, who have probably followed P. E. B. Jourdain, who refers to the translation as being 'by Lord Brougham', A. De Morgan, *Essays on the Life and Work of Newton*, ed. P. E. B. Jourdain (Chicago, IL, and London: Open Court Publishing Co., 1914), p. 3n. The British Library catalogue credits the translation to 'Sir Howard Crawfurd Elphinstone', in confusion with the major-general of that name (born 1829).
69. [H. P. Brougham], 'Lord King's Life of John Locke', *Edinburgh Review*, 50 (1829), pp. 1–31, on pp. 21–5.
70. Brewster, *Life of Newton*, p. 227.

71. Grobel, 'The Society for the Diffusion of Useful Knowledge', p. 340. Just before publishing 'Newton', the SDUK set up a Translation Committee to look into the possibility of publishing further translations, UCL, SDUK Papers, SDUK 1, Minutes of the General Committee, 1827–9, 29 April 1829. At its first meeting, the SDUK's publisher, Baldwin, advised that translations did not usually sell well, unless they were biographies. John Lubbock suggested acquiring a list of the Académie's *éloges*, UCL, SDUK Papers, SDUK 5, Minutes of Sub-Committees, 1827–9, 11 May 1829, pp. 165–6. However, at their second and last meeting, a letter from Baldwin was read opining that a series of translations would not be successful and the Committee 'resolved that it is inexpedient to undertake translations unless as a part of the Library of Useful Knowledge & by the direction of the Publication Committee', ibid., 9 June 1829, p. 190.
72. UCL, SDUK Papers, SDUK 1, Minutes of the General Committee, 1826–9, 18 June 1829, p. 148; SDUK 18, Committee Letter-Book, March 1827–June 1829, letter from Thomas Cotes to Howard Elphinstone, 18 June 1829, p. 151; SDUK 2, Minutes of the General Committee, 1830–5, 26 January 1831, p. 68. Elphinstone was proposed by Lubbock and Brougham and elected unanimously.
73. George Airy's autobiography records that he and Drinkwater had taken some of their pupils, including Elphinstone, to a cottage during the 1823 long vacation, G. B. Airy, *Autobiography of Sir George Biddell Airy*, ed. W. Airy (Cambridge: Cambridge University Press, 1896), p. 52.
74. Biographical details taken from F. Boase, *Modern English Biography: Containing Many Thousand Concise Memoirs of Persons who have Died Between the Years 1851–1900 with an Index of the Most Interesting Matter* (London: F. Cass, 1965); and W. W. Rouse Ball and J. A. Venn (eds), *Admissions to Trinity College, Cambridge, 1801–1850*, 5 vols (London: Macmillan, 1911–16).
75. See e.g. *The Times*, 19 February 1842, p. 3f; obituary, *The Times*, 4 May 1846, p. 6a.
76. Frankel, 'Jean-Baptiste Biot', pp. 158–9, 162–3.
77. Guicciardini, *Development of the Newtonian Calculus*, pp. 95–138.
78. G. N. Cantor, 'Henry Brougham and the Scottish Methodological Tradition', *Studies in History and Philosophy of Science*, 2 (1971), pp. 69–89, on p. 70. For Brougham's comments on Edinburgh education, see G. E. Davie, *The Democratic Intellect: Scotland and her Universities in the Nineteenth Century* (Edinburgh: Edinburgh University Press, 1961), p. 39.
79. On the Analytical Society see P. C. Enros, 'The Analytical Society (1812–1813): Precursor of the Renewal of Cambridge Mathematics', *Historia Mathematica*, 10 (1983), pp. 24–47; on the introduction of Continental techniques into the curriculum see A. Warwick, *The Masters of Theory: Cambridge and the Rise of Mathematical Physics* (Chicago, IL, and London: University of Chicago Press, 2003), pp. 66–84.
80. C. Babbage, *Reflections on the Decline of Science in England, and on Some of its Causes* (London: B. Fellowes, 1830). On the 'decline of science' debate see Chapter 2 below. The 'Declinists' and 'Analyticals' overlapped, with Charles Babbage and John Herschel being spokesmen for both, and, generally, both groups were made up of reformist Whigs, Foote, 'The Place of Science'.
81. Brougham to William Somerville, 27 March 1827, in M. Somerville, *Queen of Science: Personal Recollections of Mary Somerville*, ed. D. McMillan (Edinburgh: Canongate, 2001), pp. 131–2. Somerville's *Mechanism of the Heavens* (1831) was ultimately published by John Murray, not the SDUK. This was probably because the work, 'from its analytical character, could only be read by mathematicians', ibid., p. 176.

82. Elphinstone in Biot, 'Life of Newton', p. 63.
83. Ibid., p. 1. The translation was first laid before the Publication Committee on 1 November 1828, and it was resolved to return the manuscript to Elphinstone, presumably with comments, on 5 March 1829. On 20 May 1829 the committee discussed the length of the printed manuscript, UCL, SDUK Papers, SDUK 5, Minutes of Sub-Committees, 1827–9, pp. 78, 137, 180.
84. This was recorded on 1 November 1828 in the Publication Committee minutes, and two members were appointed to make enquiries, ibid., p. 78.
85. Similarly, John Herschel had told Laplace that the story 'would require much stronger evidence than that of a casual expression in a single letter from Huyghens to Leibnitz to establish such a fact in opposition to the *speaking silence* of all his other contemporaries', RSL, Herschel Papers, JFWH 19.7, letter from Herschel to Laplace, 18 March 1824.
86. A. de la Pryme, *The Diary of Abraham de la Pryme, the Yorkshire Antiquary*, ed. C. Jackson, Publications of the Surtees Society, vol. 54 (Durham, London and Edinburgh: Andrews, 1870), p. 23. There is a discrepancy in the dating of Newton's illness from the evidence of Huygens (about November 1692) and de la Pryme (before January 1692). Brewster subsequently used this to cast doubt on the story, while Biot, 'Revue de *The Life of Newton*', wrongly suggested that de la Pryme's diary was dated by the old calendar. None of these dates fit easily with the evidence from Newton's letters to Locke or Pepys in September 1693.
87. P. King, *The Life of John Locke, with Extracts from his Correspondence, Journals, and Common-Place Books* (London: Colburn, 1829), pp. 224–5.
88. [Roebuck], 'Life of Mahomet', p. 13; H. P. Brougham, *Lives of Men of Letters and Science, who flourished in the Time of George III*, 2 vols (London: C. Knight and Co., 1845, 1846), vol. 1, p. 302.
89. Ibid., p. 30.
90. See [Jeffrey], 'Memoirs of Sir James Mackintosh', pp. 208–9, for a comparable faith in the moral lessons that archive-based history or biography would teach. This point of view is different to that of De Morgan (see Chapter 4 below), who believed that historians should not make such assumptions.
91. Geoffrey Cantor has pointed to two biographical styles – the Romantic and the Smilesian – and their link to the theoretically speculative and empirical methodologies respectively, Cantor, 'The Scientist as Hero'. See also M. Shortland and R. Yeo, 'Introduction', in Shortland and Yeo (eds), *Telling Lives in Science*, pp. 1–44, p. 31.
92. Shapin and Barnes, 'Science, Nature and Control'; Biot, 'Life of Newton', p. 11; [Biot], 'Newton (Isaac)', p. 137.
93. However, in this case the Publication Committee was prepared to be flexible: The Secretary was 'directed to apply to the General Committee' to make arrangements regarding the length, 'any curtailment of the work appearing impossible', UCL, SDUK Papers, SDUK 5, Minutes of Sub-Committees, 1827–9, 20 May 1829, p. 180.
94. The University of London (founded 1826), changed its name to University College when it received its charter in 1836. I will hereafter refer to it as UCL regardless of date.
95. Quoted in P. Corsi, *Science and Religion: Baden Powell and the Anglican Debate, 1800–1860* (Cambridge: Cambridge University Press, 1988), p. 97.
96. Brewster, *Life of Newton*, p. 283.
97. Desmond, *The Politics of Evolution*, p. 406. Desmond also suggested that 'the entire Broughamite educational empire suffered from a radical-Whig ideological split, with the radicals arguing for a more materialist self-determining nature, and the Paleyites

promoting a delegated divine power of arrangement', A. J. Desmond, 'Lamarkism and Democracy: Corporations, Corruption and Comparative Anatomy in the 1830s', in J. R. Moore (ed.), *History, Humanity and Evolution: Essays for John C. Green* (Cambridge: Cambridge University Press, 1989), pp. 99–130, on p. 118.

98. Topham, 'Science and Popular Education', pp. 404–7, quotation on p. 407; H. P. Brougham, *A Discourse on Natural Theology* (London: Charles Knight, 1835), p. 6. W. Paley, *Paley's Natural Theology Illustrated*, ed. H. Brougham and C. Bell, 5 vols (London: Charles Knight, 1835–9).
99. Brougham, *A Discourse on Natural Theology*, p. 6.
100. Ibid., p. 5. Daunou's discussion of Newton's chronology was omitted from the SDUK translation of Biot's article.
101. Ibid., p. 193.
102. Ibid., p. 17.
103. [Brougham], 'King's Life of Locke', p. 23. The general regret at losing Diamond is indicated by a reference to the 'beautiful but traditional story of [Newton's] dog' in C. R. Weld, *History of the Royal Society, with Memoirs of the Presidents. Compiled from Authentic Documents*, 2 vols (London, 1848), vol. 1, p. 371.
104. [Brougham], 'King's Life of Locke', pp. 23, 22, 25.
105. See Worrall, 'Thomas Young and the "Refutation" of Newtonian Optics', pp. 108–9. This version of history was promulgated particularly by Whewell and Peacock, see e.g. Whewell, *History of the Inductive Sciences*, vol. 2, pp. 346–9. However, we are apt to be kinder to Brougham and to recognize problems in Young's theory and methodology, see Cantor, 'Henry Brougham'; Cantor, *Optics after Newton*, pp. 4–7, 144–6; Worrall, 'Thomas Young and the "Refutation" of Newtonian Optics', pp. 109–12.
106. Cantor, 'Henry Brougham', p. 85.
107. Shapiro, *Fits, Passions and Paroxysms*, p. 282.
108. D. Brewster, 'Report on the Recent Progress of Optics', *Report of the First and Second Meetings of the British Association for the Advancement of Science* (London: John Murray, 1833), pp. 308–22, on pp. 312–14.
109. Quoted in Cantor, *Optics after Newton*, p. 84. However, Brewster, who accepted the role of hypothesis in discovery, wrote of the 'beautiful theory of Fits', ibid., p. 86.
110. [H. P. Brougham], 'Bakerian Lecture on Light and Colours', *Edinburgh Review*, 1 (1803), pp. 450–6, on p. 455. Brougham saw Newton's earlier discussions of an ether as 'playful relaxations', and his interpretation of the 1675 Hypothesis was similar to Brewster's, see Chapter 5 below.
111. H. P. Brougham and E. J. Routh, *Analytical View of Sir Isaac Newton's Principia* (London: Longman, Brown, Green & Longman, 1855), p. xxviii. Brougham's interest in the history of science and philosophy was also manifest in Brougham, *Lives of Men of Letters*.
112. UCL, Brougham Papers 26,633, letter from Brewster to Brougham, 8 October 1847.
113. UCL, Brougham Papers 26,679, letter from Brewster to Brougham, 11 July 1854; Brougham Papers 26,680, Brewster to Brougham, 8 August 1854; Brougham Papers 11,248, Brewster to Brougham, 11 November 1854; Brougham Papers 31,833, Brewster to Brougham, n.d.; Brougham Papers 26,795, Brewster to Brougham, n.d.; Brougham Papers 17,101, letter from De Morgan to Brougham, 11 July 1854 and correspondence between De Morgan and Brougham in RAS, MSS De Morgan 1.

2 David Brewster's *Life of Sir Isaac Newton*

1. John Clare to Thomas Inskip, 10 August 1824, quoted in J. Clare, 'Popularity in Authorship', *The European Magazine*, 1 (1825), ed. J. Birtwhistle, http://www.johnclare.info/bitwhistle.htm (accessed 5 May 2006), n. 20.
2. M. M. Gordon, *The Home Life of Sir David Brewster* (Edinburgh: Edmonston & Douglas, 1869), pp. 260, 169.
3. Ibid., p. 260.
4. Herschel's *A Preliminary Discourse on the Study of Natural Philosophy*, in D. Lardner (ed.), *The Cabinet Cyclopaedia*, 133 vols (London: Longman, 1830–49), vol. 14 (1831), championed Bacon, but presented a more sophisticated version of his inductive methodology, informed by modern examples. See R. Yeo, 'An Idol of the Marketplace: Baconianism in Nineteenth-Century Britain', *History of Science*, 23 (1985), pp. 251–98.
5. National Library of Scotland, Add. MS 1808, f. 31, 'Advertisement for Brewster's *Life of Newton*'. In opposition to these claims, when writing the 1855 *Memoirs of Newton*, Brewster told Brougham that his *Life* 'was a mere sketch, drawn up from materials known to every body', UCL, Brougham Papers 11,248, Brewster to Brougham, 11 November 1854.
6. [Malkin], 'Brewster's *Life of Newton*', p. 3.
7. P. Wallis and R. Wallis, *Newton and Newtoniana, 1672–1975: A Bibliography* (Folkestone: Dawson, 1977), pp. 228–30; [C. R. Weld], 'Review of Brewster's *Memoirs of Sir Isaac Newton*', *Athenaeum*, 1442 (1855), pp. 697–9. Of the first edition of 12,500 copies, 6,669 were sold and the remainder went to Thomas Tegg, who presumably sold them before the second edition was printed, S. Bennett, 'John Murray's Family Library and the Cheapening of Books in Early Nineteenth Century Britain', *Studies in Bibliography*, 29 (1976), pp. 139–66, on p. 164. The volume cost 5 shillings, John Murray Archives, Ledgers.
8. Bennett, 'John Murray's Family Library', pp. 140–1. J. A. Secord, *Victorian Sensation: The Extraordinary Publication, Reception, and Secret Authorship of Vestiges of the Natural History of Creation* (Chicago, IL, and London: University of Chicago Press, 2000), pp. 50–1, also discusses the Family Library as an attempt to stop the dominance of Whigs in cheap publishing, a format which 'threatened to take over entirely' during the years of reform agitation.
9. William Whewell, quoted in D. Brewster (ed.), *Testimonials in Favour of Sir David Brewster as a Candidate for the Chair of Natural Philosophy at the University of Edinburgh* (Edinburgh, 1832), pp. 12–13.
10. Brewster received £300 each for his two contributions to the Family Library, National Library of Scotland, Accessions, Acc. 11678, letter from Brewster to Thomas Cadell, 19 March 1831; John Murray Archives, Ledgers and Brewster Correspondence, 'Assignment of Copyright for "The Life of Isaac Newton"', 1850, and Brewster to Murray, 1 September 1841.
11. Seven were devoted to discussing the history of optics, the science in which Brewster had made his name. Only two chapters discussed astronomy and the *Principia*, while mathematics and theology received one chapter apiece. The discussion of Brewster's history of science is left to Chapter 5: Brewster's second biography of Newton repeated much of the earlier historical sections but was informed by his later antagonism to the wave theory of light.

12. Brewster, *Life of Newton*, p. vi. Page references will hereafter be given in the text.
13. BLO, MSS Rigaud 60, f. 77, letter from Brewster to S. P. Rigaud, 26 September 1828. The surviving correspondence only indicates that Brewster saw Elphinstone's proofs in May 1829, but he undoubtedly already knew Biot's original biography, having been in regular correspondence since 1814, UCL, SDUK Correspondence, letter from Brewster to Thomas Coates, 16 May 1829; Gordon, *The Home Life*, pp. 80–4.
14. National Library of Scotland, Add. MS 1765, f. 84, letter from Brewster to Thomas Carlyle, 21 January 1828.
15. [D. Brewster], 'The British Scientific Association', *Edinburgh Review*, 60 (1835), pp. 363–94, on p. 380.
16. Quoted in J. Gascoigne, 'From Bentley to the Victorians: The Rise and Fall of British Newtonian Natural Theology', *Science in Context*, 2 (1988), pp. 219–56, on p. 234.
17. Yeo, 'Genius, Method and Morality', p. 266.
18. However, Brewster seems to have had a personal attachment to the apple story, for, like Biot, he apparently collected some roots from the legendary tree when visiting Woolsthorpe, Gordon, *The Home Life*, p. 259.
19. Yeo has discussed nineteenth-century Baconianism among Scottish natural philosophers, including Brewster's teachers. Except on the merits of Bacon, Brewster's position was similar to that of Robison, for both pointed to the danger of purely conjectural systems. Brewster's opinion of the wave theory of light, which assumed the existence of an ether, demonstrates the danger he saw in premature hypothesizing. Equally, Robison, despite his appreciation of Bacon, acknowledged a role for hypotheses, provided that they were tested by experiment. According to Macvey Napier, John Leslie, another of Brewster's teachers, 'denied all merit and influence to the immortal delineator of the Inductive Logic', Yeo, 'An Idol of the Marketplace', pp. 265–6.
20. Quoted in Yeo, 'An Idol of the Marketplace', p. 266.
21. Whewell, who saw a role for method, as well as what he termed Fundamental Ideas, in science, believed the history of science could teach the philosophy or methodology of science, while Brewster, who emphasized the role of genius, saw biography as the appropriate form. Shortland and Yeo, 'Introduction', p. 20. See also Yeo, *Defining Science*, pp. 116–18.
22. Ibid., pp. 139.
23. See p. 000 above; Gerard, *An Essay on Genius*.
24. Babbage, *Reflections on the Decline of Science*. See D. P. Miller, '"Into the valley of darkness": Reflections on the Royal Society in the Eighteenth Century', *History of Science*, 27 (1989), pp. 155–66, and 'Between Hostile Camps: Sir Humphry Davy's Presidency of the Royal Society of London 1820–1827', *British Journal for the History of Science*, 16 (1983), pp. 1–47.
25. BL, Babbage Correspondence, Add. MS 37,185, f. 23, letter from T. J. Hussey to Babbage, [23 January 1830].
26. See M. B. Hall, *All Scientists Now: The Royal Society in the Nineteenth Century* (Cambridge: Cambridge University Press, 1984), pp. 45–62. On Babbage's reform activities, see A. Hyman, *Charles Babbage: Pioneer of the Computer* (Oxford: Oxford University Press, 1982), pp. 82–92.
27. Babbage, *Reflections on the Decline of Science*, pp. 19, 23. On the Board of Longitude, see D. P. Miller, 'The Royal Society of London, 1800–1835: A Study of the Cultural Politics of Scientific Organization' (unpublished PhD thesis, University of Pennsylvania, 1981), pp. 313–27.

28. Babbage, *Reflections on the Decline of Science*, pp. 28, 29–34.
29. BL, Babbage Correspondence, Add. MS 37,185, f. 229, letter from Brewster to Babbage, 16 June 1830.
30. Ibid., f. 49, letter from Brewster to Babbage, 12 February 1830.
31. Ibid., ff. 72, 229, 261, letters from Brewster to Babbage, 12 February 1830, 24 February 1830, 16 June 1830, 12 July 1830.
32. UCL, Brougham Papers 26,608, letter from Brewster to Brougham, 14 March 1829.
33. BL, Babbage Correspondence, Add. MS 37,185, f. 229, letter from Brewster to Babbage, 16 June 1830.
34. Morrell and Thackray, *Gentlemen of Science*. Examples of works containing the earlier emphasis on Brewster, Babbage and professionalization are O. J. R. Howarth, *The British Association for the Advancement of Science: A Retrospect, 1831–1921* (London: BAAS, 1922); and L. Pearce Williams, 'The Royal Society and the Founding of the British Association for the Advancement of Science', *Notes and Records of the Royal Society*, 16 (1961), pp. 221–33.
35. See Barton, '"Men of Science"', on the problematic nature of the terms 'professional' and 'amateur', and alternatives, commonly used to indicate inclusion within the scientific community from the 1850s.
36. [Brewster], 'The British Scientific Association', p. 382.
37. [D. Brewster], 'Decline of Science in England', *Quarterly Review*, 43 (1830), pp. 305–42, on pp. 341–2. While Brewster clearly shared William Whewell's elevated notion of scientific discoverers, they did not, for Brewster, take the directional role of Whewell's scientific clerisy. See T. L. Alborn, 'The Business of Induction: Industry and Genius in the Language of British Scientific Reform, 1820–1840', *History of Science*, 34 (1996), pp. 91–121, on pp. 94–5.
38. [Brewster], 'Decline of Science', p. 327. Brewster gave Young as an example of men of science who wasted their abilities on literary work, but was obviously thinking of himself.
39. Ibid., p. 325.
40. Brewster's view was similar to that of E. L. Bulwer-Lytton, who outlined three classes of scientific mind, two of which might benefit from government patronage. The highest, and rarest, minds were those to whom we owe 'the discovery, or the full establishment of the primary and general principles' and had 'habits of mind and modes of inquiry only obtained by long years of profound thought and abstract meditation'. It was doubtful whether the number or productivity of these minds could be increased, Bulwer-Lytton, *England and the English* (1833), quoted in Foote, 'The Place of Science', p. 203.
41. But see Miller, '"Puffing Jamie"', pp. 16–17, for Brewster's opinion of Watt, whom he viewed as a philosopher and an original discoverer.
42. [Brewster], 'The British Scientific Association', pp. 380, 381–2. Brewster believed that men might be directed to collect data of various kinds, but thought this was better done by existing national, local or specialized societies than the intermittent, peripatetic BAAS.
43. [Brewster], 'The British Scientific Association', p. 382.
44. See Morrell and Thackray, *Gentlemen of Science*, pp. 267–9, 517, and M. Ruse, 'William Whewell: Omniscientist', in M. Fisch and S. Schaffer (eds), *William Whewell: A Composite Portrait* (Oxford: Clarendon Press, 1991), pp. 87–116, on pp. 95–8.
45. See ibid. and W. J. Ashworth, 'The Calculating Eye: Baily, Herschel, Babbage and the Business of Astronomy', *British Journal for the History of Science*, 27 (1994), pp. 409–41.
46. Schaffer, 'Scientific Discoveries', p. 413.

47. The very title of the collection of essays on Brewster suggests this: A. D. Morrison-Low and J. R. R. Christie (eds), *Martyr of Science: Sir David Brewster 1781–1868* (Edinburgh: The Royal Scottish Museum, 1984).
48. S. Shapin, 'Brewster and the Edinburgh Career in Science', in Morrison-Low and Christie (eds), *Martyr of Science*, pp. 17–23, on p. 19.
49. Gordon, *The Home Life*, p. 368. See also A. D. Morrison-Low, 'Brewster and Scientific Instruments', in Morrison-Low and Christie (eds), *Martyr of Science*, pp. 59–65.
50. Gordon, *The Home Life*, p. 95.
51. UCL, Brougham Papers 26,215, letter from Brewster to Brougham, 21 January 1832.
52. UCL, Brougham Papers 15,728 and 26,616, letters from Brewster to Brougham, 9 May 1832, 28 May 1832.
53. [Brewster], 'The British Scientific Association', p. 381.
54. Gordon, *The Home Life*, pp. 55, 56, 57.
55. Shapin, 'Brewster and the Edinburgh Career', p. 18.
56. Quoted in W. Cochran, 'Sir David Brewster: An Outline Biography', in Morrison-Low and Christie (eds), *Martyr of Science*, pp. 11–14, on p. 13. A similar opinion is stated in [Brewster], 'Decline of Science', p. 326, but the Edinburgh chair, contested in 1832, seems to have offered sufficient remunerative reward without pupils.
57. UCL, Brougham Papers 15,477, letter from Brewster to Brougham, 6 November 1832.
58. Gordon, *The Home Life*, p. 154.
59. BL, Babbage Correspondence, Add. MS 37,187, ff. 408–11, letter from Brewster to Babbage, 3 February 1833.
60. J. B. Morrell, 'Brewster and the Early British Association for the Advancement of Science', in Morrison-Low and Christie (eds), *Martyr of Science*, pp. 25–9.
61. Brewster told Macvey Napier that it 'was stated and perhaps justly that I might be able to lecture under an excitement; and yet not be able to lecture when the excitement was removed', BL, Napier Correspondence, Add. MS 34,616, f. 12, letter from Brewster to Macvey Napier, 23 June 1833.
62. Quoted in Shapin, 'Brewster and the Edinburgh Career', p. 19.
63. See Davie, *The Democratic Intellect*, pp. 107–86, quotation on p. 174; Morrell and Thackray, *Gentlemen of Science*, p. 480. Brewster was incensed that Forbes had been 'backed, by the testimonials of *Cambridge & Oxford Professors*, & other English Philosophers, who certified what was not true[?], and what was not within their knowledge', BL, Babbage Correspondence, Add. MS 37,187, ff. 408–11, letter from Brewster to Babbage, 3 February 1833. It seems likely that this contest was one of the reasons, if not *the* reason, behind Brewster's subsequent antagonism towards Cambridge and overt hostility regarding the wave theory. Prior to this Brewster had openly admired Cambridge, and emphasized Newton's debt to his education there, Brewster, *Life of Newton*, p. 13.
64. Quoted in Morrison-Low, 'Brewster and Scientific Instruments', p. 63.
65. Gordon, *The Home Life*, pp. 289–91.
66. [T. Galloway], 'French and English Biographies', p. 90.
67. D. Brewster, *The Martyrs of Science; or the Lives of Galileo, Tycho Brahe, and Kepler* (London: John Murray. 1841). In addition, this book revisited his opposition to Baconianism, particularly where he highlighted the importance of imagination to Kepler.
68. BLO, MSS Rigaud 60, f. 80, letter from Brewster to Rigaud, 15 September 1830.
69. TCL, Whewell Papers, Add.MS a.201/80, letter from Brewster to Lord Braybrooke (forwarded to Whewell), 14 April 1831. Richard Griffin, third Lord Braybrooke (1783–1858), whose brother George was Master of Magdalene College, Cambridge,

edited Pepys's diary and correspondence. He provided Brewster with letters between Pepys, Newton and John Millington, which he had not published in S. Pepys, *Memoirs of Samuel Pepys, Comprising his Diary from 1659 to 1669, deciphered by the Rev. J. Smith, from the Original Short-Hand MS. in the Pepysian Library, and a Selection from his Private Correspondence*, ed. R. Griffin, 2 vols (London: Henry Colburn, 1825). See *DNB*. These letters, which presumably once belonged to Magdalene College, have been in the British Library since 1992 (BL, Charnwell Autographs, Add. MS 70,951).

70. [Malkin], 'Brewster's *Life of Newton*', p. 17.
71. Newton to Bentley, 10 December 1692, 17 January 1693, 11 February 1693, 25 February 1693, in Turnbull et al., *The Correspondence of Newton*, vol. 3, pp. 233–6, 238–40, 244, 253–6.
72. Ibid., vol. 3, pp. 282–3, cites the Millington–Pepys correspondence from Brewster's 1855 biography and includes his emphasis, which does not exist in the original letter, BL, Charnwell Autographs, Add. MS 70,951, f. 296, John Millington to Samuel Pepys, 30 September 1693.
73. This is plausible as a contributory factor to Newton's illness, see Iliffe, 'Newton', pp. 141–3.
74. King, *The Life of Locke*, pp. 224–5. King printed the note by Stewart, which pointed to 'the ingenuous and almost infantine simplicity of Newton's letters', ibid., p. 223. Stewart quoted from these letters in his 'Dissertation on the Progress of Metaphysical, Ethical, and Political Philosophy', in M. Napier (ed.), *Supplement to the fourth, fifth, and sixth editions of the Encyclopaedia Britannica*, 6 vols (Edinburgh: A. Constable and Co., 1824).
75. J. C. Gregory, 'Notice concerning an Autograph Manuscript by Sir Isaac Newton, Containing Some Notes upon the Third Book of the Principia, and Found Among the Papers of Dr David Gregory ... (Read March 2. 1829)', *Transactions of the Royal Society of Edinburgh*, 12 (1834), pp. 66–76.
76. Biot, 'Revue de *The Life of Newton*', p. 265n.
77. Ibid., p. 194; 'less difficult to exercise, and more detailed'. See Paul, *Science and Immortality*, pp. 88–9, on the influence of Plutarchian biography on the *éloges* of the *Académie des Sciences*.
78. G. Pomata, 'Versions of Narrative: Overt and Covert Narrators in Nineteenth Century Historiography', *History Workshop Journal*, 27 (1989), pp. 1–17, on p. 1; Thierry quoted ibid., p. 3.
79. T. B. Macaulay, 'Hallam's *Constitutional History*' (1828), in T. B. Macaulay, *Essays, Critical and Miscellaneous* (New York, D. Appleton and Co., 1861), pp. 67–99, on p. 67. See M. Phillips, 'Macaulay, Scott, and the Literary Challenge to Historiography', *Journal of the History of Ideas*, 50 (1989), pp. 117–33, on pp. 117–20.
80. [D. Brewster], 'Macaulay's *History of England*', *North British Review*, 10 (1849), pp. 367–424, on pp. 421–2.
81. Biot, 'Revue de *The Life of Newton*', p. 195; 'The compliment is flattering to Cambridge, but it is completely at variance with the truth of his character, and with the very details that Dr. Brewster, like all biographers, records of Newton's childhood'.
82. Ibid., p. 196; 'everything testifies that the birth of this solitary genius was a gift of nature and not a work of art'.
83. Ibid., pp. 264–5; 'so completely breaks the logical order in which Newton's ideas were successively to develop'.
84. Ibid., p. 264; 'The succession of the inventions of these two great men, and their epistolary communications, are related with a clever dexterity, the characteristics of their

methods are presented as so analogous, the differences of their process of calculation as so slight, the irritation of the one as so quick compared to the meekness of the other, that all the faults, all the injustices seem to be on Leibnitz's side, if even we cannot blame him for anything else'.

85. Ibid., pp. 266, 267; 'an intention of infidelity'.
86. Ibid., p. 323; 'above all the memory of Laplace'.
87. Ibid., pp. 338–9.
88. RSL, Herschel Papers, JFWH 11.102, letter from Pierre-Simon Laplace to John Herschel, 18 April 1823.
89. Biot, 'Revue de *The Life of Newton*', p. 323; 'anti-religious mission'; 'very philosophical interest'.
90. Ibid., pp. 325–31, pp. 333, 339; 'the letters of Newton to Bentley are not of this class of philosophy'; 'Here is a type of argument which greatly illuminates literary debates, and scholars of the nineteenth century [will be] grateful to Dr. Brewster for having resurrected it for their use'.
91. [Malkin], 'Brewster's *Life of Newton*', p. 29.
92. The Wellesley Periodicals Index attributes authorship to B. H. Malkin senior, from the evidence of Malkin's letters to Napier in 1832. These letters, however, are clearly from his son as he refers to Arthur Malkin as his brother. See BL, Napier Correspondence, Add. MS 34,615, ff. 346, 354, 386, letters from B. H. Malkin to Macvey Napier, 14 June 1832, 28 June 1832, 14 August 1832.
93. See J. A. Venn, *Alumni Cantabrigienses*, Part 2 (Cambridge: Cambridge University Press, 1951).
94. [Malkin], 'Brewster's *Life of Newton*', pp. 7, 8.
95. Ibid., p. 3.
96. BL, Napier Correspondence, Add. MS 34,615, f. 346, letter from B. H. Malkin to Macvey Napier, 14 June 1832.
97. [Malkin], 'Brewster's *Life of Newton*', pp. 4, 6.
98. Ibid., pp. 8–9.
99. Ibid., pp. 11, 16.
100. [Brougham], 'King's Life of Locke', p. 23, [Malkin], 'Brewster's *Life of Newton*', pp. 17, 19.
101. Ibid., pp. 20, 22.
102. Ibid., pp. 22, 29.
103. Ibid., p. 29.
104. Hays, 'Science and Brougham's Society', pp. 234–5.
105. [B. H. Malkin], 'Astronomy', in *Natural Philosophy*, Library of Useful Knowledge (London: Baldwin & Cradock, 1829–38), vol. 3 (1834), pp. 1–2.
106. [B. H. Malkin?], 'Newton', in A. T. Malkin (ed.), *Gallery of Portraits: with Memoirs*, 7 vols (London: Charles Knight, 1833–7), vol. 1 (1833), pp. 79–88, on p. 83. This short article was not detailed or controversial, but comparable in outline to other short biographies of Newton. It contained Malkin's claim that Newton had experienced a period of mental illness and, although he recovered, was forced to change his lifestyle.

107. BL, Napier Correspondence, Add. MS 34,615, f. 346, B. H. Malkin to Macvey Napier, 14 June 1832. Malkin seems to have been happy with Napier's additions, for he saw the proofs and returned them without changes. However, he was due to leave the country the following day and would have had little time for revisions, ibid., f. 386, Malkin to Napier, 14 August 1832. Editorial additions must have been common: Whewell commented on additions to a *Quarterly Review* article he had written, which were 'interpolations of the worthy editors', Whewell to Richard Jones, 15 July 1831, quoted in I. Todhunter, *William Whewell, D.D., Master of Trinity College, Cambridge. An Account of his Writings with Selections from his Literary and Scientific Correspondence*, 2 vols (London: Macmillan, 1876), vol. 2, p. 123.
108. Yeo, 'An Idol of the Marketplace', pp. 259–6, Napier quoted on p. 261. Yeo suggested that Brewster anticipated a later reassessment of Bacon's legacy, and uses Malkin's review as evidence that such views were seen as eccentric in 1831. In fact, the explicit rejection of Bacon's methodology, as antithetical to a positive role for hypothesis and imagination in scientific thinking, was already current.
109. [Malkin], 'Brewster's *Life of Newton*', pp. 29, 32–3.
110. Ibid., pp. 36, 34.
111. Malkin's father also positioned himself against hero-worship in his inaugural lecture at UCL. He condemned as unhistorical 'the reading of history in the spirit of a romance; adopting a hero ... and taking a deep interest in his personal success'. He felt 'The vices of Alexander are too often overlooked by a young reader for the gallantry of his spirit; and Caesar enslaved Rome to be sure; but then what a complete gentleman he was!', B. H. Malkin, *An Introductory Lecture on History, delivered in the University of London on Thursday, March 11, 1830* (London, 1830), p. 7.
112. [Malkin], 'Brewster's *Life of Newton*', pp. 3, 2.
113. BL, Napier Correspondence, Add. MS 34,615, ff. 346, 386, letters from Malkin to Napier, 14 June 1832, 14 August 1832.
114. [Malkin], 'Brewster's *Life of Newton*', pp. 19, 21.
115. [A. De Morgan?], 'Newton', in Long (ed.), *Penny Cyclopaedia*, vol. 16 (1840), pp. 197–203, on p. 202. An annotated copy of the *Penny Cyclopaedia* at the British Library gives the author as 'Lecappelain'. I have found no other reference to such an individual and Jourdain credits the article to De Morgan in De Morgan, *Essays*, p. 3.
116. UCL, SDUK Papers, SDUK 9, Publication Committee Book, 16 October 1832, pp. 106–7.
117. [Galloway], 'French and English Biographies', p. 67.
118. Ibid., pp. 71, 89.
119. Ibid., p. 90.
120. [A. De Morgan], 'Review of *The Works of Francis Bacon*', *Athenaeum*, 1612 (1858), pp. 367–8. See Yeo, 'Genius, Method and Morality', pp. 268–9.
121. J. Davy, *Memoirs of the Life of Sir Humphry Davy*, 2 vols (London: Longman, Rees, Orme, Brown, Green & Longman, 1836), vol. 2, p. 410.
122. Gordon, *The Home Life*, pp. 260–1.
123. B. Powell, *History of Natural Philosophy, from the Earliest Periods to the Present Time*, in D. Lardner (ed.), *The Cabinet Cyclopaedia*, 133 vols (London: Longmans, 1830–49), vol. 51 (1834), pp. 358–9.

124. S. E. De Morgan, *Memoir of Augustus De Morgan ... With Selections from his Letters* (London: Longmans, Green & Co., 1882), p. 256.

3 Francis Baily's *Account of the Revd. John Flamsteed*

1. CUL, RGO MSS, Baily Papers, RGO 60/3, letter from Caroline Herschel to Baily, 15 February 1836.
2. Yeo, 'Genius, Method and Morality', p. 278. See also the similar, but extended, discussion in Yeo, *Defining Science*, pp. 129–34.
3. W. J. Ashworth, '"Labour harder than *thrashing*": John Flamsteed, Property and Intellectual Labour in Nineteenth-Century England', in F. Willmoth (ed.), *Flamsteed's Stars: New Perspectives on the Life and Work of the First Astronomer Royal* (Woodbridge: Boydell Press, 1997), pp. 199–216, on p. 199.
4. [J. Herschel], 'Memoir of Francis Baily', *Monthly Notices of the Royal Astronomical Society*, 6 (1844), pp. 89–128, on p. 119; A. Johns, *The Nature of the Book: Print and Knowledge in the Making* (Chicago, IL, and London: University of Chicago Press, 1998), p. 550. See also Fara, *Newton*, p. 253.
5. See accounts of the dispute by R. S. Westfall, *Never at Rest: A Biography of Isaac Newton* (Cambridge: Cambridge University Press, 1980), pp. 541–50, 655–66; and Johns, *The Nature of the Book*, pp. 543–62. Flamsteed's writings on Newton, mostly first published by Baily, are re-transcribed in Iliffe et al. (eds), *Early Biographies*, vol. 1, pp. 15–54.
6. Baily, *Account of Flamsteed*, p. xxii. Page references will hereafter be given in the text.
7. Flamsteed spent the rest of his life working on his version of the catalogue, a task continued after his death by his assistants Abraham Sharp and Joseph Crosthwait. It was published by his wife Margaret and J. Hodgson.
8. Quoted in Johns, *The Nature of the Book*, p. 549.
9. Ibid.
10. [A. De Morgan], 'Flamsteed, John', in Long (ed.) *Penny Cyclopaedia*, vol. 10 (1838), pp. 296–7, on p. 297.
11. Caroline Herschel had already corrected many of Flamsteed's errors in *Catalogue of Stars, taken from Mr Flamsteed's Observations contained in the Second Volume of the Historia Cœlestis, and not Inserted in the British Catalogue* (London, 1798).
12. [Herschel], 'Memoir of Baily', p. 110, see pp. 121–8 for a full bibliography.
13. Baily 'has had so much experience in the examination of Catalogues, that his authority will have great weight', letter from Thomas Maclear to Herschel, 4 March 1836, quoted in B. Warner and N. Warner, *Maclear and Herschel: Letters and Diaries at the Cape of Good Hope, 1834–1838* (Cape Town: A. A. Balkema, 1984), pp. 129–30. Baily's high status in the scientific community is clear from the fact that he was, in 1835, Vice-President of the RAS, Treasurer of the Royal Society and General Secretary of the BAAS.
14. Ashworth, '"Labour harder than *thrashing*"', and 'The Calculating Eye'; W. J. Ashworth, 'Memory, Foresight and Production: The Work of Analysis in Early 19th-Century England' (unpublished PhD thesis, 1996), pp. 13–35. On Baily see [Herschel], 'Memoir of Baily', [A. De Morgan], 'Baily, Francis', in Long (ed), *Penny Cyclopaedia*, Supplement 1 (1845), pp. 166–8; and L. G. H. Horton-Smith, *Francis Baily, the Astronomer, 1774–1844* (Newbury: Blacket, Turner & Co., 1938), and *The Baily Family of Thatcham and Later of Speen and of Newbury, all in the County of Berkshire* (Leicester: W. Thornley & Sons, 1951), pp. 69–79.
15. [Herschel], 'Memoir of Baily', p. 118.

16. [De Morgan], 'Baily, Francis', p. 168.
17. RSL, Herschel Papers, JFWH 6.183, letter from Augustus De Morgan to John Herschel, 20 February 1837.
18. Letter from Augustus De Morgan to William Frend, n.d., quoted in S. E. De Morgan, *Threescore Years and Ten: Reminiscences of the Late Sophia Elizabeth De Morgan, to which are Added Letters to and from her Husband the Late Augustus De Morgan, and Others*, ed. M. A. De Morgan (London: Richard Bentley & Son, 1895), p. 125.
19. Airy and others noted that the original observations of James Bradley and Nevil Maskelyne, third and fourth Astronomers Royal, 'were considered private property', causing severe delays in publication, G. B. Airy, 'Report on the Progress of Astronomy during the Present Century', *Report of the British Association for the Advancement of Science*, p. 127; Baily, *Account of Flamsteed*, pp. 731–3.
20. See C. MacLeod, 'Concepts of Invention and the Patent Controversy in Victorian Britain', in R. Fox (ed.), *Technological Change*, Studies in the History of Science, Technology and Medicine (Amsterdam: Harwood Academic Publishers, 1996), pp. 137–53, for the connection between patent laws and ideas about genius versus co-operative effort. However, Baily, the champion of co-operative science, apparently supported the claim to individual copyright.
21. C. Rumker, 'Astronomical Observations made at the Observatory at Paramatta in New South Wales', *Philosophical Transactions*, 119 (1829), pp. 1–152.
22. RAS, MSS Baily 5, 'Letters and Papers relating to Rumker's observations 1830', 'Statement of Sir Thomas Brisbane', ff. 13–14.
23. Ibid., 'Notice', f. 4.
24. Ibid., 'Report on the Paramatta Observations', f. 5; ibid., ff. 95–6, draft letter from Francis Baily to Thomas Brisbane, 13 February 1830.
25. Ibid., 'Report on the Paramatta Observations', f. 5.
26. Ibid., ff. 93–4, draft letter from Baily to Brisbane, 28 January 1830. Baily saw this episode as proof of the poor judgment of some members of the Royal Society Council, who had accepted and printed Rumker's copied observations, and he offered it to Babbage as material for his *Reflections on the Decline of Science*, BL, Babbage Correspondence, Add. MS 37,185, f. 55, letter from Baily to Charles Babbage, [16 February 1830].
27. RSL, Herschel Papers, JFWH 3.122, Baily to Herschel, 18 August 1834.
28. [De Morgan], 'Baily, Francis', p. 166; Ashworth, '"Labour harder than *thrashing*"', p. 203.
29. [Herschel], 'Memoir of Baily', pp. 104, 94.
30. Quoted in Ashworth, '"Labour harder than *thrashing*"', p. 203.
31. F. Baily, 'Report on the Pendulum Experiments made by the Late Captain Henry Foster, R. N., in his Scientific Voyage in the Years 1828–1831 with a View to Determine the Figure of the Earth', *Memoirs of the Royal Astronomical Society*, 7 (1834), p. 19. Baily's admiration of Foster's techniques – which included the use of pre-printed forms and the maintenance of books and papers 'in the most neat and scientific order' – echoes that of Flamsteed's record-keeping, especially since Baily was again using these records to publish results posthumously, ibid., p. 2.
32. RAS, MSS Baily 11, 'Letters from Mr. John Flamsteed [124] and Mr Joseph Crosthwait [60] addressed to Mr. Abraham Sharp during the years 1702–30, Copied from the Original Manuscripts; and presented to The Royal Astronomical Society', f. 4.
33. The amanuenses for the *Account* seem to have left something to be desired, for 'Baily's transcriptions were replete with errors and omissions', Johns, *The Nature of the Book*, p.

550. Some of these were flagged by critics, causing Baily embarrassment and forcing him to include two pages of errata in his *Supplement* to the *Account of Flamsteed*, pp. 750–1.
34. Levine, *The Amateur and the Professional*, p. 74. I agree with A. N. L. Munby's view that Baily's 'energetic search for manuscript materials both in private and institutional hands and in his meticulous listing and citation of these sources' filled a 'pioneer role', A. N. L. Munby, *The History and Bibliography of Science in England: The First Phase, 1833–1845* (Berkeley and Los Angeles: School of Librarianship and Graduate School of Library Service, 1968), p. 4.
35. F. Baily, 'Some Particulars relative to the Life and Writings of the Late Mr. Flamsteed, Never Yet Published', *Monthly Notices of the Royal Astronomical Society*, 3 (1833), pp. 4–10, on p. 10.
36. BL, Babbage Correspondence, Add. MS 37,188, f. 87, letter from Baily to Babbage, 9 November 1833.
37. RAS, MSS Baily 11, 'Letters from Flamsteed to Crosthwait'; *Monthly Notices of the Royal Astronomical Society*, 3 (1834), p. 29, entry for 14 February 1834.
38. RAS, MSS Baily 1, letter from Brewster to Baily, 5 February 1834.
39. RSL, Herschel Papers, JFWH 3.121, Baily to Herschel, 12 February 1834.
40. National Archives, Admiralty Papers, ADM 1/4282, letter from Baily to the Duke of Sussex, 3 June 1834, quoted in 'Resolutions passed at the Annual Visitation of the Royal Observatory 7 June 1834'.
41. Baily, 'Report on the Pendulum Experiments'.
42. Ashworth, '"Labour harder than *thrashing*"', p. 203.
43. [Herschel], 'Memoir of Baily', pp. 118–19.
44. This was seen as a good way of dealing with controversies in biography, Altick, *Lives and Letters*, pp. 191–8. Robert Southey commented that autobiographical materials should only be used 'where the biographer is conscious of a paucity of materials for his own share of the work, or of some nice and delicate points in the story, upon which he does not choose to express himself with the responsibility of an author', [Southey], 'Hayley's *Life*' p. 457.
45. Baily only had a maximum of eight days to both view and consider the Portsmouth Papers: on 24 April 1835 Henry Fellowes wrote in anticipation of his visit and on 2 May 1835 Airy gave his opinion of the Preface, CUL, RGO MSS, Baily Papers, RGO 60/4, 60/3.
46. See [Herschel], 'Memoir of Baily', p. 95.
47. Astronomical Society, *Address and Regulations of the Astronomical Society of London* (London, 1821), p. 6.
48. R. Grant, *History of Physical Astronomy, from the Earliest Ages to the Middle of the Nineteenth Century: Comprehending a Detailed Account of the Establishment of the Theory of Gravitation by Newton, and its Development by his Successors; with an Exposition of the Progress of Research on all the Other Subjects of Celestial Physics* (London: Baldwin, 1852), p. 477.
49. BLO, MSS Rigaud 60/22, Baily to S. P. Rigaud, 17 March 1836.
50. Brewster, *Life of Newton*, pp. 243–4.
51. F. Baily, 'Some Account of the Astronomical Observations made by Dr. Edmund Halley, at the Royal Observatory at Greenwich', *Memoirs of the Royal Astronomical Society*, 8 (1835), pp. 169–90, on pp. 170, 188, (read 14 November 1834).
52. National Archives, Admiralty Papers, ADM 1/4282, 'Resolutions Passed at the Annual Visitation of the Royal Observatory 7 June 1834', annotation dated 11 June 1834.

53. This study provides a contrast to other studies on publications of this period, where increasing literacy and cheap book-production means that achieving precise data on readership is difficult, see Secord, *Victorian Sensation*, pp. 42–68; J. R. Topham, 'Scientific Publishing and the Reading of Science in Nineteenth-Century Britain: A Historiographical Survey and Guide to Sources', *Studies in the History and Philosophy of Science*, 31 (2000), pp. 559–612.
54. Letter from De Morgan to William Frend, September 1835, in De Morgan, *Threescore Years and Ten*, p. 130. Baily did ensure that copies were made available at libraries and societies.
55. Baily's list was first drawn up in June 1835, but was checked twice by the Admiralty, CUL, RGO MSS, Baily Papers, RGO 60/4, letters from A. B. Becher to Baily, 11 September 1835, 29 September 1835.
56. James Epps, W. Robertson and Beaufort also gave suggestions for the list, CUL, RGO MSS, Baily Papers, RGO 60/4.
57. See F. Baily, *Journal of a Tour in Unsettled Parts of North America in 1796 and 1797*, ed. A. De Morgan (London: Baily Bros, 1856).
58. [Herschel], 'Memoir of Baily', p. 97.
59. Those who were not Fellows of the RAS were mostly included because they had given help to Baily. These included the librarian of the British Museum, Sir Henry Ellis; the keeper of Public Records, Joseph Hunter; the keeper of State Papers, Robert Lemon; and others with antiquarian expertise, including Nicholas Harris Nicolas and Dawson Turner, or who owned relevant documents, such as Henry and Newton Fellowes.
60. See D. P. Miller, 'The Revival of the Physical Sciences in Britain, 1815–1840', *Osiris*, 2nd series, 2 (1986), pp. 107–34, on these three groups and their importance in the revival of the physical sciences and the foundation of the RAS.
61. BL, Babbage Correspondence, Add. MS 37,182, f. 291, letter from Baily to Babbage, 11 November 1820.
62. Anon., *An History of the Instances of Exclusion from the Royal Society ... with Strictures on the Formation of the Council, and Other Instances of the Despotism of Sir J. Banks* (London: Debrett, 1784), p. 20; Miller, 'The Royal Society of London', p. 10.
63. J. L. Heilbron, 'A Mathematicians' Mutiny with Morals', in P. Horwich (ed.), *World Changes: Thomas Kuhn and the Nature of Science* (Cambridge, MA, and London: MIT Press, 1993), pp. 81–129.
64. Anon., *An History of the Instances*, p. 17; S. Horsley et al., *An Authentic Narrative of the Dissensions and Debates in the Royal Society, Containing the Speeches at Large of Dr. Horsley, Dr. Maskelyne, Mr. Maseres, Mr. Poore, Mr. Glennie, Mr. Watson, and Mr. Maty* (London, 1784), p. 66. See also J. Gascoigne, *Joseph Banks and the English Enlightenment: Useful Knowledge and Polite Culture* (Cambridge: Cambridge University Press, 1994), pp. 62–3.
65. Dr Lort, quoted in Weld, *History of the Royal Society*, vol. 2, p. 169.
66. D. P. Miller, 'Method and the "Micropolitics" of Science: The Early Years of the Geological and Astronomical Societies of London', in J. A. Schuster and R. R. Yeo (eds), *The Politics and Rhetoric of Scientific Method: Historical Studies* (Dordrecht: Reidel, 1986), pp. 227–57, on p. 246.
67. Airy, *Autobiography*, pp. 92–3. Miller, 'The Royal Society of London', p. 328.
68. E.g. F. Baily, *Further Remarks on the Present Defective State of the Nautical Almanac. To which is added an Account of the New Astronomical Ephemeris published at Berlin* (London: C. J. G. & F. Rivington, 1829), [J. South], *Refutation of the Numerous Mistate-*

ments and Fallacies contained in a Paper Presented to the Admiralty by Dr. Thomas Young (Superintendent of the Nautical Almanac) and Printed by Order of the House of Commons (London, 1829). Quotation in Hall, *All Scientists Now*, p. 43.

69. Letter from De Morgan to William Frend, 21 September [1835], in De Morgan, *Threescore Years and Ten*, p. 131.
70. Miller, 'The Revival of the Physical Sciences in Britain', p. 109. Flamsteed's £100 salary remained the payment for the Astronomer Royal until Airy's appointment in 1835. This sum was, however, usually increased by the grant of an ecclesiastical living.
71. RSL, Herschel Papers, JFWH 6.261, letter from De Morgan to Herschel, 25 August 1852.
72. See extract of letter from Herschel to Francis Beaufort, 3 July 1836, copied by Baily, in the De Morgan Collection copy of the *Account of Flamsteed*, University of London Library, Senate House.
73. CUL, RGO MSS, Baily Papers, RGO 60/3, letter from Caroline Herschel to Baily, 15 February 1836. This suggests that she did not, as Johns claims, ultimately 'declare for Whewell' and his defence of Newton, Johns, *The Nature of the Book*, p. 619.
74. RSL, Herschel Papers, JFWH 3.128, letter from Herschel to Baily, 17 August 1835.
75. CUL, RGO MSS, Baily Papers, RGO 60/3, letter from George Airy to Baily, 2 May 1835. Baily also sent his Preface to Galloway and Hunter; De Morgan claimed that Baily 'was one of the few men who ask advice to get the means of coming to a decision', [De Morgan], 'Baily, Francis', p. 168.
76. Airy, 'Report on the Progress of Astronomy', p. 182; Airy quoted in Morrell and Thackray, *Gentlemen of Science*, p. 481. On Airy see A. Chapman, 'Sir George Airy (1801–1892) and the Concept of International Standards in Science, Timekeeping and Navigation', *Vistas in Astronomy*, 28 (1985), pp. 321–8.
77. CUL, RGO MSS, Baily Papers, RGO 60/3, letter from Adam Sedgwick to Baily, 5 November 1835.
78. Ibid., letter from John Rickman to Baily, 15 October 1835.
79. Ibid., letter from John Bostock to Baily, 17 September 1835.
80. CUL, RGO MSS, Baily Papers, RGO 60/4, letter from William Wallace to Baily, 19 April 1834.
81. CUL, RGO MSS, Baily Papers, RGO 60/3, letter from George Innes to Baily, 22 February 1836.
82. Letter from Thomas Maclear to John Herschel, 4 March 1836, in Warner and Warner, *Maclear and Herschel*, p. 130. It is perhaps not coincidental that Pond had apparently pressed for the printing of Flamsteed's autobiographical writings, as well as the catalogue, at public expense, BLO, MSS Rigaud 60/9, letter from Baily to Rigaud, 29 October 1833.
83. CUL, RGO MSS, Baily Papers, RGO 60/3, letter from John Britton to Baily, 30 August 1835.
84. Ibid., letter from John Britton to Baily, 25 December 1835.
85. Ibid., letter from Richard Phillips to Baily, 21 November 1835. Phillips had been imprisoned for selling Paine's *Rights of Man*, see *DNB*.
86. A. De Morgan, *A Budget of Paradoxes ... Reprinted, with the Author's Additions, from the Athenaeum*, ed. S. E. De Morgan (London: Longmans, Green & Co., 1872), p. 242. Phillips came up with an alternative to the Newtonian system.
87. Phillips claimed his 'impressions about Newton' were 'derived from the late Dr. Hutton who had access to his papers at Hurtsbourne, & also from conversations with old

inhabitants of Woolstrope [sic]', CUL, RGO MSS, Baily Papers, RGO 60/3, letter from Richard Phillips to Baily, 21 November 1835. Hall (ed.), *Newton*, p. 178, states that in his *Mathematical and Philosophical Dictionary* (1795–6) Hutton 'makes no claim to have examined the Portsmouth papers himself', but Phillips's comment suggests that he did.

88. CUL, RGO MSS, Baily Papers, RGO 60/3, letter from Thomas Brisbane to Baily, 26 January 1836.
89. Miller, 'The Revival of the Physical Sciences in Britain', p. 109.
90. Baily was Anglican, but many of his friends were Nonconformist, including Priestley, Hunter, De Morgan and the Unitarian Minister James Martineau, Horton-Smith, *The Baily Family*, p. 68. These friendships also indicate his political views, as does his visit to republican America less than a decade after the French Revolution, Baily, *Journal of a Tour*. Because of the difficulty of collecting such information, I am only sure that seven of Baily's recipients were non-Anglican, but my general impression is of a group that rejected links between science and established religion, and accompanying political and moral ideas.
91. CUL, RGO MSS, Baily Papers, RGO 60/3, letter from J. Wood to Baily, 21 November 1835.
92. Secord, *Victorian Sensation*.
93. CUL, RGO MSS, Baily Papers, RGO 60/3, letter from Thomas Galloway to Baily 6 May 1835, [B. Powell], 'Sir Isaac Newton and his Contemporaries', *Edinburgh Review*, 78 (1843), pp. 402–37, on p. 404, A. De Morgan, *Newton: His Friend: And His Niece*, ed. S. E. De Morgan and A. Cowper Ranyard (London: Elliot Stock, 1885; reprinted London: Dawson, 1968), p. 105.
94. E.g. Phillips and a correspondent who had 'reason to believe that there is not a single copy in Liverpool', CUL, RGO MSS, Baily Papers, RGO 60/3, letter from R. Prescot to Baily, 31 March 1836.
95. *The Times*, 5 February 1836, p. 6e.
96. CUL, RGO MSS, Baily Papers, RGO 60/3, letter from Basil Hall to Baily, 4 April 1836; see also ibid., Hall to Baily, 11 April 1836. Of course, 109 copies did not go to individuals.
97. Letter from W. H. Smyth to M. Somerville, 3 October 1835, quoted in Somerville, *Queen of Science*, pp. 169–70.
98. J. Barrow, *An Autobiographical Memoir of Sir John Barrow, Bart., Late of the Admiralty; including Reflections, Observations, and Reminiscences at Home and Abroad, from Early Life to Advanced Age* (London: John Murray, 1847), pp. 503–4.
99. [J. Barrow], 'Account of the Rev. John Flamsteed', *Quarterly Review*, 55 (1835), pp. 96–128, on pp. 97, 121.
100. Ibid., pp. 120, 128, 112.
101. Ibid., p. 108. Banks had told Barrow 'that all these new-fangled associations will finally dismantle the Royal Society, and not leave the old lady a rag to cover her', Barrow, *An Autobiographical Memoir*, p. 10.
102. Ibid., p. 37.
103. Ashworth, '"Labour harder than *thrashing*"', pp. 207–8, my emphasis.
104. Miller, 'Between Hostile Camps', p. 40.
105. Ashworth, '"Labour harder than *thrashing*"', p. 208.
106. See C. C. Lloyd, *Mr. Barrow of the Admiralty: A Life of Sir John Barrow 1764–1848* (London: Collins, 1970). Barrow's description of the disorder in which he found the

Admiralty's archives is very similar to Baily's description of those in the Royal Observatory, ibid., p. 109.
107. W. Whewell, *Newton and Flamsteed: Remarks on an Article in Number CIX of the Quarterly Review* (Cambridge and London: J. & J. J. Deighton, 1836), reprinted in Iliffe et al. (eds), *Early Biographies*, vol. 2, pp. 133–43. Whewell told Richard Jones that the pamphlet was 'scribbled off at a sitting after reading Barrow's article', Todhunter, *Whewell*, vol. 1, p. 96.
108. Whewell, *Newton and Flamsteed*, p. 133.
109. Ibid., p. 133.
110. Ibid., pp. 134, 139.
111. Ibid., p. 141.
112. [S. P. Rigaud], 'Review of *Newton and Flamsteed*. Remarks on an Article in Number CIX of the Quarterly Review by the Rev. William Whewell', *Philosophical Magazine*, 8 (1836), pp. 139–47, on pp. 139, 147.
113. Ibid., p. 143.
114. CUL, RGO MSS, Baily Papers, RGO 60/3, letter from W. S. Stratford to Baily, 5 February 1836. A letter to the *Philosophical Magazine* described the 'Note' as 'revolting on account of its coarseness', C. S., 'On Whiston, Halley, and the Quarterly Reviewer of the "Account of Flamsteed"', *Philosophical Magazine*, 8 (1836), pp. 225–6, on p. 225.
115. [J. Barrow?], 'Note on a Pamphlet Entitled "Newton and Flamsteed, by the Rev. Wm. Whewell, M. A. ...", *Quarterly Review*, 55 (1836), pp. 568–72.
116. [S. P. Rigaud], 'Observations on a Note respecting Mr. Whewell, which is Appended to No. CX, of the Quarterly Review', *Philosophical Magazine*, 8 (1836), pp. 218–25; W. Whewell, 'Remarks on a Note on a Pamphlet Entitled "Newton and Flamsteed"', *Philosophical Magazine*, 8 (1836), pp. 211–18; TCL, Whewell Papers, Add. MS a.211/81, letter from Rigaud to Whewell, 7 February 1836.
117. [Rigaud], 'Observations on a Note', p. 225.
118. TCL, Whewell Papers, Add. MS a.200/207, letter from Baily to Whewell, 27 January 1837.
119. TCL, Whewell Papers, Add. MS a.201/8, letter from Francis Beaufort to Whewell, 2 January 1836.
120. CUL, RGO MSS, Baily Papers, RGO 60/4, letters from S. P. Rigaud to Baily, 9 February 1836, 8 November 1835; S. P. Rigaud, 'Some Particulars respecting the Principal Instruments at Greenwich in the Time of Dr. Halley', *Memoirs of the Royal Astronomical Society*, 9 (1836), pp. 205–27. TCL, Whewell Papers, Add. MS a.211/82, letter from Rigaud to Whewell, 25 February 1836.
121. BLO, MSS Rigaud 60/21, letter from Baily to Rigaud, 11 March 1836; CUL, RGO MSS, Baily Papers, RGO 60/3, letter from Rigaud to Baily, 14 March 1836.
122. BLO, MSS Rigaud 60/22, letter from Baily to Rigaud, 17 March 1836.
123. Ashworth, '"Labour harder than *thrashing*"', pp. 108–10, Yeo, 'Genius, Method and Morality', pp. 271–2; Whewell, 'Remarks on a Note', p. 215.
124. TCL, Whewell Papers, Add. MS a.211/79, letter from Rigaud to Whewell, 25 January 1836.
125. BLO, MSS Rigaud 62/483, letter from Whewell to Rigaud, 25 May 1836; TCL, Whewell Papers, Add. MS a.211/84, letter from Rigaud to Whewell, 24 June 1836.
126. His paper on Halley's instruments and observational methods suggested that he was more meticulous than Baily allowed, while S. P. Rigaud, *Historical Essay on the First Pub-*

lication of Sir Isaac Newton's Principia (Oxford, 1838) revealed Halley's vital role in that undertaking.
127. BLO, MSS Rigaud 62/484, letter from Whewell to Rigaud, 7 January 1839 (also at TCL, Whewell Papers, O.15.47). The defence of Halley was completed by his son: S. J. Rigaud, *A Defence of Halley Against the Charge of Religious Infidelity* (Oxford: Ashmolean Society, 1844).
128. TCL, Whewell Papers, Add. MS a.211/88, letter from Rigaud to Whewell, 13 October 1836.
129. TCL, Whewell Papers, Add. MS a.211/84, letter from Rigaud to Whewell, 24 June 1836.
130. [T. Galloway], 'Life and Observations of Flamsteed – Newton, Halley, and Flamsteed', *Edinburgh Review*, 62 (1836), pp. 359–97, reprinted in Iliffe et al. (eds), *Early Biographies*, vol. 2, pp. 93–132.
131. TCL, Whewell Papers, Add. MS a.211/80, letter from Rigaud to Whewell, 31 January 1836.
132. CUL, RGO MSS, Baily Papers, RGO 60/3, letter from Thomas Galloway to Baily, 14 March 1836.
133. [T. Galloway], 'Astronomical Society of London – Recent History of Astronomical Science', *Edinburgh Review*, 51 (1830), pp. 81–114, on pp. 104, 103–4.
134. [Galloway], 'Life and Observations of Flamsteed', pp. 99, 101, 103.
135. Ibid., pp. 126, 127.
136. Ibid., pp. 127, 131; CUL, RGO MSS, Baily Papers, RGO 60/3, letter from Galloway to Baily, 11 July 1836.
137. [Galloway], 'Life and Observations of Flamsteed', p. 97.
138. Biot, 'Revue de *An Account*'; CUL, RGO MSS, Baily Papers, RGO 60/3, letter from B. Lailland[?] to Baily, 1 October [1836]; letter from W. S. Stratford to Baily, 30 December 1836.
139. ibid., letters from W. S. Stratford to Baily, 22 April 1836, 30 December 1836.
140. Biot revisited the issue of Newton's breakdown, noting that communications with Flamsteed were apparently interrupted in the vital period, and suggested that Newton's rising irritability may have been connected to a fragile state of mind. His other main point was that the published correspondence provided an important insight into the development of the theory of refraction and showed Newton to be 'le créateur de la théorie des refractions astronomiques commes il l'est de la théorie de la gravitation' ('the creator of the theory of astronomical refraction as he is of the theory of gravitation'). He believed that Baily had demonstrated that the generally held opinion of Flamsteed 'était injuste' ('was unjust'), but appreciated that his foremost concern had been to restore 'un grand monument astronomique' ('a great astronomical monument'), in the form of the *British Catalogue*, Biot, 'Revue de *An Account*', pp. 655, 158, 159.
141. Baily, *Supplement to the Account*, p. 676.
142. Ibid., p. 677. As similar points were raised by correspondents before publication these can hardly have been 'unexpected'.
143. See ibid., pp. 680–707, 685.
144. Ibid., pp. 709, 717, 716.
145. Ibid., p. 720.
146. Ibid., pp. 731–3.
147. Ibid., pp. 728–9.
148. Yeo, *Defining Science*, p. 134.

149. CUL, RGO MSS, Baily Papers, RGO 60/3, letter from George Peacock to Baily, 3 February 1837.
150. Ibid., letter from John Britton to Baily, 29 January 1837; letter from William Frend to Baily, n.d.
151. Whewell skirted around the controversy, but presented Newton as a remarkable and morally courageous thinker, Whewell, *History of the Inductive Sciences*, vol. 2, p. 141.
152. CUL, RGO MSS, Baily Papers, RGO 60/3, letter from Joseph Hunter to Baily, 19 June 1835.
153. TCL, Whewell Papers, Add. MS a.211/88, letter from Rigaud to Whewell, 13 October 1836. See S. P. Rigaud (ed.), *Correspondence of Scientific Men of the Seventeenth Century; Including Letters of Barrow, Flamsteed, Wallis, and Newton, Printed from the Originals in the Collection of the Right Hon. the Earl of Macclesfield*, 2 vols (Oxford, 1841, 1862).
154. See CUL, RGO MSS, Baily Papers, RGO 60/3, letter from Henry Fellowes to Baily, 24 April 1835; Newton Fellowes to Baily, 24 November 1835.
155. Ibid., letter from Walker Skirrow to Baily, 26 November 1835.
156. Ibid., letter from W. S. Stratford to Baily, 30 December 1836.
157. [Galloway], 'Life and Observations of Flamsteed', pp. 131–2.
158. BLO, MSS Rigaud 60/86, letter from Brewster to Rigaud, 21 May 1837.
159. *The Times*, 28 June 1837, p. 5b.

4 Newtonian Studies and the History of Science

1. [B. Powell], 'Sir Isaac Newton', *Edinburgh Review*, 103 (1856), pp. 499–534, on p. 500. This article is reprinted in Iliffe et al. (eds), *Early Biographies*, vol. 2, pp. 253–87.
2. BLO, MSS Rigaud 61/267, letter from Lee to Rigaud, 26 September 1831.
3. [Powell], 'Newton and his Contemporaries', p. 403. Rigaud, *Historical Essay*; Rigaud (ed.), *Correspondence of Scientific Men*.
4. A. De Morgan, 'References for the History of the Mathematical Sciences', *Companion to the Almanac* (1843), pp. 40–65, on p. 58. Listed before this comment were the biographies produced by the SDUK, including [Drinkwater], 'Galileo' and 'Kepler', considered 'both of the best kind of biography'.
5. Weld, *History of the Royal Society*, J. Edleston, *Correspondence of Sir Isaac Newton and Professor Cotes, including Letters of Other Eminent Men, now first Published from the Originals in the Library of Trinity College, Cambridge; together with an Appendix, containing other Unpublished Letters and Papers by Newton; with Notes, Synoptical View of the Philosopher's Life, and a Variety of Details Illustrative of his History* (Cambridge and London: John Deighton and J. W. Parker, 1850; reprinted London: Frank Cass, 1969); Brewster, *Memoirs of Newton*; Grant, *History of Physical Astronomy*.
6. H. F. Cohen, *The Scientific Revolution: A Historiographical Inquiry* (Chicago, IL, and London: University of Chicago Press, 1994) moves from Kant to Whewell to Comte. I. B. Cohen, *Revolution in Science* (Cambridge, MA, and London: Belknap Press, 1985), pp. 521–32, gives extracts from Playfair and Leslie but makes Whewell the last word on nineteenth-century historiography of science. D. C. Lindberg, 'Conceptions of the Scientific Revolution from Bacon to Butterfield: A Preliminary Sketch', in D. C. Lindberg and R. S. Westman (eds), *Reappraisals of the Scientific Revolution* (Cambridge: Cambridge University Press, 1990), pp. 1–26, points to Whewell and Comte before crediting the 'rehabilitation of medieval science' to Pierre Duhem. Laudan, 'Histories of the Sci

ences and their Uses' takes a wide view, chronologically and geographically, but stresses the 'rival accounts of progress' of Comte and Whewell.
7. Cohen, *The Scientific Revolution*, pp. 30–9.
8. A. Rice, 'Augustus De Morgan: Historian of Science', *History of Science*, 34 (1996), pp. 201–40, and P. A. Maccioni Ruju and M. Mostert, *The Life and Times of Guglielmo Libri (1802–1869): Scientist, Patriot, Scholar, Journalist and Thief: A Nineteenth-Century Story* (Hilversum: Verloren, 1995) do something to correct the balance by contextualizing the historical achievements of De Morgan and Libri. Nothing has been written on Edleston, who edited only one book, and Rigaud has been under-studied, with the exception of his contribution to the Flamsteed debates, but see J. Fauvel, R. Flood and R. Wilson (eds), *Oxford Figures: 800 Years of the Mathematical Sciences* (Oxford: Oxford University Press, 2000), pp. 160–1. Chasles's historical work has only received brief treatment in English, see e.g. *Dictionary of Scientific Biography*.
9. On Whewell's *History of the Inductive Sciences* (1837), see the essays in Fisch and Schaffer (eds), *William Whewell*, especially G. Cantor, 'Between Rationalism and Romanticism: Whewell's Historiography of the Inductive Sciences', pp. 67–86; Yeo, *Defining Science*; J. H. Brooke, 'Joseph Priestley (1733–1804) and William Whewell (1794–1866): Apologists and Historians of Science. A Tale of Two Stereotypes', in Anderson and Lawrence (eds), *Science, Medicine and Dissent*, pp. 11–27; R. E. Butts, 'Whewell on Newton's Rules of Philosophizing', in R. E. Butts and J. W. Davis (eds), *The Methodological Heritage of Newton* (Oxford: Blackwell, 1970), pp. 132–49; Y. Elkana, 'William Whewell, Historian', *Rivista di storia della scienza*, 1 (1984), pp. 149–97; J. Losee, 'Whewell and Mill on the Relation between Philosophy of Science and History of Science', *Studies in the History and Philosophy of Science*, 14 (1983), pp. 113–26; and D. B. Wilson, 'Herschel and Whewell's Version of Newtonianism', *Journal of the History of Ideas*, 35 (1974), pp. 79–97. The *History* was structured to reveal Whewell's understanding of the pattern of development of scientific theories.
10. Munby, *History and Bibliography of Science*, p. 5. Accounts of Galileo and research among his archives are chronicled in M. A. Finocchiaro, *Retrying Galileo, 1633–1992* (Berkeley, CA, and London: University of California Press, 2005).
11. Addition by Philip Bliss to an obituary, possibly from the *Oxford Herald*, BLO, Rigaud Family Papers, MS Eng.misc.c.807.
12. R. Yeo, 'Introduction', W. Whewell, *Collected Works of William Whewell*, ed. R. Yeo, 16 vols (Bristol: Thoemmes, 2001), vol. 1, p. xxii.
13. See e.g. S. P. Rigaud, *Miscellaneous Works and Correspondence of The Rev. James Bradley, D.D. F.R.S., Astronomer Royal, Savilian Professor of Astronomy in the University of Oxford &c. &c. &c.* (Oxford: Oxford University Press, 1832), *Supplement to Dr. Bradley's Miscellaneous Works: with an Account of Harriot's Astronomical Papers* (Oxford: Oxford University Press, 1833), 'Biographical Account of John Hadley, Esq. V.P.R.S., the Inventor of the Quadrant ...', *Nautical Magazine*, 4 (1835), pp. 12–22, 137–46, and 'Some Particulars respecting the Principal Instruments at Greenwich'. Rigaud, *A Defence of Halley* relied on the research completed by Rigaud senior.
14. BLO, MSS Rigaud 60/6, letter from Baily to Rigaud, 31 December 1830.
15. Addition by Phillip Bliss to BLO, Rigaud Family Papers, MS Eng.misc.c.807; J. Rigaud, *Stephen Peter Rigaud: A Memoir* (Oxford, 1883), p. 18.
16. Addition by Bliss to BLO, Rigaud Family Papers, MS Eng.misc.c.807.
17. Fauvel et al. (eds), *Oxford Figures*, p. 12.
18. BLO, Rigaud Family Papers, MS Eng.misc.c.807, letter from Rigaud to J. W. Jordan, 18 August 1833; Rigaud, *Miscellaneous Works and Correspondence of Bradley*, Preface, p. [6].
19. Munby, *History and Bibliography of Science*, p. 5.

20. It is interesting to note the similarity in the painstaking neatness of the historical and astronomical manuscripts of Baily and Rigaud. Rigaud, however, has left a great deal more evidence of his approach to reading, note-taking and historical composition in the many notebooks held in the Bodleian Library.
21. Rigaud, *Miscellaneous Works and Correspondence of Bradley*, Preface, p. [4].
22. Several important items were printed by Rigaud for the first time, most notably Newton's *De Motu* (1684) and his account of his early notes on the fluxions, Rigaud, *Historical Essay*, Appendix, pp. 1–24.
23. [Biot], 'Newton (Isaac)', p. 154, attributes this story to Robison.
24. Rigaud, *Historical Essay*, p. 2.
25. Fara, *Newton*, p. 223.
26. Rigaud, *Historical Essay*, p. 18.
27. Ibid., pp. 47, 49, 51, 65.
28. Ibid., pp. 96, 65.
29. S. J. Rigaud in Rigaud (ed.), *Correspondence of Scientific Men*, p. vii. Rigaud's son noted that copying the letters was necessary because 'the printers declared themselves unable to work from the originals'. This indicates the extent to which manuscripts might be moved around in an age before easy methods of copying. Rigaud modernized and standardized the spelling, which Rigaud junior presented as a bonus.
30. [Powell], 'Newton and his Contemporaries', p. 407. Because Powell had succeeded to the chair of the recently deceased Rigaud, was a member of the Ashmolean Society and promoted science and its history within and outside Oxford, it is unsurprising that this review was positive.
31. Grafton has credited Ranke – a man who abandoned his early love of Scott after convincing himself that 'the historical tradition is more beautiful, and certainly more interesting, than the romantic fiction' – with raising the footnote to a new level of importance, A. Grafton, *The Footnote: A Curious History* (London: Faber and Faber, 1997), pp. 24, 33, 37.
32. Bann, *The Clothing of Clio*, p. 35.
33. Addition by Bliss to BLO, Rigaud Family Papers, MS Eng.misc.c.807.
34. BLO, MSS Rigaud 36, ff. 27–36, 'Mrs Catharine Barton'; ff. 25–6, letter from Whewell to Rigaud, 2 July 1836. This essay is discussed further in Chapter 6. Rigaud did not publish because he wished to avoid treading on Brewster's toes, since his *Memoirs of Newton*, which eventually appeared sixteen years after Rigaud's death, was expected imminently.
35. TCL, Whewell Papers, Add. MS a.211/80, letter from Rigaud to Whewell, 31 January 1836.
36. Rigaud's books relating to Newton were reviewed but *Miscellaneous Works and Correspondence of Bradley* was not. A friend did prepare a review, but it was put aside to make way for '"Melton Mowbray" which was thought to be a more taking subject', BLO, Rigaud Family Papers, MS Eng.misc.c.807, letter from Rigaud to J. W. Jordan, 18 August 1833.
37. [Powell], 'Newton and his Contemporaries', pp. 408, 409.
38. BLO, Rigaud Family Papers, MS Eng.misc.c.807, Obituary.
39. See BLO, MSS Rigaud 61/203, draft letter from Rigaud to E. Goodenough, 3 May 1831, and Goodenough to Rigaud, 5 May 1831; MSS Rigaud 62/452, petition and letter from William Wallis to Rigaud, 23 April 1831; [S. P. Rigaud], *Defence of the Resolution for Omitting Mr. Panizzi's Bibliographical Notes from the Catalogue of the Royal Society* (London: Richard & John E. Taylor, 1838).

40. In some cases these individuals were clearly included on the list in recognition of their assistance, but some were closer acquaintances, especially Hunter (who before coming to London in 1833 had been a Dissenting minister), and Nicolas, who was De Morgan's legal advisor, De Morgan, *Memoir of De Morgan*, pp. 70–3.
41. Levine, *The Amateur and the Professional*, p. 2.
42. Miller, 'The Royal Society of London', p. 283, N. H. Nicolas, *Observations on the State of Historical Literature, and on the Society of Antiquaries, and Other Institutions for its Advancement in England; with Remarks on Record Offices, and on the Proceedings of the Record Commission* (London, 1830). Nicolas wrote to Babbage to declare himself a 'fellow labourer … in an attempt to rouse the Government and the Public to attend to subjects which, far more than any military glory, promote the renown of a great country', BL, Babbage Correspondence, Add. MS 37,185, f. 366, N. H. Nicolas to Babbage, 3 November 1830.
43. J. Evans, *A History of the Society of Antiquaries* (Oxford: Oxford University Press, 1956), pp. 263–4, Levine, *The Amateur and the Professional*, ch. 2.
44. Ibid., p. 44.
45. Weld, *History of the Royal Society*.
46. Levine, *The Amateur and the Professional*, p. 3.
47. T. Wright, 'Antiquarianism in England' (1847), quoted in Levine, *The Amateur and the Professional*, pp. 91–2.
48. On Turner and his enthusiasm for manuscripts see A. N. L. Munby, *The Cult of the Autograph Letter in England* (London: Athlone Press, 1962), pp. 33–60.
49. D. Turner, *Thirteen Letters from Sir Isaac Newton to John Covel D.D. – from Original Manuscripts in the Library of Dawson Turner, Esq., Yarmouth* (Norwich, 1848), p. 18.
50. Ibid., pp. 7, 3, 8.
51. Rice, 'De Morgan', p. 223.
52. M. Spevack, *James Orchard Halliwell-Phillipps: The Life and Works of the Shakespearian Scholar and Bookman* (New Castle, DE, and London: Oak Knoll Press, 2001), pp. 28, 29. Halliwell testified, 'As early as 1835 I was in the habit of purchasing MSS generally of a scientific character and relating chiefly to Mathematics, Geometry, Physics, Astrology, &c.', ibid., p. 12.
53. On Libri see Maccioni Ruju and Mostert, *The Life and Times of Libri*, and P. A. Maccioni, 'Guglielmo Libri and the British Museum: A Case of Scandal Averted', *British Library Journal*, 17 (1991), pp. 36–60.
54. See *DNB*; Spevack, *Halliwell-Phillipps*, pp. 124–43; and Munby, *History and Bibliography of Science*, pp. 28–9.
55. The forgeries of the previous age, such as those by Brewster's father-in-law James 'Ossian' Macpherson and Thomas 'Rowley' Chatterton, were imaginatively realized literary epics, while forgers and detectors of the nineteenth century increasingly relied on knowledge of handwriting, paper, bindings, etc., as well as literary knowledge.
56. Spevack, *Halliwell-Phillipps*, p. 14.
57. See e.g. J. O. Halliwell (ed.), *The Private Diary of Dr. John Dee, and the Catalogue of his Library of Manuscripts, from the Original Manuscripts in the Ashmolean Museum at Oxford, and Trinity College Library, Cambridge* (London: Camden Society, 1842).
58. Prospectus of the Historical Society of Science, June 1840, quoted in H. W. Dickinson, 'J. O. Halliwell and the Historical Society of Science', *Isis*, 18 (1932), pp. 127–32, on p. 128.

59. The council and members were listed in J. O. Halliwell (ed.), *A Collection of Letters Illustrative of the Progress of Science in England from the Reign of Queen Elizabeth to that of Charles the Second* (London: Historical Society of Science, 1841).
60. University of Edinburgh Library, Halliwell-Phillipps Collection, L.O.A. 7/48, letter from Whewell to Halliwell, 12 June 1840.
61. Another twelve works were suggested, covering fourteenth- to sixteenth-century science, but none appeared.
62. T. Hornberger, 'Halliwell-Phillipps and the History of Science', *Huntingdon Library Quarterly*, 12 (1949), pp. 391–9, on p. 394; T. Wright, *Popular Treatises on Science Written During the Middle Ages in Anglo-Saxon, Anglo-Norman and English* (London: Historical Society of Science, 1841).
63. Hornberger, 'Halliwell-Phillips', p. 397.
64. Munby, *History and Bibliography of Science*, p. 29; University of Edinburgh Library, Halliwell-Phillipps Collection, L.O.A. 34/46, letter from Baden Powell to Halliwell, 7 March 1841.
65. Munby, *History and Bibliography of Science*, p. 21; University of Edinburgh Library, Halliwell-Phillipps Collection, L.O.A. 160/5, letter from De Morgan to Halliwell, [annotated '20 or 23 Jul 1840']; [A. De Morgan], 'Review of Halliwell's *A Collection of Letters, Illustrative of the Progress of Science in England*, published by the Historical Society of Science', *Athenaeum*, 719 (1841), pp. 588–9.
66. Halliwell, *A Collection of Letters*, p. ix.
67. Spevack, *Halliwell-Phillipps*, p. 10.
68. [De Morgan], 'Review of Halliwell's *A Collection*', p. 589.
69. Ibid.
70. Such a response had been called for both to compete with Oxford and to make use of the rich collections in Cambridge. See e.g. statements in BLO, MSS Rigaud 61/302, letter from J. O. Halliwell to Rigaud, 20 December 1838; and [Powell], 'Newton and his Contemporaries', p. 403.
71. Edleston is not included in the *DNB*, and information about his life is taken from the introduction to the catalogue of Edleston Papers at Durham County Record Office, from the contents of those papers, Rouse Ball and Venn, *Admissions to Trinity College*, and Venn, *Alumni Cantabrigienses*.
72. Edleston was involved with a number of improvement campaigns, including a scheme to light Senate House with gas in 1847, Durham County Record Office, Edleston Papers, D/Ed/12/2/26, letter from C. B. Broadley, to Edleston, 19 June 1847.
73. Durham County Record Office, Edleston Papers, D/Ed/11/1/53, letter from W. H. Bull to Edleston, 26 February 1845.
74. Proof, enclosed in Durham County Record Office, Edleston Papers, D/Ed/12/2/32, letter from *Cambridge Chronicle* to Edleston, 13 November 1861. Edleston's relatively humble origins and conservative politics are reminiscent of Whewell.
75. Whewell accepted the proposal, offering 'every assistance which I can give', Durham County Record Office, Edleston Papers, D/Ed/12/2/679, letter from Whewell to Edleston, 21 October 1851. The cataloguing of manuscripts at Trinity has previously been ascribed to Halliwell, see Spevack, *Halliwell-Phillipps*, p. 20.
76. Durham County Record Office, Edleston Papers, D/Ed/12/2/326, letter from Frederick Martin to Edleston, 2 December 1857. If the date on this letter is correct, this was old news, for the marriage took place in 1851. Turner sold his paintings and books in 1853,

but the auction of manuscripts took place in 1859, Munby, *The Cult of the Autograph Letter*, pp. 49–50, see pp. 52–7 on the sale.
77. RAS, MSS De Morgan 1, letter from Whewell to De Morgan, 29 August[?] 1855.
78. The letters of Georgina Gregory to a Mr Walton, which frequently mention Edleston's assistance and judgments, are among Durham County Record Office, Edleston Papers, D/Ed/11/2/40/48.
79. There are several letters from Thomas Winter, secretary to the statue committee, to Edleston, Durham County Record Office, Edleston Papers, D/Ed/11/2/35–9.
80. See the Historical Science Collection, http://www.dur.ac.uk/Library/asc/print/science.html. Edleston also corresponded with Samuel Crompton, who planned to publish a book on the portraits of Newton, Durham County Record Office, Edleston Papers, D/Ed/11/2/85, letter from Samuel Crompton to Edleston, 30 June 1882.
81. Warwick, *The Masters of Theory*, p. 58. See also A. R. Hall, 'Cambridge: Newton's Legacy', *Notes and Records of the Royal Society*, 55 (2001), pp. 205–26.
82. Edleston, *Correspondence of Newton and Cotes*, p. ix.
83. Ibid., pp. x–xi. While Edleston felt that 'the infirmities of Flamsteed's temper and bodily health' were the principal cause of delay, he admitted that Newton's appointment to the Mint in 1696 may have contributed.
84. [C. R. Weld], 'Review of Edleston's *Correspondence of Sir Isaac Newton and Professor Cotes*', *Athenaeum*, 1217 (1851), pp. 211–12, on p. 212.
85. TCL, Whewell Papers, Add. MS a.202/123, letter from De Morgan to Whewell, 8 January 1851.
86. A. De Morgan, 'A Short Account of Some Recent Discoveries in England and Germany relative to the Controversy on the Invention of the Fluxions', *Companion to the Almanac* (1852), pp. 5–20, reprinted in De Morgan, *Essays*, pp. 67–101, on p. 74. Edleston had objected to De Morgan's earlier essays, A. De Morgan, 'On a Point Connected with the Dispute between Keil and Leibnitz, *Philosophical Transactions*, 136 (1846), pp. 107–9, and [A. De Morgan], 'Review of Weld's *History of the Royal Society*', *Athenaeum*, 1078 (1848), pp. 621–2, and 1079 (1848), pp. 651–3, see below.
87. Edleston, *Correspondence of Newton and Cotes*, pp. lxiv–lxvii.
88. Ibid., p. xxxiv. Later in the notes Edleston tackled the second dispute, over the publication of the catalogue, in a similar manner. He suggested that reading Baily's *Account of Flamsteed* would in itself be enough to convince the reader of the fallacy of Flamsteed's complaints, pp. lxxi–lxxiii.
89. Ibid., pp. lxi, lxiii.
90. Durham County Record Office, Edleston Papers, D/Ed/12/2/444, letter from George Pryme to Edleston, 5 April 1850; D/Ed/12.2.439, letter from Charles de la Pryme to Edleston, 28 March 1870. Charles was keen to distance his ancestor from the dispute, saying that he wished 'to state in a Note to the Edition of Abraham De la Pryme's Diary that the best judges are of opinion that the (*supposed*) derangement of Sir Isaac Newton is not warranted by the known facts, & such clearly is Sir David Brewster's view of it'.
91. John Wallis, quoted in Edleston, *Correspondence of Newton and Cotes*, p. lxii.
92. Ibid., pp. lxx–lxxi.
93. Ibid., p. lxxv.
94. Ibid., p. lxxvi; [Biot], 'Newton (Isaac)', p. 193.
95. De Morgan, *Memoir of De Morgan* includes a list of De Morgan's writings. See also Rice, 'De Morgan', pp. 208–9. Rice's article, which is more descriptive than analytical, divides De Morgan's interests into the following categories: astronomy, the calculus controversy,

Newton and his niece, arithmetic and bibliography. My account naturally focuses on the second and third areas, putting these writings, their historical style and content fully into context.

96. J. L. Richards, 'Augustus De Morgan, the History of Mathematics, and the Foundations of Algebra', *Isis*, 78 (1987), pp. 7–30, on p. 7. De Morgan's 'openness' included signing many journal articles, which would normally remain anonymous. His controversial 1846 biography of Newton was signed, 'For reasons which will be easily understood', A. De Morgan, 'Newton', in C. Knight (ed.), *The Cabinet Portrait Gallery of British Worthies*, 12 vols (London: Charles Knight & Co., 1845–7), vol. 11 (1846), pp. 78–117, reprinted in De Morgan, *Essays*, pp. 3–63, and in Iliffe et al. (eds), *Early Biographies*, vol. 2, pp. 183–211, on p. 211. Rice, 'De Morgan'.
97. Quoted in Richards, 'De Morgan', p. 9.
98. A. De Morgan, 'On the Earliest Printed Almanacs', *Companion to the Almanac* (1846), pp. 1–31, on p. 1.
99. [A. De Morgan], 'Review of Whewell's *History of the Inductive Sciences*', *Athenaeum*, 541 (1838), pp. 179–81. This is omitted from the Athenaeum index, http://athenaeum.soi.city.ac.uk/athall.html, but internal evidence suggests that De Morgan is the author.
100. See Elkana, 'William Whewell, Historian'; Lindberg, 'Conceptions of the Scientific Revolution', pp. 12–13; and Yeo, *Defining Science*, pp. 150–1 for such claims about Whewell's history. [De Morgan], 'Review of Whewell's *History*', p. 179. De Morgan also criticized Whewell for not considering the interaction between science and the 'arts' or technology.
101. Ibid., p. 180. Brewster's review of Whewell's *History* considered the section on medieval science 'occupying 180 pages ... extended beyond the importance of the subjects which it embraces', [D. Brewster], 'Whewell's *History of the Inductive Sciences*', *Edinburgh Review*, 66 (1837), pp. 110–51, on p. 122.
102. [A. De Morgan], '*Novum Organum Renovatum*. By W. Whewell', *Athenaeum*, 1628 (1859), pp. 42–4, and '*The Philosophy of Discovery, Chapters Historical and Critical*. By W. Whewell', *Athenaeum*, 1694 (1860), pp. 501–3. Richards refers to his 'essential agreement with Whewell's view of mathematical and scientific history as a process of conceptual unfolding', but notes that De Morgan insisted that there were no infallible rules for discovery, Richards, 'De Morgan', pp. 19, 20.
103. De Morgan, 'References for the History', pp. 48, 44.
104. Ibid., pp. 56, 53.
105. A. De Morgan, *Arithmetical Books from the Invention of Printing to the Present Time, being Brief Notices of a Large Number of Works Drawn up from Actual Inspection* (London: Taylor and Walton, 1847; reprinted London: H. K. Elliott, 1967), p. i. For a discussion of the impact of Peacock's mathematical and historical ideas on De Morgan see Richards, 'De Morgan', pp. 13–18. The appreciation was mutual, for Peacock called De Morgan 'the most accurate & learned of all modern writers on the History of Mathematics', quoted in Rice, 'De Morgan', p. 201. De Morgan also considered Peacock's *Life of Thomas Young* (1855) 'one of the best scientific biographies', [A. De Morgan], 'Sir David Brewster's *Life of Newton*', *North British Review*, 23 (1855), pp. 307–38, reprinted in De Morgan, *Essays*, pp. 119–82, and in Iliffe et al. (eds), *Early Biographies*, vol. 2, pp. 213–44, on p. 222.
106. A. De Morgan, 'The Progress of the Doctrine of the Earth's Motion, Between the Times of Copernicus and Galileo; being Notes on the AnteGalilean Copernicans', *Companion to the Almanac* (1855), pp. 5–25; [De Morgan], 'Review of *The Works of Bacon*'.

107. BL, Add. MS 22,786, 'Tractatus de mundi sphæra', pp. 12–13, letter from De Morgan to Anthony Panizzi, 18 April 1859. This manuscript was bought by the British Museum in 1859 from a sale of Libri's manuscripts. Suspicions regarding the handwriting were aroused after purchase. De Morgan was not familiar with Galileo's hand, and deferred to Libri, who trumpeted the manuscript as 'an AUTOGRAPH AND UNKNOWN WORK of Galileo'. De Morgan thought the internal evidence did not argue against this conclusion. Ultimately the Museum catalogue concluded it was not by Galileo, but the trustees did not demand a refund.
108. TCL, Whewell Papers, Add. MS a.202/138, letter from De Morgan to Whewell, 10 October 1858; and De Morgan, *Memoir of De Morgan*, p. 296.
109. See for example TCL, Whewell Papers, Add. MS a.202/100, 143, letters from De Morgan to Whewell, 30 April 1844, 3 March 1860.
110. See J. L. Richards, '"In a rational world all radicals would be exterminated": Mathematics, Logic and Secular Thinking in Augustus De Morgan's England', *Science in Context*, 15 (2002), pp. 137–64, on pp. 145–6.
111. De Morgan, *Newton*, pp. 127, 128. Joshua Milne was an actuary at the Sun Life Assurance Society. The historical work mentioned by De Morgan probably refers to his articles in the *Encyclopaedia Britannica* on 'Annuities' and 'Bills of Mortality'. See *DNB*.
112. Richards, 'De Morgan', p. 28.
113. [A. De Morgan], Notices of English Mathematical Writers Between the Norman Conquest and the Year 1600', *Companion to the Almanac* (1837), pp. 21–44, on p. 24.
114. [De Morgan], '*Novum Organum*', p. 43.
115. A. De Morgan, 'Fly-Leaves of Books: Reuben Burrow', *Notes and Queries*, 12 (1855), pp. 142–3, on p. 143, and 'Publication of Diaries', *Notes and Queries*, 2nd series, 5 (1864), pp. 107–8. T. T. Wilkinson, a Fellow of the RAS, the editor of the diary, was a friend but responded vigorously in the same journal, T. T. Wilkinson, 'Publication of Diaries', *Notes and Queries*, 2nd series, 5 (1864), pp. 215–16. It is difficult to gauge the seriousness of the subsequent dispute: although apparently rancorous, they may have enjoyed the literary jousting.
116. A. De Morgan, 'Newton's Nephew, the Rev. B. Smith', *Notes and Queries*, 2nd series, 3 (1857), pp. 41–2. De Morgan's informant was Anne Sheepshanks, sister of Richard Sheepshanks, De Morgan's friend and Fellow of the RAS. See RAS, MSS De Morgan 1, letters from A. S. to De Morgan, 7 December [1856], 4 January 1857, and a piece subsequently published in *Notes and Queries*, in MSS De Morgan 4.
117. De Morgan, *Newton*, pp. 5–7.
118. De Morgan, 'On a Point Connected with the Dispute', and 'On the Additions Made to the Second Edition of the *Commercium Epistolicum*, *Philosophical Magazine*, 3rd series, 23 (1848), pp. 446–56, on p. 447.
119. Ibid., p. 447.
120. [De Morgan], 'Brewster's *Life of Newton*', p. 227.
121. University of London Library, De Morgan Papers, MS 913/A/2/5, letter from De Morgan to W. H. Smyth, 26 June 1848.
122. De Morgan's version puts the Royal Society's motto back into the context from which it was stripped: Horace's 'Ac ne forte roges que me duce, quo lare tuter, nullius addictus iurare in verba magistri, quo me cumque rapit tempestas, deferor hospes' (first Epistle, ll. 13–15). It is likely that most Victorians knew this context and did not, as in recent times, mistranslate the Royal Society motto as 'nothing in words' or 'don't take anybody's word for it'. S. J. Gould, 'Royal Shorthand', *Science*, 251 (1991), p. 142, pointed out this

'canonical mistranslation', which privileges empiricism over authority, or words themselves. In his reading the motto indicates a plea for tolerance (i.e. 'not bound to swear allegiance to anyone's doctrine') in scientific, religious or political terms. If this is how it was read by De Morgan, his version takes on additional resonance.

123. De Morgan to William Frend, n.d. [1835], in De Morgan, *Threescore Years and Ten*, p. 128.
124. [De Morgan?], 'Newton'; De Morgan, 'Newton', 'A Short Account of some Recent Discoveries', 'On the Authorship of the Account of the Commericum Epistolicum', *Philosophical Magazine*, 4th series, 3 (1852), pp. 440–4, and 'On the Early History of Infinitesimals in England', *Philosophical Magazine*, 4th series, 4 (1852), pp. 321–30.
125. A. De Morgan, 'Notes on the History of the English Coinage', *Companion to the Almanac* (1856), pp. 5–21, which is dated 20 October 1855; [De Morgan] 'Brewster's *Life of Newton*'.
126. De Morgan believed that between 1700 and 1750 the Royal Society had 'a strange deficiency in their controversial library ... as if an expurgatorial visit had been paid, for the purpose of expelling everything which might be grating to a strong Newtonian, even to works which use the infinitesimal principle or the differential notation', A. De Morgan, 'Book Dust', *Notes and Queries*, 2nd series, 4 (1857), pp. 301–2, on p. 301.
127. De Morgan, *Memoir of De Morgan*, p. 256.
128. De Morgan, 'Newton', p. 198; [De Morgan], 'Brewster's *Life of Newton*', p. 230.
129. De Morgan, 'Newton', p. 203.
130. Ibid., p. 204; see P. Theerman, 'Unaccustomed Role: The Scientist as Historical Biographer – Two Nineteenth-Century Portrayals of Newton', *Biography*, 8 (1985), pp. 145–62, on p. 154.
131. De Morgan, 'Newton', pp. 203, 205.
132. BL, Napier Correspondence, Add. MS 34,615, f. 346, letter from B. H. Malkin to Macvey Napier, 14 June 1832.
133. [DeMorgan?] 'Newton', p. 201.
134. De Morgan, 'Newton', p. 201.
135. L. Daston and P. Galison, 'The Image of Objectivity', *Representations*, 40 (1992), pp. 81–128, p. 81; P. Galison, 'Objectivity is Romantic', in J. Freidman, P. Galison and S. Haack, *The Humanities and the Sciences* (New York: American Council of Learned Societies, 1999), pp. 15–43; L. Daston, 'Objectivity and the Escape from Perspective', *Social Studies of Science*, 22 (1992), pp. 597–618.
136. J. Herschel, 'Outlines of Astronomy', (1849), reprinted in I. B. Cohen and H. M. Jones (eds), *Science Before Darwin: An Anthology of British Scientific Writing in the Early Nineteenth Century* (London: Andre Deutsch, 1963), pp. 97–121, on p. 98.
137. Daston and Galison, 'The Image of Objectivity', p. 118.
138. Quoted in D. Forbes, *The Liberal Anglican Idea of History* (Cambridge: Cambridge University Press, 1952), p. 125.
139. G. Levine, *Dying to Know: Scientific Epistemology and Narrative in Victorian England* (Chicago, IL, and London: University of Chicago Press, 2002), p. 4.
140. De Morgan, 'References for the History', p. 50; [Brewster], 'Whewell's *History*', p. 112.
141. Forbes, *The Liberal Anglican Idea*, p. 124.
142. Yeo, *Defining Science*, pp. 155–7.
143. Brooke, 'Priestley and Whewell', p. 15.
144. [J. P. Collier], 'Lectures on the History of Rome', *Athenaeum*, 1058 (1848), pp.139–40, on p. 140.

5 David Brewster's *Memoirs of Sir Isaac Newton*

1. Anon., 'The House of Fame', *Punch*, 25 (1853), pp. 106–7, on p. 107.
2. Brewster, *Memoirs of Newton*, vol. 2, p. 406. Page references will hereafter be given in the text. Compare Brewster, *Life of Newton*, p. 337.
3. John Murray Archives, letter from Brewster to John Murray, 5 May 1831.
4. BLO, MSS Rigaud 60/86, letter from Brewster to Rigaud, 21 May 1837; John Murray Archives, letter from Brewster to Murray, 14 November 1838.
5. University of Edinburgh Library, Halliwell-Phillipps Collection, L.O.A. 8/30, letter from Brewster to Halliwell-Phillips, 7 September 1841.
6. [Powell], 'Newton and his Contemporaries', p. 402n.
7. See UCL, Brougham Papers 26,672 and 26,679, letters from Brewster to Brougham, 15 August 1853, 11 July 1854; and Durham County Record Office, Edleston Papers, D/Ed/11/2/76, letter from Brewster to Edleston, 31 March 1855.
8. See R. Anderson, 'Brewster and the Reform of the Scottish Universities', in Morrison-Lowe and Christie (eds), *Martyr of Science*, pp. 31–4.
9. P. Baxter, 'Brewster, Evangelism and the Disruption of Scotland', in Morrison-Lowe and Christie (eds), *Martyr of Science*, pp. 45–50, on p. 46.
10. Gordon, *The Home Life*, pp. 428–35.
11. [D. Brewster], 'Newton, Sir Isaac', in M. Napier (ed.), *Encyclopaedia Britannica*, 7th edn, 21 vols (Edinburgh: A. and C. Black, 1842), pp. 175–81, reprinted in Iliffe et al. (eds), *Early Biographies*, vol. 2, pp. 159–73. This article is interesting as a relatively raw response to Baily's *Account of Flamsteed*, not tempered by the subsequent work of Rigaud, Edleston and De Morgan, perhaps partly written before his visit to Hurtsbourne Park.
12. John Murray Archives, letter from Brewster to Murray, 2 April 1842; letter from Murray to Brewster, 20 March 1847. This, Murray said, was the arrangement made with Charles Lyell for his *Principles of Geology*.
13. John Murray Archives, letter from Brewster to Murray, 2 April 1842. Brewster had published *Martyrs of Science* under a different scheme, taking all the risk and the profit himself, John Murray Archives, Ledgers. On these various means of publishing see A. Fyfe, 'Conscientious Workmen or Bookseller's Hacks? The Professional Identities of Science Writers in the Mid-Nineteenth Century', *Isis*, 96 (2005), pp. 192–223, on pp. 211–12.
14. BL, Napier Correspondence, Add. MS 34,618, f. 38, letter from Brewster to Napier, 20 February 1837.
15. TCL, Whewell Papers, O.15.47/32–93, letter from Whewell to Forbes, 10 October 1849; Rigaud, *Historical Essay*, Preface.
16. BLO, MSS Rigaud 60/88, letter from Brewster to Rigaud, 23 December 1837; Durham County Record Office, Edleston Papers, D/Ed/11/2/63, letter from Brewster to Edleston, 30 July 1853.
17. [Brewster], 'Newton, Sir Isaac', p. 172.
18. BLO, MSS Rigaud 60/86, letter from Brewster to Rigaud, 21 May 1837; BL, Napier Correspondence, Add. MS 34,618, f. 161, letter from Brewster to Napier, 14 June 1837; RAS, MSS De Morgan 1, letter from Brewster to De Morgan, 6 August 1842; Edinburgh University Library, Halliwell-Phillipps Collection, L.O.A. 8/30, letter from Brewster to Halliwell, 7 September 1841; Edinburgh University Library, AAF Brewster, letter from Brewster to Mrs Liddle, 12 September 1838. Quotations are taken from the letters to

Mrs Liddle and Halliwell. Since Turnor had published part of the Conduitt collection in his *Collections* of 1806, I am at a loss to explain why it was 'supposed not to exist'.

19. BL, Napier Correspondence, Add. MS 34,618, f. 161, letter from Brewster to Napier, 14 June 1837. The papers Brewster saw on this visit should not be confused with the selection 'examined and arranged' by Fellowes (Brewster, *Memoirs of Newton*, vol. 1, p. x), which were sent to Brewster in 1854. I believe that Brewster at least glanced at most of the collection.
20. BL, Napier Correspondence, Add. MS 34,618, f. 161, letter from Brewster to Napier, 14 June 1837.
21. BLO, MSS Rigaud 60/86, letter from Brewster to Rigaud, 21 May 1837; Durham County Record Office, Edleston Papers, D/Ed/11/2/64, letter from Brewster to Edleston, 10 November 1853 (latter emphasis mine). Brewster was more generous to Hooke than other writers, although he was suspicious of his methodology and morality (vol. 1, pp. 144–5). None of his reviewers referred to this uncharacteristic sympathy for an enemy of Newton (and a proponent of the wave theory of light). Hooke to Newton, 20 January 1676, and Newton to Hooke, 5 February 1676, Turnbull et al. (eds), *The Correspondence of Newton*, vol. 1, pp. 412–16.
22. BLO, MSS Rigaud 60/86, letter from Brewster to Rigaud, 21 May 1837.
23. UCL, Brougham Papers 26,633, letter from Brewster to Brougham, 8 October 1847.
24. UCL, Brougham Papers 26,679, letter from Brewster to Brougham, 11 July 1854.
25. UCL, Brougham Papers 26,680 and 11,248, letters from Brewster to Brougham, 8 August 1854, 11 November 1854.
26. UCL, Brougham Papers 26,680, letter from Brewster to Brougham, 8 August 1854.
27. RAS, MSS De Morgan 1, letter from Brewster to De Morgan, 6 August 1842, and draft letter De Morgan to Brewster[?], reprinted in Iliffe et al. (eds), *Early Biographies*, vol. 2, pp. 175–82, on p. 176. The draft is annotated by Sophia De Morgan, 'No date. Was this to Brewster?' It was a reply to Brewster's letter, but there is no proof that it was sent.
28. BLO, MSS Rigaud 60/88, letter from Brewster to Rigaud, 23 December 1837.
29. Durham County Record Office, Edleston Papers, D/Ed/11/2/70, letter from Brewster to Edleston, 11 January 1855.
30. Gordon, *The Home Life*, p. 262.
31. R. S. Westfall, 'Introduction', in Brewster, *Memoirs of Newton*, reprinted in The Sources of Science, vol. 14 (New York and London: Johnson Reprint Corporation, 1965); Theerman, 'Unaccustomed Role'; and Yeo, 'Genius, Method and Morality'.
32. J. R. R. Christie, 'Sir David Brewster as an Historian of Science', in Morrison-Low and Christie (eds), *Martyr of Science*, pp. 53–6, on p. 56.
33. Westfall, 'Introduction', pp. xxxvi–xxxvii.
34. Brewster, *Life of Newton*, p. 196.
35. [De Morgan], 'Brewster's *Life of Newton*', pp. 214–15.
36. Ibid., pp. 215–16.
37. Brewster complained that the reviewer in the *Edinburgh Review* made up his article by 'abstracting the unscientific portion of the book', UCL, Brougham Papers, 26,692, letter from Brewster to Brougham, 30 May 1856. [Powell], 'Newton'.
38. BL, Napier Correspondence, Add. MS 34,623, f. 636, letter from Baden Powell to Macvey Napier, 20 June 1843.
39. [De Morgan], 'Brewster's *Life of Newton*', pp. 220, 221.
40. Ibid.; J.-B. Biot, 'Revue de *Memoirs of the Life, Writings, and Discoveries of Sir Isaac Newton*', *Journal des savants* (1855), pp. 589–606, 662–77, on pp. 596, 605.

41. [De Morgan], 'Brewster's *Life of Newton*', p. 220. Christie has described the passages on the history of optics as the real strength of the first volume of the *Memoirs*, containing, unlike Whewell's *History*, a strong appreciation of technique, the design of experiments and the importance of instruments. He also notes their vindication of the corpuscular theory of light, Christie, 'Brewster as an Historian', p. 55.
42. Whewell, *History*, vol. 2, pp. 346–54.
43. Cantor, *Optics after Newton*, pp. 9–15.
44. UCL, Brougham Papers, 26,674, letter from Brewster to Brougham, 12 September 1853.
45. Chen and Barker, 'Cognitive Appraisal and Power', p. 76.
46. UCL, Brougham Papers 26,632, letter from Brewster to Brougham, 26 September 1847.
47. Chen and Barker, 'Cognitive Appraisal and Power', p. 80; UCL, Brougham Papers 26,636, letter from Brewster to Brougham, 21 February 1849.
48. H. P. Brougham, 'Experiments and Observations upon the Properties of Light', *Philosophical Transactions*, 140 (1850), pp. 235–59, on pp. 235–6; Chen and Barker, 'Cognitive Appraisal and Power', p. 86. See also Cantor, *Optics after Newton*, p. 177.
49. UCL, Brougham Papers 26,638, letter from Brewster to Brougham, 21 February 1849. Brewster subsequently withdrew his support for Brougham because he departed from the agreed strategy by initiating direct attacks against the principle of interference and the undulatory theory, Chen and Barker, 'Cognitive Appraisal and Power', p. 93.
50. Ibid., pp. 98, 95.
51. Brewster, *Life of Newton*, p. 113. The 1855 additions are given in angled brackets and deletions are struck through.
52. [De Morgan], 'Brewster's *Life of Newton*', p. 224.
53. Christie, 'Brewster as an Historian', p. 55.
54. [De Morgan], 'Brewster's *Life of Newton*', p. 225.
55. Gordon, *The Home Life*, p. 260–1.
56. UCL, Brougham Papers 26,633, letter from Brewster to Brougham, 8 October 1847.
57. Brewster had charged the deceased Baily 'with fiction, calumny, and misrepresentation'; terms, according to De Morgan, 'he would not have dared to use if Baily had been able to answer', De Morgan, *Newton*, pp. 100, 102.
58. UCL, Brougham Papers 19,955 and 26,686, letters from Brewster to Brougham, 14 August 1855, 30 August 1855.
59. Durham County Record Office, Edleston Papers, D/Ed/11/2/75, letter from Brewster to Edleston, 19 March 1855.
60. Baily, *Account of Flamsteed*, p. xx.
61. [De Morgan], 'Brewster's *Life of Newton*', p. 225.
62. Brewster apparently felt secure in attacking the *Account*, remembering that there 'was a strong feeling against [Baily] at the time' and that he 'never heard of any scientific individual holding Baily's views about Flamsteed', UCL, Brougham Papers 26,686, letter from Brewster to Brougham, 30 August 1855. This is not borne out by the correspondence described in Chapter 3.
63. [De Morgan], 'Brewster's *Life of Newton*', p. 225.
64. Draft letter from De Morgan to Brewster [1842], in Iliffe et al. (eds), *Early Biographies*, vol. 2, p. 176.
65. De Morgan, 'A Short Account of Some Recent Discoveries', p. 75.

66. RSL, Herschel Papers, JFWH 6.314, letter from De Morgan to Herschel, 26 August 1858.
67. [De Morgan], 'Brewster's *Life of Newton*', pp. 226, 227.
68. Durham County Record Office, Edleston Papers, D/Ed/11/2/68, letter from Brewster to Edleston, 22 December 1854.
69. [De Morgan], 'Brewster's *Life of Newton*', p. 230.
70. Quoted in De Morgan, *Essays*, p. 183.
71. [De Morgan], 'Brewster's *Life of Newton*', p. 239.
72. On the eighteenth-century reading of Newton's theological writings, as both for and against orthodox Christianity, see L. Stewart, 'Seeing through the Scholium: Religion and Reading Newton in the Eighteenth Century', *History of Science*, 34 (1996), pp. 123–65. When still under the influence of his High-Church upbringing, Baden Powell had attempted to counter Unitarian claims regarding Newton. Powell did not present Newton as a great theologian but sought to demonstrate 'the orthodoxy of so eminent a man' and to take 'from the Socinian cause its claim to the countenance of such a powerful auxiliary', [B. Powell], 'An Examination into the Charge of Heterodoxy Brought Against Eminent Men. In a Letter to the Editor of the Christian Remembrancer', *Christian Remembrancer*, 7 (1825), pp. 566–75, on p. 575. See also B. Powell, *Rational Religion Examined: Or, Remarks on the Pretensions of Unitarianism; Especially as Compared with those Systems which Professedly Discard Reason* (London, 1826); and Corsi, *Science and Religion*, pp. 27–8.
73. Burgess sent his article to Brewster but, despite Brewster's protests and Burgess's acknowledgment that he had not read the whole of Brewster's account, he published it unchanged. Brewster included this exchange in an appendix to the *Memoirs*, and defended his decision to publish an account of Newton's essay, which was already in the public domain and not, in his opinion (or Burgess's), anti-Trinitarian, Brewster, *Memoirs of Newton*, vol. 2, pp. 523–5.
74. UCL, Brougham Papers 17,101, letter from De Morgan to Brougham, 11 July 1854.
75. De Morgan, 'Newton', pp. 205–10, 208.
76. UCL, Brougham Papers 17,101, letter from De Morgan to Brougham, 11 July 1854.
77. De Morgan, 'Newton', p. 210.
78. Ibid., p. 210.
79. Ibid., p. 207 with note. De Morgan did not take his Cambridge MA because of his inability to subscribe to the thirty-nine articles of the Anglican Church.
80. Whewell, *Newton and Flamsteed*, p. 134.
81. Ekins possessed a number of theological manuscripts that Catherine Conduitt intended to publish, but which passed into the hands of Catherine's daughter's executor, Jeffrey Ekins senior. Ekins donated the manuscripts to New College, Oxford, in 1872. They are now in BLO, New College MSS 361.
82. Gordon, *The Home Life*, p. 326.
83. Ibid., p. 262.
84. UCL, Brougham Papers 26,679, letter from Brewster to Brougham, 11 July 1854. Brewster also told Brougham that he had consulted with Robert Smith Candlish, Principal of the Free Church College in Edinburgh, who believed Newton to be 'a *Semi-Arian* which I suppose means that Christ was not merely the highest of created beings, but *above* all created beings', UCL, Brougham Papers 31,833, letter from Brewster to Brougham, n.d. [1854].
85. Christie, 'Brewster as an Historian', p. 56.

86. National Library of Scotland, Add. MS 4949, ff. 24–6, letter from Brewster to Col. William Mure, 30 January 1844. On Mure (1799–1860), see *DNB*.
87. [De Morgan], 'Brewster's *Life of Newton*', p. 242.
88. Ibid., p. 239, 241. He suggested that Newton was a Humanitarian, p. 242.
89. Ibid., p. 242.
90. Richards, '"In a rational world"', p. 143.
91. Letter from De Morgan to his mother, 1836, quoted in De Morgan, *Memoir of De Morgan*, pp. 139–44; and University of London Library, Senate House, De Morgan Papers, MS 913/A/2/12, letter from De Morgan to his mother, n.d.; De Morgan, *Memoir of De Morgan*, p. 86. See also Richards, '"In a rational world"', pp. 142–3.
92. Gordon, *The Home Life*, pp. 311, 313, See Baxter, 'Brewster, Evangelism and the Disruption', pp. 45–6 and J. H. Brooke, 'Natural Theology and the Plurality of Worlds: Observations on the Brewster-Whewell Debate', *Annals of Science*, 34 (1977), pp. 221–86, on pp. 231–2, 271–2. Powell had taken the opposite journey, from High Church to political and religious liberalism, which explains his later very different views on Newton and friendship with De Morgan, see Corsi, *Science and Religion*; and Iliffe et al. (eds), *Early Biographies*, vol. 2, pp. xxx–xxxiii.
93. Baxter, 'Brewster, Evangelism and the Disruption', p. 47. On *Vestiges* see Secord, *Victorian Sensation*.
94. Gordon, *The Home Life*, p. 324, Rev. Cousins quoted on pp. 323–34. This list was the 1865 'Scientists' Declaration', which Brewster signed, see W. H. Brock and R. M. MacLeod, 'The Scientists' Declaration: Reflexions on Science and Belief in the Wake of *Essays and Reviews*, 1864–5', *British Journal of the History of Science*, 9 (1976), pp. 39–66.
95. Gordon, *The Home Life*, p. 311.
96. On Law, Boehme and Newton in the context of the debates over electricity, see S. Schaffer, 'The Consuming Flame: Electrical Showmen and Tory Mystics in the World of Goods', in J. Brewer and R. Porter (eds), *Consumption and the World of Goods* (London and New York: Routledge, 1993), pp. 489–526, on pp. 502–6, 507–8; and A. Wormhoudt, 'Newton's Natural Philosophy in the Behmenistic Works of William Law', *Journal of the History of Ideas*, 10 (1949), pp. 411–29. Law also referred to Humphrey Newton's letter to Conduitt, which described Newton's toils in his laboratory and an alchemical text. Law's informant had evidently actually seen the Portsmouth Papers.
97. King, *The Life of Locke*, pp. 410, 413.
98. Brewster, *Life of Newton*, p. 302, and *Martyrs of Science*, pp. 154–5, 109, 151.
99. See Christie, 'Brewster as an Historian', p. 55.
100. [Powell], 'Newton', pp. 268–9.
101. J. M. Keynes, 'Newton, the Man', in J. M. Keynes, *Essays in Biography … New Edition with Three Additional Essays*, ed. G. Keynes (London: Rupert Hart-Davis, 1951), pp. 310–23. By the 1850s, alchemy was beginning to receive more attention from historians of chemistry, especially from Ferdinand Hoefer and Hermann Kopp, but Newton as alchemist was rarely mentioned.
102. See letters from Brewster to Brougham in UCL, Brougham Papers: 19,955, 14 August 1855 (on De Morgan); 26,692, 30 May 1856 (on the *Edinburgh Review* [Powell]); 26,709, 18 November 1858 (on *The Times* and Biot).
103. UCL, Brougham Papers 26,692, letter from Brewster to Brougham, 30 May 1856.
104. UCL, Brougham Papers 26,708, letter from Brewster to Brougham, 13 November 1858.
105. See Westfall, *Never at Rest*, pp. 671–7.

106. Baily, *Account of Flamsteed*, p. 294.
107. Brewster in principle agreed with Edleston as to the dating of de la Pryme's story, Edleston, *Correspondence of Newton and Cotes*, pp. lxi–lxiii.
108. Durham County Record Office, Edleston Papers, D/Ed/11/2/77, letter from Brewster to Edleston, 31 August 18??.
109. As Westfall says, the sins are 'neither salacious nor sensational', but it is likely that they would have been unpleasant to Brewster, R. S. Westfall, 'Short-Writing and the State of Newton's Conscience, 1662 (I)', *Notes and Records of the Royal Society of London*, 18 (1963), pp. 10–16, on p. 10. He may have avoided deciphering it after the warning provided by the recent publication of Pepys's shorthand diary in 1825. Newton's shorthand was certainly deciphered in 1876 by John Couch Adams when cataloguing the Portsmouth Papers, and he reported some of the lesser misdemeanours as dinner-party conversation, M. E. G. Duff, *A Victorian Vintage: Being a Selection of the Best Stories from the Diaries of the Right Hon. Sir Mountstuart E. Grant Duff*, ed. A. T. Bassett (London: Methuen & Co., 1930), p. 35.
110. [Weld], 'Review of Brewster's *Memoirs of Newton*', p. 698; UCL, Brougham Papers 26,697, letter from Brewster to Brougham, 18 March 1857.
111. Letter from Ada Lovelace to Lady Byron, 21 July 1837, quoted in B. A. Toole, *Ada, the Enchantress of Numbers: Prophet of the Computer Age* (Mill Valley, CA: Strawberry Press, 1998), p. 70.
112. [De Morgan], 'Brewster's *Life of Newton*', p. 231.
113. King, *The Life of Locke*, p. 259.
114. UCL, Brougham Papers 26,633, letter from Brewster to Brougham, 8 October 1847.
115. [Powell], 'Newton', p. 276.
116. [De Morgan], 'Brewster's *Life of Newton*', pp. 244, 243.
117. [Powell], 'Newton', p. 279, my emphasis.
118. UCL, Brougham Papers 26,633, letter from Brewster to Brougham, 8 October 1847.
119. [De Morgan], 'Brewster's *Life of Newton*', p. 242.
120. Anon. 'Memoirs of Sir Isaac Newton', *The Times*, 21 September 1855, pp. 8e–9a, reprinted in Iliffe et al. (eds), *Early Biographies*, vol. 2, pp. 245–52, on p. 248. The author may have been E. S. Dallas, see ibid., p. xxxiii.
121. [Weld], 'Review of Brewster's *Memoirs of Newton*', p. 698.
122. The apocryphal story illustrated in Figure 8 was also the subject of the poem used as the epigraph to Chapter 1, which described Newton absentmindedly using the finger of a lady whom he was courting as a pipe-stopper. In both cases it clearly refers to the successful, London-based Newton. See Fara, *Newton*, pp. 202–3.
123. [De Morgan], 'Brewster's *Life of Newton*', p. 242.
124. Anon., 'Memoirs of Newton', p. 249.
125. [Powell], 'Newton', p. 275.
126. [Weld], 'Review of Brewster's *Memoirs of Newton*', p. 698. The 1888 catalogue of the Portsmouth Papers did not credit this letter to Newton, [H. R. Luard, G. G. Stokes, J. C. Adams and G. D. Living], *A Catalogue of the Portsmouth Collection of Books and Papers Written by or Belonging to Sir Isaac Newton, the Scientific Portion of which has been Presented by the Earl of Portsmouth to the University of Cambridge, drawn up by the Syndicate Appointed the 6th November 1872* (Cambridge: Cambridge University Press, 1888), p. 41.
127. However, De Morgan suggested that this and his own review both 'illustrate the reaction against hero-worship, which is sure to take place', De Morgan, *Newton*, p. 148.

128. Anon., 'Memoirs of Newton', p. 245.
129. Fara, *Newton*, pp. 30–1, 215–16.
130. S. Crompton, 'On the Portraits of Sir Isaac Newton; and Particularly on one of him by Kneller, Painted about the Time of the Publication of the *Principia*, and Representing him as he was in the Prime of Life', *Proceedings of the Literary and Philosophical Society of Manchester* (2 October 1866), pp. 1–7, on p. 2.
131. Edleston, *Correspondence of Newton and Cotes*, p. xix.
132. Crompton, 'On the Portraits of Newton' pp. 2–3.
133. Edleston, *Correspondence of Newton and Cotes*; Durham County Record Office, Edleston Papers, D/Ed/11/2/75, letter from Brewster to Edleston, 19 February 1855. Edleston received other compliments on this portrait: Dawson Turner considered it 'in every point of view exquisite, & cannot be looked at without the deepest admiration', Durham County Record Office, Edleston Papers, D/Ed/11/2/61, letter from D. Turner to Edleston, 26 February 1851, see also D/Ed/11/2/60, letter from T. Ebor[?] to Edleston, 8 February 1851, and D/Ed/11/2/24, letter from James Stothart to Edleston, 3 January 1864.
134. B. Bensaude-Vincent, 'Between History and Memory: Centennial and Bicentennial Images of Lavoisier', *Isis*, 87 (1996), pp. 481–99, on pp. 483, 493.
135. [De Morgan], 'Brewster's *Life of Newton*', p. 216.

6 The 'Mythical' and the 'Historical' Newton

1. [De Morgan], 'Brewster's *Life of Newton*', p. 244.
2. RSL, Miscellaneous Manuscripts, MM.14.10, extract from minute of Grantham Town Council meeting, 25 May 1853.
3. Royal Society, Council Minutes, 16 May 1853, pp. 205–51.
4. Fara has described how the event became a forum for various individuals and groups to represent their concerns. She notes the tensions between local and national interests, suggesting that the latter became dominant, P. Fara, 'Isaac Newton Lived Here: Sites of Memory and Scientific Heritage', *British Journal for the History of Science*, 33 (2000), pp. 407–26, on pp. 416–19; see also Fara, *Newton*, pp. 243–6.
5. E. F. King, *A Biographical Sketch of Sir Isaac Newton. To which are Added Authorized Reports of the Oration of Lord Brougham (with his Lordship's Notes) at the Inauguration of the Statue at Grantham; and of Several of the Speeches Delivered on that Occasion*, 2nd edn (Grantham and London, 1858), p. 109.
6. Royal Society, Council Minutes, 28 October 1858, p. 449.
7. Brougham and Routh, *Analytical View of Newton's Principia*. Routh was Senior Wrangler in 1854, but had previously been a pupil of De Morgan in London, where he had probably become acquainted with Brougham, Warwick, *The Masters of Theory*, p. 231, see also pp. 227–64.
8. The inauguration was described and pictured in detail in the *Illustrated London News*, 33 (1858), pp. 288, 299 (see Figure 14) and the *Illustrated News of the World* (2 October 1858), pp. 212–13, 220–1. *The Times* printed Brougham's speech, 22 September 1858, p. 7b–e.
9. See letters from Winter to Edleston, Durham County Record Office, Edleston Papers, D/Ed/11/2/35–9.
10. UCL, Brougham Papers 26,705, letter from Brewster to Brougham, 15 August 1858.
11. Fara, 'Newton Lived Here', p. 418. Brewster's daughter records this illness as occurring in July 1858, but she also reports that Brewster attended the September BAAS meeting

in Leeds, Gordon, *The Home Life*, p. 286. He wrote the letter to Brougham cited above in August, which suggests that other circumstances made the trip to Grantham impossible.
12. *Ilustrated London News*, 33 (1858), pp. 315–16.
13. UCL, Brougham Papers 26,705 and 26,707, letters from Brewster to Brougham, 15 August 1858, 18 September 1858.
14. UCL, Brougham Papers, 26,712, letter from Brewster to Brougham, 15 June 1859.
15. King, *A Biographical Sketch*; Durham County Record Office, Edleston Papers, D/Ed/11/2/38, letter from Thomas Winter to Edleston, 4 March 1858. Winter passed on King's request for an introduction to Edleston to aid him in writing his biographical account of Newton.
16. King, *A Biographical Sketch*, p. 101. A draft of Whewell's speech survives: TCL, Whewell Papers, R.18.7/18. See Fara, 'Newton Lived Here'. Whewell told De Morgan that he gave the audience 'something from [Isaac] Barrow about our Cambridge philosophy of that time' but did not, of course, tell them that Newton's 'discoveries were according to your anagram, *Not New*', letter from Whewell to De Morgan, 10 October 1858, quoted in Todhunter, *Whewell*, vol. 2, p. 415.
17. Fara, 'Newton Lived Here', p. 419.
18. King, *A Biographical Sketch*, p. 113.
19. *The Times*, 21 January 1854, p. 6d.
20. King, *A Biographical Sketch*, p. 32.
21. [A. De Morgan], 'Review of King's *Biographical Sketch of Newton* and Brougham's *Address on Popular Literature*', *Athenaeum*, 1621 (1858), pp. 641–2. De Morgan probably also contributed sections on the statue to the 'Weekly Gossip' column in *Athenaeum*, 1339 (1853), p. 773; 1337 (1853), pp. 505–6; and 1360 (1853), pp. 1389–90. The latter two columns were spliced together and kept in the Royal Society Archive (MS 657.9). This is the 'unidentified' article referred to by Fara, 'Newton Lived Here', p. 417.
22. [De Morgan], 'Review of King's *Biographical Sketch*', p. 641.
23. Ibid., pp. 642, 641. UCL, Brougham Papers 5,500, letter from De Morgan to Brougham, 6 October 1862.
24. Will of Lord Halifax, quoted in De Morgan, *Newton*, p. 58. Excerpts from this work relevant to these debates are reprinted in Iliffe et al. (eds), *Early Biographies*, pp. 289–337.
25. A. De Morgan, 'Lord Halifax and Mrs. Catherine Barton', *Notes and Queries*, 8 (1853), pp. 429–33, and 'Lord Halifax and Mrs. Catherine Barton', *Notes and Queries*, 2nd series, 2 (1856), pp. 161–3; Brewster, *Memoirs of Newton*, vol. 2, pp. 276–81.
26. De Morgan, *Newton*. Charles Knight asked De Morgan to alter or cut the article. De Morgan, who had contributed many abstruse articles to the *Companion* felt this altered the arrangement they had enjoyed and withdrew his contribution: RAS, MSS De Morgan 1, draft letter from De Morgan to Knight, 17 October 1857. This was apparently the reason why De Morgan did not offer further contributions to the *Companion*, De Morgan, *Memoir of De Morgan*, p. 264.
27. Rice, 'De Morgan', p. 217; Fara, *Newton*, p. 273.
28. Rice, 'De Morgan', p. 217; Theerman, 'Unaccustomed Role', p. 155; Richards, 'De Morgan', p. 17.
29. Manuel, *A Portrait of Newton*, pp. 260, 262; More, *Newton*, pp. 465–74.
30. E. A. Osborne, 'Introduction', in De Morgan, *Newton*, pp. v–ix, on pp. v, vii.
31. [De Morgan], 'Brewster's *Life of Newton*', p. 214.
32. De Morgan, *Memoir of De Morgan*, p. 264.

33. De Morgan, *Newton*, p. v.
34. Quoted De Morgan, 'Lord Halifax' (1853), p. 430.
35. The longest account of Bartica's role in this book appears in Manuel, *Portrait of Newton*, pp. 249–51. It suggested that Bartica's 'parent' connived at the relationship.
36. De Morgan, *Newton*, p. 3.
37. Baily, *Account of Flamsteed*, pp. 314, 72, Flamsteed's emphasis. In addition, the 'slander' of Voltaire was current knowledge, see e.g. [Galloway], 'French and English Biographies', p. 91.
38. Weld, *History of the Royal Society*, vol. 1, p. 371; [De Morgan], 'Review of Weld's *History*', p. 622.
39. Ibid., p. 622.
40. CUL, RGO MSS, Baily Papers, RGO 60/4, letters from Rigaud to Baily, 19 December 1836, 23 July 1838; letter from Dawson Turner to Baily, 3 December 1836; letter from Joseph Hunter to Baily, 4 March 1837.
41. De Morgan, 'Lord Halifax' (1853), p. 430. See RSL, Herschel Papers, JFWH 6.261–2, letters from De Morgan to John Herschel, 25 August 1852, 29 August 1852, on his examination of Baily's correspondence.
42. De Morgan, 'Lord Halifax' (1853), p. 429.
43. Ibid., p. 429; De Morgan, *Newton*, p. 5.
44. De Morgan, 'Lord Halifax' (1853), pp. 431, 433.
45. Ibid., p. 431. See also De Morgan, *Newton*, pp. 22–5.
46. Brewster, *Memoirs of Newton*, vol. 2, p. 271.
47. BLO, New College MSS, Jeffrey Ekins Papers, MSS 361/4, f. 176, letter from Brewster to Ekins, 18 April 1855.
48. [De Morgan], 'Brewster's *Life of Newton*', pp. 234. De Morgan's whole approach to this topic, which lacked positive proof, was led by reason and logic, the subject of his academic interest at this time.
49. De Morgan, 'Lord Halifax' (1856), p. 161. Of course Brewster had also claimed that he had written more about Newton's faults than any biographer.
50. RSL, Herschel Papers, JFWH 6.286, letter from De Morgan to Herschel, 15 August 1856.
51. Letter from Newton to Sir John [Newton], 23 May 1715, quoted in De Morgan. 'Lord Halifax' (1856), p. 162. The emphasis is De Morgan's.
52. De Morgan, 'Lord Halifax' (1856), p. 161; UCL, Brougham Paper 10,302, letter from De Morgan to Brougham, 15 August 1856.
53. RSL, Herschel Papers, JFWH 6.286, letter from De Morgan to Herschel, 15 August 1856.
54. De Morgan, *Newton*, pp. 1–2.
55. TCL, Whewell Papers, Add. MS a.211/84, letter from Rigaud to Whewell, 24 June 1836. See also BLO, MSS Rigaud 60/26, letter from Baily to Rigaud, 13 December 1836; MSS Rigaud 62/487, letter from Newton B. Young to Rigaud, 13 July 1838.
56. BLO, MSS Rigaud 36, ff. 27–36, 'Mrs Catharine Barton' (June 1836), transcribed in Iliffe et al. (eds), *Early Biographies*, vol. 2, pp. 145–58, on p. 145.
57. De Morgan, *Newton*, p. 43; Iliffe et al. (eds), *Early Biographies*, vol. 2, pp. 146, 156, 151.
58. BLO, MSS Rigaud 36, f. 24, draft letter from Rigaud to Cuthbert Barton, 22 July 1837.
59. In 1861 he retired from the Council of the RAS and in 1866 he resigned his chair for a second time when a professorial candidate was rejected on the grounds of his religious beliefs, De Morgan, *Memoir of De Morgan*, pp. 270, 366, The letter of resignation, in

which De Morgan declared 'the college has left me', is printed in ibid., pp. 339–45. De Morgan did take on the presidency of the Mathematical Society founded by his son George and Arthur Cowper Ranyard in 1864.
60. De Morgan, *Newton*, pp. 102, 107.
61. Ibid., pp. 109, 108. The letter quoted, for which I have been unable to find the original, is dated 14 March 1854.
62. Letter from Brewster to De Morgan, 13 March 1855, in De Morgan, *Newton*, p. 110, the emphasis is De Morgan's.
63. De Morgan, *Newton*, pp. 130, 137.
64. Richards, '"In a Rational World"', pp. 139–43.
65. S. Shapin, *Science as a Vocation: Scientific Authority and Personal Virtue in Late Modernity* (forthcoming; unpublished draft, 2004), ch. 3: 'The Moral Equivalence of the Scientist: The History of an Idea', p. 24.
66. De Morgan, *Newton*, pp. 154–5, 111.
67. See [A. De Morgan], 'Newton Ousted', *Athenaeum*, 2077 (1867), pp. 209–10, reprinted in J. Rosenblum, *Prince of Forgers* (New Castle, DE: Oak Knoll Press, 1998), pp. 149–53. A version of this section appears in R. Higgitt, '"*Newton dépossédé!*" The British Response to the Pascal Forgeries of 1867', *British Journal for the History of Science*, 36 (2003), pp. 437–53.
68. His reputation as a historian rested mainly on his *Histoire de l'arithmétique* ([Paris]: Bachelier, [1843]).
69. M. Chasles, *Rapport sur le progrès de la géométrie* (Paris: Imprimerie nationale, 1870). Details of Chasles's life are taken from the *Dictionary of Scientific Biography*.
70. Details of the affair are taken from the translation of H. L. Bordier and É. Mabille, *Une fabrique de faux autographes, or récit de l'affaire V. Lucas* (Paris, 1870), trans. J. Rosenblum, in Rosenblum, *Prince of Forgers*, pp. 10–143.
71. No one, however, was more credulous than Chasles. The hundreds of letters made public were dwarfed by the thousands that he had bought from one individual over two decades, which included letters from Cleopatra to Caesar, Mary Magdalene to Saint Peter and Joan of Arc to the Parisians. Chasles appeared to believe that, if not actually ancient, the letters had at least once existed in the collection of Tours monastery, where they had been transcribed the sixteenth century. Such at least was one of the explanations arising from yet more forged documents.
72. Bordier and Mabille, *Une fabrique*, trans. Rosenblum, p. 16.
73. 'Weekly Gossip', *Athenaeum*, 2097 (1868), p. 22; *The Times*, 2 October 1867, p. 8d.
74. Bordier and Mabille, *Une fabrique*, trans. Rosenblum, pp. 11–18, M. P. Faugère, *Défense de B. Pascal, et accessoirement de Newton, Galilée, Montesquieu, &c. Contre les faux documens présentés par M. Chasles à l'Académie des Sciences* (Paris: Hachette, 1868); U. Le Verrier, 'Examen de la discussion soulevée au sein de l'Académie des Sciences, au sujet de la découverte de l'attraction universelle', *Comptes rendus*, 68 (1869), pp. 1425–33.
75. Bordier and Mabille, *Une fabrique*, trans. Rosenblum, p. 59.
76. See for example the account in J. A. Farrer, *Literary Forgeries* (London: Longmans, Green and Co., 1907), pp. 202–14; and Rosenblum, *Prince of Forgers*.
77. Bordier and Mabille, *Une fabrique*, trans. Rosenblum, p. 56.
78. Farrer, *Literary Forgeries*, pp. 202–3.
79. See e.g. *Notes and Queries*, 4th series, 1 (1868), p. 444, and 4th series, 4 (1868), p. 248. Rosenblum, *Prince of Forgers*, pp. 149–78, reprints articles from the *Penn Monthly*.

80. See L. A. Marchand, *The Athenaeum: A Mirror of Literature* (Chapel Hill, NC: University of North Carolina Press, 1941).
81. *The Times*, 2 October 1867, p. 8d–e.
82. Letters from Brewster were printed in *Comptes rendus*, 65 (1867), p. 262; *Athenaeum*, 2078 (1867), p. 243, and 2085 (1867), pp. 467–8; *The Times*, 12 October 1867, p. 9a, 13 November 1867, p. 3f, and 21 November 1867, p. 8d. See also Gordon, *The Home Life*, pp. 385–8, for an account of Brewster's objections.
83. [De Morgan], 'Newton Ousted'.
84. [A. De Morgan], 'Review of Faugère's *Défence de B. Pascal*', *Athenaeum*, 2135 (1868), pp. 398–400, 'The Pascal Forgeries', *Athenaeum*, 2090 (16 November 1867), pp. 648–9, and 'Sir D. Brewster and the *Athenaeum*', *Athenaeum*, 2092 (1867), pp. 724–5.
85. R. Grant, 'Letter to the Editor', *The Times*, 20 September 1867, p. 9c–d, reprinted in Rosenblum, *Prince of Forgers*, pp. 156–62; *Comptes rendus*, 65 (1867), pp. 571–7; T. A. Hirst, 'Letter to the Editor', *The Times*, 1 October 1867, p. 7c.
86. [De Morgan], 'Newton Ousted', p. 152.
87. Grant. 'Letter to the Editor', p. 157.
88. Gordon, *The Home Life*, p. 385.
89. Ibid., p. 386.
90. Ibid., p. 384.
91. Ibid., p. 385.
92. Letter from B. J., *Athenaeum*, 2098 (11 January 1868), p. 58, commenting on an article in the *Figaro*.
93. The Society sent a polite refusal, feeling it was 'not expedient' that they should act, 'considering the obvious want of authenticity in the documents'. Once again it kept a dignified distance from matters of popular rather than strictly scientific concern, RSL, Miscellaneous Correspondence, MC.8.91, letter from Brewster to Edward Sabine, 10 October 1867; letter from Sabine to Brewster, quoted in Royal Society, Council Minutes, 31 October 1867, p. 383.
94. Letter from Brewster to the editor, *Athenaeum*, 2078 (1867), p. 243, 2083 (1867), p. 401, and 2085 (1867), p. 467.
95. Letter from Brewster to the editor, *The Times*, 13 November 1867, p. 3f.
96. *Athenaeum*, 2085 (1867), p. 468, and 2090 (1867), p. 648. He also commented in his regular column: e.g. *Athenaeum*, 2084 (1867), reprinted in De Morgan, *A Budget of Paradoxes*, p. 29.
97. [De Morgan], 'Brewster and the *Athenaeum*', p. 724.
98. RSL, Herschel Papers, JFWH 6.398, letter from De Morgan to Herschel, 9 September 1867.
99. De Morgan, *A Budget of Paradoxes*, p. 29.
100. UCL, Brougham Papers 10,301, letter from De Morgan to Brougham, 9 November 1855.
101. UCL, Brougham Papers 10,299, letter from De Morgan to Brougham, 16 May 1855.
102. UCL, Brougham Papers 10,301, letter from De Morgan to Brougham, 9 November 1855.
103. RSL, Herschel Papers, JFWH 6.412, letter from De Morgan to Herschel, 8 November 1869.
104. De Morgan, 'Newton Ousted', pp. 151–2.
105. Ibid., p. 153.
106. Ibid.

107. [A. De Morgan], 'Weekly Gossip', *Athenaeum*, 2081 (1867), p. 334.
108. Grant, 'Letter to the Editor', p. 156.
109. De Morgan had been impressed by Grant's *History of Physical Astronomy*, probably being instrumental in ensuring it won the RAS Gold Medal, and called it a '*standard historical work*', University of London Library, Senate House, De Morgan Papers, MS 913/A/2/10, letter from De Morgan to Brougham, 18 June 1852 (De Morgan's emphasis).
110. Grant, 'Letter to the Editor', pp. 156, 156–7.
111. Ibid., p. 157.
112. Ibid., p. 160.
113. *Athenaeum*, 2082 (1867), p. 367.
114. T. A. Hirst, *Natural Knowledge in Social Context: The Journals of Thomas Archer Hirst*, ed. W. H. Brock and R. M. MacLeod (London: Mansell, 1980), 31 December 1867, p. 1824, 29 July 1869–2 August 1869, p. 1851.
115. RSL, Herschel Papers, JFWH 6.399, letter from De Morgan to Herschel, 8 September 1867.
116. Report of communications of Hirst and Brewster to Section A in *Dundee Advertiser*, copied in Hirst, *Natural Knowledge*, pp. 1813–16, 1813. See also *The Times*, 12 September 1867, p. 8f; and BAAS, 'Notices and Abstracts of Miscellaneous Communications to the Sections', *Report of the 37th Meeting of the British Association for the Advancement of Science* (London: John Murray, 1868), pp. 1–3.
117. Hirst, 'Letter to the Editor'.
118. RSL, Miscellaneous Correspondence, MC.8.114, letter from Hirst to William Sharpey, 26 November 1867. The minutes of the Royal Society confirm that the Council resolved to put a sum of not more than £25 at Hirst's disposal on 30 November 1867. Hirst records that he received £20, Hirst, *Natural Knowledge*, 5 December 1867, p. 1823.
119. Hirst, *Natural Knowledge*, 2 November 1867, 7 November 1867, pp. 1822–3.
120. Ibid., 30 November 1867, p. 1823.
121. Ibid., 29 June 1869, p. 1851.
122. *Athenaeum*, 2090 (1867), pp. 648, 649.
123. D. Brewster, 'Letter to the Editor', *The Times*, 21 November 1867, p. 8d.
124. 'Weekly Gossip', *Athenaeum*, 2093 (1867), pp. 768–9.
125. RSL, Herschel Papers, JFWH 6.396, letter from De Morgan to Herschel, 29 August 1867.
126. [De Morgan], 'Brewster's *Life of Newton*', p. 214.
127. Discussing the exposure of forgeries, Stephen Bann suggests that the 1860s 'was certainly [a period] in which long-established but unfounded myths about the past were finally exposed', Bann, *The Clothing of Clio*, p. 2.
128. *The Times*, 21 February 1854, p. 6d; King, *A Biographical Sketch*, p. 117.
129. Manuel, *A Portrait of Newton*, p. viii.
130. TCL, Whewell Papers, O.15.47/85, letter from Whewell to J. D. Forbes, 17 September 1861.
131. [G. W. Hemming], 'Newton as a Scientific Discoverer', *Quarterly Review*, 110 (1861), pp. 401–35, on pp. 401, 402–3.
132. Ibid., pp. 401, 403.
133. J. Tyndall, *Address Delivered Before the British Association Assembled at Belfast, With Additions* (London: Longmans, Green & Co., 1874), p. 13.
134. J. Tyndall, 'On the Scientific Use of the Imagination', in J. Tyndall, *Fragments of Science: A Series of Detached Essays, Lectures, and Reviews*, 3rd edn (London: Longmans, Green and Co., 1871), pp. 125–62.

135. J. Tyndall, *Faraday as a Discoverer* (London: Longmans, Green and Co., 1868). See Cantor, 'The Scientist as Hero', pp. 173–7, on his similar treatment of Faraday's discoveries.
136. See Turner, *Contesting Cultural Authority*, pp. 270–83; and F. M. Turner, 'John Tyndall and Victorian Scientific Naturalism', in W. H. Brock, N. D. McMillan and R. C. Mollan (eds), *John Tyndall: Essays on a Natural Philosopher* (Dublin: Royal Dublin Society, 1981), pp. 169–80.
137. De Morgan, 'Newton', p. 183.

Conclusion

1. Reviews of the most recent biography of Newton, J. Gleick, *Isaac Newton* (London: Fourth Estate, 2003), have concentrated on its depiction of Newton as an alchemist, e.g. J. Banville, 'The Magus', *The Guardian*, 'Review', 30 August 2003, p. 9. The BBC's *Newton: The Dark Heretic* (2003) likewise presented Newton's alchemy and theology as less well known and remarkable. Such depictions have also emphasized Newton's treatment of Hooke and other rivals.
2. Jann, *The Art and Science*, p. xxv, is typical in placing the claims for a scientific methodology in history in the period when the field was beginning to professionalize. See also Goldstein, 'History at Oxford and Cambridge'.
3. [De Morgan], 'Brewster's *Life of Newton*', p. 215.
4. See above, n. 72, p. 236.
5. Theerman, 'Unaccustomed Role', p. 152.
6. [A. De Morgan], '*History of the Rise and Influence of the Spirit of Rationalism in Europe.* By W. E. H. Lecky', *Athenaeum*, 1960 (1865), pp. 676–7, on p. 677; Brock and MacLeod, 'The Scientists' Declaration', pp. 49–51. De Morgan was referring to the Broad Church *Essays and Reviews* (1860), formally condemned by both Houses of Convocation in 1864. John William Colenso was Bishop of Natal, and his critical studies of the Bible led to his excommunication, but an appeal held in 1864 found him legally still in possession of his see. This decision was considered by some the 'final triumph of the secular over the ecclesiastical jurisdiction', P. B. Hinchcliffe, *John William Colenso, Bishop of Natal* (1963), quoted in ibid., p. 40.
7. Shapin, *Science as a Vocation*, pp. 19–24. See R. K. Merton, *The Sociology of Science: Theoretical and Practical Investigations* (Chicago, IL, and London: University of Chicago Press, 1973), pp. 275–7.
8. Yeo, *Defining Science*, p. 117.
9. Shapin, *Science as a Vocation*, p. 24.
10. Daston and Gallison, 'The Image of Objectivity'.
11. Baily, *Account of Flamsteed*, p. xxi; [Galloway], 'Life and Observations of Flamsteed', pp. 131–2.
12. Iliffe, 'A "connected system"?', p. 145. In the same year Jeffrey Ekins gave the papers in his possession to New College, Oxford, see ibid., p. 146.
13. Ibid., p. 146.
14. W. W. Rouse Ball, *An Essay on Newton's "Principia"* (London: Macmillan & Co., 1893); Iliffe, 'A "connected system"?', p. 150.
15. Ibid., p. 148, and see pp. 148–9 for an account of the dispersal of the papers at this date. Keynes, 'Newton', p. 310. The Newton Project includes a number of important documents relating to Keynes's Newtonian interests, see http://www.newtonproject.imperial.ac.uk/jmk.html.

APPENDIX: TRANSLATIONS OF QUOTATIONS FROM BIOT'S 'NEWTON' IN CHAPTER 1

Translations of [J.-B. Biot], 'Newton (Isaac)', in L. G. Michaud (ed.), *Biographie universelle, ancienne et moderne*, 83 vols (Paris: Michaud Frères, 1811–53), vol. 31 (1822), pp. 127–94', as rendered in J. B. Biot, 'Life of Newton', trans. H. Elphinstone (1829), in *Lives of Eminent Persons*, Library of Useful Knowledge (London: Baldwin & Cradock, 1833), reprinted in R. Iliffe, M. Keynes and R. Higgitt (eds), *Early Biographies of Isaac Newton, 1660–1885*, 2 vols (London, Pickering & Chatto, 2006), vol. 2: R. Higgitt (ed.), *Nineteenth-Century Biography of Isaac Newton: Public Debate and Private Controversy*, pp. 1–63, unless otherwise stated. Translations of other texts, or in different chapters, are given in the notes.

Where Elphinstone's translation has significantly altered the meaning, or where he did not translate a phrase at all, a modern translation has been provided. Elphinstone's translation was neither faithful nor very skilful. His alterations tended to moderate Biot's heightened language and the image of the Romantic genius.

- p. 23, 'le créateur ... jamais existé' (p. 169): 'almost the creator of Natural Philosophy, as one of the chief promoters of mathematical analysis' (Elphinstone, p. 43). The translation adds the 'almost' and misses out 'foremost among the physicists that have ever existed'.
- p. 23, 'la prééminence ... de l'esprit humain' (p. 165): 'a lasting pre-eminence over all other productions of the human mind' (Elphinstone, p. 39). The translation adds 'lasting'.
- p. 23, 'un homme qui ... à Newton même' (p. 139): 'a man of extensive acquirements, and of an original turn of thought, with great activity of mind' (Elphinstone, p. 13). 'A man who, for genius of invention and extent of insight was scarcely inferior to Newton himself' (translation Caroline Higgitt).
- p. 23, 'une excessive ambition de renommée' (p. 139): 'an excessive desire of renown' (Elphinstone, p. 13).

p. 23, 'le grand avantage ... dans les sciences' (p. 139): 'the great advantage possessed by Newton, and which assured to his researches a precision and a certainty hitherto unknown in science' (Elphinstone, p. 14).

p. 23, 'seulement dans leurs ... autrefois imaginée' (p. 140): 'only in relation to an hypothesis which he had formerly imagined' (Elphinstone, p. 14).

p. 23, 'pour pouvoir être ... éprouvé par le calcul' (p. 140): 'in order to place such an hypothesis on an equal footing with another hypothesis, shown by calculation to be consistent with experiment and observation, it ought to be detailed with exactness, and to be rigorously accordant with mathematical calculation' (Elphinstone, p. 14). 'In order to be accepted today as true and certain, it would be necessary first for it to be defined in its details, and then that it be open to rigorous demonstration through calculation' (translation Caroline Higgitt).

p. 24, 'n'a jamais mis en doute' (p. 144): 'has never treated as doubtful' (Elphinstone, p. 18).

p. 24, 'ne puissent s'appliquer qu'à des particules matérielles' (p. 143): 'is applicable ... to material particles only' (Elphinstone, p. 18).

p. 24, 'sont si rigidement ... des ondulations propagées ...' (p. 144): 'are described in such exact conformity with experiment, that they would exist without any change, even were it discovered that light is constituted in any other manner – that it consists, for instance, in the propagation of undulations ...' (Elphinstone, p. 18).

p. 24, 'hypothèse physique très hardie' (p. 144): 'very bold physical hypothesis' (Elphinstone, p. 18).

p. 24, 'non pas dans l'intention ... moins explicite' (p. 144): 'without the intention of either defending or combating it, but in order that the reader may see precisely in what the general views of Newton from this time forward consisted, and how, while they continued unchanged by lapse of time, he made a more or less explicit declaration of them according to circumstances' (Elphinstone, p. 19).

p. 25, 'onze ans plus tard ... aujourd'hui' (p. 133): 'eleven years later, Leibnitz again discovered [the fluxions], and presented to the world in a different form, that, namely, of the modern *Differential calculus*' (Elphinstone, p. 7).

p. 25, 'gardé long-temps et obstinément le secret de ces découvertes' (p. 173): 'obstinately guarded the secret of his discoveries' (Elphinstone, p. 47).

p. 25, 'noble loyauté de Leibnitz' (p. 174): 'noble frankness of Leibniz' (Elphinstone, p. 48).

p. 25, 'le signal de l'attaque de la part des écrivains anglais' (p. 176): 'the signal for attack, on the part of the English writers' (Elphinstone, p. 50).

p. 25, 'qui ne furent point ... nullement consulté' (p. 176): 'who were not known, and about whose appointment Leibnitz was *not* consulted' (Elphinstone, p. 50).

p. 25, 'il faut dire que ... ni moins injuste' (p. 177): 'it is necessary to say that Leibnitz, on his side, had neither been less passionate nor less unjust' (Elphinstone, p. 52).

p. 26, 'Le 29 mai *1694* ... son livre des PRINCIPES' (p. 168): 'On the 29th May, 1694, a Scotchman of the name of Colin informed me, that Isaac Newton, the celebrated mathematician, eighteenth months previously, had become deranged in his mind, either from too great application to his studies, or from excessive grief at having lost, by fire, his chemical laboratory and some papers. Having made observations before the Chancellor of Cambridge, which indicated the alienation of his intellect, he was taken care of by his friends, and being confined to his house, remedies were applied, by means of which he has lately so far recovered his health as to begin to again understand his own Principia' (Elphinstone, p. 42).

p. 26, 'On raconte que ... le tort que m'as fait"' (p. 168): 'It is said, that on first perceiving this great loss, he contented himself by exclaiming, "Oh Diamond! Diamond! thou little knowest the mischief thou hast done"' (Elphinstone, p. 42). 'It is said that, in the first anguish of this great loss, he confined himself to saying: "Oh! Diamond, Diamond, you do not know the harm you have done me"' (translation Caroline Higgitt).

p. 26, 'la douleur qu'il ... pendant quelque temps (p. 168): 'the grief caused by this circumstance, grief which reflection must have augmented, instead of alleviating, injured his health, and, if we may venture to say so, for some time impaired his understanding' (Elphinstone, p. 42). The words 'instead of alleviating' are added.

p. 26, 'cette tête qui ... raison humaine' (p. 168): 'this mind which, during many years was applied continually to meditations so deep that they seemed to stand at the ultimate limit of human reason' (translation Caroline Higgitt).

p. 27, 'Quelle vive et naïve ... moment d'inspiration!' (p. 156): 'What a lifelike and simple portrait of genius awaiting the moment of inspiration!' (translation Caroline Higgitt).

p. 27, 'dérangement d'esprit' (p. 169): 'the derangement in his intellect' (Elphinstone, p. 43).

p. 27, 'il paraîtrait ... commerce du monde' (p. 193): 'it would appear that he was very ignorant of the habits of society' (Elphinstone, p. 61).

p. 27, 'presque puérile' (p. 193): 'almost puerile' (Elphinstone, p. 61).

p. 27, 'l'effet d'une timidité ... retirée et méditative' (p. 193): 'the effect of excessive shyness, produced by the retired and meditative habits of his life' (Elphinstone, p. 61).

p. 27, 'on est tenté ... sert le génie' (p. 179): 'we are disposed to compassionate the occasional weaknesses of the finest intellects, and to deplore the petty passions which tarnish the splendour of genius' (Elphinstone, pp. 53–4). 'One is tempted to feel sorry for our feeble human intellect and to wonder what is the point of genius' (translation Caroline Higgitt).

p. 28, 'les lectures religieuses ... unique délassement' (p. 190): 'the reading of religious works had become one of his most habitual occupations; and after he had performed the duties of his office, they formed, along with the conversation of his friends, his only amusement' (Elphinstone, p. 58).

p. 28, 'Sa tête ... dans les affaires' (p. 192): 'His mind, fatigued by long and painful efforts, had need of complete and entire repose. At least we know, that thenceforward he only occupied his leisure with religious studies, or sought relief in literature or in business' (Elphinstone, p. 61). 'Religious' is substituted for 'serious'.

p. 28, 'il n'y a réellement ... méthode d'interprétation' (p. 187): 'we find, in fact, nothing new, except the precise and, in some degree, systematic explanation of the method of interpretation' (Elphinstone, p. 55).

pp. 28–9, 'On demandera ... il les établit' (pp. 188–9): 'It will, doubtless, be asked, how a mind of the character and force of Newton's, so habituated to the severity of mathematical considerations, so accustomed to the observation of real phenomena, so methodical, and so cautious, even at his boldest moments in physical speculation, and consequently so well aware of the conditions by which alone truth is to be discovered, could put together such a number of conjectures, without noticing the extreme improbability that is involved in all of them, from the infinite number of arbitrary postulates on which he endeavours to establish his system' (Elphinstone, p. 57).

p. 29, 'ce système est ... science chronologique' (p. 186): 'this system is a very important fact in the history of chronological science' (translation Caroline Higgitt).

p. 30, 'certes, soit que ... intimement convaincue' (p. 179): 'even those who might dispute the arguments which he gives for such an existence, must still recognize, in this passage, the sentiments of a mind deeply imbued with religious feelings, and convinced of their true foundation' (Elphinstone, p. 53). 'Religious soul' is omitted.

p. 30, 'l'esprit de prévention' (p. 187): 'the spirit of prejudice of which it unhappily bears the stamp' (Elphinstone, p. 56).

p. 30, 'comme chez d'autres ... démonstration évidente' (p. 188): 'as in those of some other protestant writers, dictated by any sectarian or party feeling; he states it with all the calm of entire conviction, and with all the simplicity of an evident demonstration.' (Elphinstone, p. 57). 'Spirit of resentment or hate' is omitted.

p. 38, 'appliqué continûment ... raison humaine' (p. 168): 'applied continually to meditations so deep that they seemed to stand at the ultimate limit of human reason' (translation Caroline Higgitt).

p. 38, 'Si la chose était vraie ... un prodige' (pp. 129–30): 'If it were true, it would indeed be something extraordinary' (translation Caroline Higgitt).

WORKS CITED

Manuscript Sources

Bodleian Library, Modern Manuscripts
    New College Manuscripts, Jeffrey Ekins Papers (MSS 361/1-4).
    Rigaud Family Papers (MS Eng.misc.c.807).
    Savile Collection, Rigaud Manuscripts (MSS Rigaud 3-65).

British Library, Manuscripts
    Additional MS 22,786, 'Tractatus de mundi sphæra'.
    Babbage Correspondence (Add. MS 37182-201).
    Charnwell Autographs, Add. MS 70,951.
    Napier Correspondence (Add. MS 34611-26).

Cambridge University Library, Manuscripts and Archives, Royal Greenwich Observatory, Baily Papers (RGO 60-1).

Durham County Record Office, Joseph Edleston Papers (D/Ed/11-12).

John Murray (Publishers) Archives, Ledgers and Correspondence.

National Archives, Admiralty Papers (ADM).

National Library of Scotland, Manuscripts, Additional Manuscripts and Accessions.

Royal Astronomical Society, Library and Archives
    Baily Manuscripts (MSS Baily).
    De Morgan Manuscripts (MSS De Morgan).

Royal Society of London, Library and Archives
    Herschel Papers (JFWH).
    Miscellaneous Correspondence (MC).
    Miscellaneous Manuscripts (MM).

Trinity College Library, Cambridge, Whewell Papers (Add. MS a.51-83, Add. MS a.200-24, Add. MS c.87-8, O.15.45-50).

University College London Library, Rare Books and Manuscripts
    Brougham Papers.

SDUK Papers and Correspondence.

University of Edinburgh Library, Special Collections, Halliwell-Phillipps Collection (L.O.A. 1–160).

University of London Library, Senate House, Archives and Manuscripts, De Morgan Papers (MS913).

Electronic Resources

Athenaeum index, http://athenaeum.soi.city.ac.uk/reviews/home.html.

Oxford Dictionary of National Biography, http://www.oxforddnb.com/.

The Times, Palmer's complete online edition, http://historynews.chadwyck.com/.

Printed Sources

Airy, G. B., 'Report on the Progress of Astronomy during the Present Century', *Report of the First and Second Meetings of the British Association for the Advancement of Science* (London: John Murray, 1833), pp. 125–89.

—, *Autobiography of Sir George Biddell Airy*, ed. W. Airy (Cambridge: Cambridge University Press, 1896).

Alborn, T. L., 'The Business of Induction: Industry and Genius in the Language of British Scientific Reform, 1820–1840', *History of Science*, 34 (1996), pp. 91–121.

Altick, R. D., *The English Common Reader: A Social History of the Mass Reading Public 1800–1900* (Chicago, IL: Chicago University Press, 1957).

—, *Lives and Letters: A History of Literary Biography in England and America* (New York: Alfred A. Knopf, 1966).

Anderson, R., 'Brewster and the Reform of the Scottish Universities', in A. D. Morrison-Low and J. R. R. Christie (eds), *Martyr of Science: Sir David Brewster 1781–1868* (Edinburgh: The Royal Scottish Museum, 1984), pp. 31–4.

Anon., *An History of the Instances of Exclusion from the Royal Society ... with Strictures on the Formation of the Council, and Other Instances of the Despotism of Sir J. Banks* (London: J. Debrett, 1784).

—, *The Consequences of a Scientific Education to the Working Classes of this Country Pointed Out; and the Theories of Mr. Brougham on the Subject Confuted; in a Letter to the Marquess of Lansdown, by a Country Gentleman* (London: T. Cadell, 1826).

—, 'The House of Fame', *Punch*, 25 (1853), pp. 106–7.

—, 'Memoirs of Sir Isaac Newton', *The Times*, 21 September 1855, pp. 8e–9a, reprinted in R. Iliffe, M. Keynes and R. Higgitt (eds), *Early Biographies of Isaac Newton, 1660–1885*, 2 vols (London, Pickering & Chatto, 2006), vol. 2, pp. 245–52.

—, 'Inauguration of the Statue of Sir Isaac Newton' *Illustrated London News*, 33 (1858), p. 299.

—, 'The British Association', *Punch*, 49 (23 September 1865), p. 113.

Ashworth, W. J., 'The Calculating Eye: Baily, Herschel, Babbage and the Business of Astronomy', *British Journal for the History of Science*, 27 (1994), pp. 409–41.

—, 'Memory, Foresight and Production: The Work of Analysis in Early 19th-Century England' (unpublished PhD thesis, University of Cambridge, 1996).

—, '"Labour harder than *thrashing*": John Flamsteed, Property and Intellectual Labour in Nineteenth-Century England', in F. Willmoth (ed.), *Flamsteed's Stars: New Perspectives on the Life and Work of the First Astronomer Royal* (Woodbridge: Boydell Press, 1997), pp. 199–216.

Astronomical Society, *Address and Regulations of the Astronomical Society of London* (London, 1821).

Ault, D. D., *Visionary Physics: Blake's Response to Newton* (Chicago, IL, and London: University of Chicago Press, 1974).

[Austin, S], 'Life of Carsten Niebuhr', in *Lives of Eminent Persons*, Library of Useful Knowledge (London: Baldwin & Cradock, 1833).

BAAS, 'Notices and Abstracts of Miscellaneous Communications to the Sections', *Report of the 37th Meeting of the British Association for the Advancement of Science* (London: John Murray, 1868), pp. 1–3.

Babbage, C., *Reflections on the Decline of Science in England, and on Some of its Causes* (London: B. Fellowes, 1830).

Baily, F., *Further Remarks on the Present Defective State of the Nautical Almanac. To which is added an Account of the New Astronomical Ephemeris published at Berlin* (London: C. J. G. & F. Rivington, 1829).

—, 'Some Particulars relative to the Life and Writings of the Late Mr. Flamsteed, Never Yet Published', *Monthly Notices of the Royal Astronomical Society*, 3 (1833), pp. 4–10.

—, 'Report on the Pendulum Experiments made by the Late Captain Henry Foster, R.N., in his Scientific Voyage in the Years 1828–1831 with a View to Determine the Figure of the Earth', *Memoirs of the Royal Astronomical Society*, 7 (1834).

—, *An Account of the Revd. John Flamsteed, the First Astronomer-Royal; Compiled from his own Manuscripts, and other Authentic Documents, Never Before Published, and Supplement to the Account of the Revd. John Flamsteed* (London: 1835, 1837; London: Dawsons of Pall Mall, 1966).

—, 'Some Account of the Astronomical Observations made by Dr. Edmund Halley, at the Royal Observatory at Greenwich', *Memoirs of the Royal Astronomical Society*, 8 (1835), pp. 169–90.

—, *Journal of a Tour in Unsettled Parts of North America in 1796 and 1797*, ed. A. De Morgan (London: Baily Bros, 1856).

Bann, S., *The Clothing of Clio: A Study of the Representation of History in Nineteenth-Century Britain and France* (Cambridge: Cambridge University Press, 1984).

Banville, J., 'The Magus', *The Guardian*, 'Review', 30 August 2003, p. 9.

Barrow, J., *An Autobiographical Memoir of Sir John Barrow, Bart., Late of the Admiralty; including Reflections, Observations, and Reminiscences at Home and Abroad, from Early Life to Advanced Age* (London: John Murray, 1847).

[Barrow, J.], 'Account of the Rev. John Flamsteed', *Quarterly Review*, 55 (1835), pp. 96–128.

[Barrow, J.?], 'Note on a Pamphlet Entitled "Newton and Flamsteed, by the Rev. Wm. Whewell, M.A. ..."', *Quarterly Review*, 55 (1836), pp. 568–72.

Barton, R., '"Men of science": Language, Identity and Professionalization in the Mid-Victorian Scientific Community', *History of Science*, 41 (2003), pp. 73–119.

Baxter, P., 'Brewster, Evangelism and the Disruption of Scotland', in A. D. Morrison-Low and J. R. R. Christie (eds), *Martyr of Science: Sir David Brewster 1781–1868* (Edinburgh: The Royal Scottish Museum, 1984), pp. 45–50.

Bennett, J. A., 'Museums and the Establishment of the History of Science at Oxford and Cambridge', *British Journal for the History of Science*, 30 (1997), pp. 29–46.

Bennett, S., 'John Murray's Family Library and the Cheapening of Books in Early Nineteenth Century Britain', *Studies in Bibliography*, 29 (1976), pp. 139–66.

Bensaude-Vincent, B., 'Between History and Memory: Centennial and Bicentennial Images of Lavoisier', *Isis*, 87 (1996), pp. 481–99.

Biot, J.-B., 'Life of Newton', trans. H. Elphinstone (1829), in *Lives of Eminent Persons*, Library of Useful Knowledge (London: Baldwin & Cradock, 1833), reprinted in R. Iliffe, M. Keynes and R. Higgitt (eds), *Early Biographies of Isaac Newton, 1660–1885*, 2 vols (London: Pickering & Chatto, 2006), vol. 2, pp. 1–63.

—, 'Revue de *The Life of Isaac Newton*', *Journal des savants*, (1832), pp. 199–203, 263–74, 321–39.

—, 'Revue de *An Account of the Rev. John Flamsteed*', *Journal des savants* (1836), pp. 156–66, 205–23, 641–58.

—, 'Revue de *Memoirs of the Life, Writings, and Discoveries of Sir Isaac Newton*', *Journal des savants* (1855), pp. 589–606, 662–77.

[Biot, J.-B.], 'Galilée Galilei', in Louis Gabriel Michaud (ed.), *Biographie universelle, ancienne et moderne*, 83 vols (Paris: Michaud Frères, 1811–53), vol. 16 (1816), pp. 318–31.

—, 'Newton (Isaac)', in L. G. Michaud (ed.), *Biographie universelle, ancienne et moderne*, 83 vols (Paris: Michaud Frères, 1811–53), vol. 31 (1822), pp. 127–94.

[Biot, J.-B., F. P. G.[?] Duvau, P. Maine de Biran and P. A.[?] Stapfer], 'Leibnitz, Godfroi-Guillaume, baron de', in Louis Gabriel Michaud (ed.), *Biographie universelle, ancienne et moderne*, 83 vols (Paris: Michaud Frères, 1811–53), vol. 23 (1819), pp. 594–642.

Birch, T., 'Sir Isaac Newton', in P. Bayle (ed.), *A General Dictionary, Historical and Critical*, 10 vols (London, 1734–41), vol. 7 (1738), pp. 776–802, reprinted in A. R. Hall, *Isaac Newton: Eighteenth-Century Perspectives* (Oxford: Oxford University Press, 1999), pp. 83–95.

Boase, F. (ed.), *Modern English Biography: Containing Many Thousand Concise Memoirs of Persons who have Died Between the Years 1851–1900 with an Index of the Most Interesting Matter* (London: F. Cass, 1965).

Bordier, H. L., and E. Mabille, *Une fabrique de faux autographes, ou récit de l'affaire V. Lucas* (Paris, 1870), trans. J. Rosenblum, in J. Rosenblum, *Prince of Forgers* (New Castle, DE: Oak Knoll Press, 1998), pp. 10–143.

Bowler, P. J, *The Invention of Progress: The Victorians and the Past* (Oxford: Basil Blackwell, 1989).

Brewster, D., *The Life of Sir Isaac Newton*, The Family Library, vol. 24 (London: John Murray, 1831).

— (ed.), *Testimonials in Favour of Sir David Brewster as a Candidate for the Chair of Natural Philosophy at the University of Edinburgh* (Edinburgh, 1832).

—, 'Report on the Recent Progress of Optics', *Report of the First and Second Meetings of the British Association for the Advancement of Science* (London: John Murray, 1833), pp. 308–22.

—, *The Martyrs of Science; or the Lives of Galileo. Tycho Brahe, and Kepler* (London: John Murray, 1841).

—, *Memoirs of the Life, Writings and Discoveries of Sir Isaac Newton*, 2 vols (Edinburgh: Constable & Co., 1855), reprinted in The Sources of Science, vol. 14 (New York and London: Johnson Reprint Corporation, 1965).

—, 'Letter to the Editor', *The Times*, 21 November 1867, p. 8d.

[Brewster, D.], 'Decline of Science in England', *Quarterly Review*, 43 (1830), pp. 305–42.

—, 'The British Scientific Association', *Edinburgh Review*, 60 (1835), pp. 363–94.

—, 'Whewell's History of the Inductive Sciences', *Edinburgh Review*, 66 (1837), pp. 110–51.

—, 'Newton, Sir Isaac', in M. Napier (ed.), *Encyclopaedia Britannica*, 7th edn, 21 vols (Edinburgh: A. and C. Black, 1842), vol. 16, pp. 175–81, reprinted in R. Iliffe, M. Keynes and R. Higgitt (eds), *Early Biographies of Isaac Newton, 1660–1885*, 2 vols (London, Pickering & Chatto, 2006), vol. 2, pp. 159–73.

—, 'Macaulay's *History of England*', *North British Review*, 10 (1849), pp. 367–424.

Brock, W. H., and R. M. MacLeod, 'The Scientists' Declaration: Reflexions on Science and Belief in the Wake of *Essays and Reviews*, 1864–5', *British Journal of the History of Science*, 9 (1976), pp. 39–66.

Brooke, J. H., 'Natural Theology and the Plurality of Worlds: Observations on the Brewster-Whewell Debate', *Annals of Science*, 34 (1977), pp. 221–86.

—, 'Joseph Priestley (1733–1804) and William Whewell (1794–1866): Apologists and Historians of Science. A Tale of Two Stereotypes', in R. G. W. Anderson and C. Lawrence (eds), *Science, Medicine and Dissent: Joseph Priestley (1733–1804)* (London: Wellcome Trust, 1987), pp. 11–27.

Brooke, J. H., and G. Cantor, *Reconstructing Nature: The Engagement of Science and Religion* (Edinburgh: T. & T. Clark, 1998).

Brougham, H. P., *Discourse on the Objects, Advantages, and Pleasures of Science*, Library of Useful Knowledge (London: Baldwin & Cradock, 1827).

—, *A Discourse on Natural Theology* (London: Charles Knight, 1835).

—, *Lives of Men of Letters and Science, who flourished in the time of George III*, 2 vols (London: C. Knight and Co., 1845, 1846).

—, 'Experiments and Observations upon the Properties of Light', *Philosophical Transactions*, 140 (1850), pp. 235–59.

—, *The Life and Times of Henry Lord Brougham, Written by Himself*, 3 vols (Edinburgh and London: W. Blackwood & Sons, 1871).

[Brougham, H. P.], 'Bakerian Lecture on Light and Colours', *Edinburgh Review*, 1 (1803), pp. 450–6.

—, 'Lord King's Life of John Locke', *Edinburgh Review*, 50 (1829), pp. 1–31.

Brougham, H. P., and E. J. Routh, *Analytical View of Sir Isaac Newton's Principia* (London: Longman, Brown, Green & Longmans, 1855).

Burke, P., 'Ranke the Reactionary', in G. G. Iggers and J. M. Powell (eds), *Leopold von Ranke and the Shaping of the Historical Discipline* (Syracuse NY: Syracuse University Press, 1990), pp. 36–44.

Butterfield, H., 'Delays and Paradoxes in the Development of Historiography', in K. Bourne and D. C. Watt (eds), *Studies in International History: Essays Presented to W. Norton Medlicott* (London: Longmans, 1967), pp. 1–15.

Butts, R. E., 'Whewell on Newton's Rules of Philosophizing', in R. E. Butts and J. W. Davis (eds), *The Methodological Heritage of Newton* (Oxford: Blackwell, 1970), pp. 132–49.

Cannon, S. F., *Science in Culture: The Early Victorian Period* (Folkestone and New York: Dawson and Science History, 1978).

Cantor, G. N., 'Henry Brougham and the Scottish Methodological Tradition', *Studies in History and Philosophy of Science*, 2 (1971), pp. 69–89.

—, *Optics after Newton: Theories of Light in Britain and Ireland, 1704–1840* (Manchester: Manchester University Press, 1983).

—, 'Between Rationalism and Romanticism: Whewell's Historiography of the Inductive Sciences', in M. Fisch and S. Schaffer (eds), *William Whewell: A Composite Portrait* (Oxford: Clarendon Press, 1991), pp. 67–86.

—, 'The Scientist as Hero: Public Images of Michael Faraday', in M. Shortland and R. Yeo (eds), *Telling Lives in Science: Essays on Scientific Biography* (Cambridge: Cambridge University Press, 1996), pp. 171–93.

Chadwick, O., *The Secularisation of the European Mind in the Nineteenth Century* (Cambridge: Cambridge University Press, 1975).

Chapman, A., 'Sir George Airy (1801–1892) and the Concept of International Standards in Science, Timekeeping and Navigation', *Vistas in Astronomy*, 28 (1985), pp. 321–8.

Chasles, M., *Histoire de l'arithmétique* ([Paris]: Bachelier, [1843]).

—, *Rapport sur le progrès de la géométrie* (Paris: Imprimerie nationale, 1870).

Chen, X., and P. Barker, 'Cognitive Appraisal and Power: David Brewster, Henry Brougham, and the Tactics of the Emission-Undulatory Controversy during the Early 1850s', *Studies in the History and Philosophy of Science*, 23 (1992), pp. 75–101.

Christie, J. R. R., 'Sir David Brewster as an Historian of Science', in A. D. Morrison-Low and J. R. R. Christie (eds), *Martyr of Science: Sir David Brewster 1781–1868* (Edinburgh: The Royal Scottish Museum, 1984), pp. 53–6.

[Christie, R. H.], 'Biographical Dictionaries', *Quarterly Review*, 157 (1884), pp. 187–230, reprinted in I. B. Nadel (ed.), *Victorian Biography: A Collection of Essays from the Period* (New York and London, Garland, 1986), unpaginated.

Clare, J., 'Popularity in Authorship', *The European Magazine*, 1 (1825), ed. J. Birtwhistle, http://www.johnclare.info/birtwhistle.htm (accessed 5 May 2006).

Cochran, W., 'Sir David Brewster: An Outline Biography', in A. D. Morrison-Low and J. R. R. Christie (eds), *Martyr of Science: Sir David Brewster 1781–1868* (Edinburgh: The Royal Scottish Museum, 1984), pp. 11–14.

Cockshut, A. O. J., *Truth to Life: The Art of Biography in the Nineteenth Century* (London: Collins, 1974).

Cohen, H. F., *The Scientific Revolution: A Historiographical Inquiry* (Chicago, IL, and London: University of Chicago Press, 1994).

Cohen, I. B., *Revolution in Science* (Cambridge, MA, and London: Belknap Press, 1985).

Cohen, I. B., and R. S. Westfall (eds), *Newton: Texts, Backgrounds, Commentaries* (New York and London: W. W. Norton, 1995).

[Collier, J. P.], 'Lectures on the History of Rome', *Athenaeum*, 1058 (1848), pp. 139–40.

Collins, J., *Commercium epistolicum ... ou Correspondance ... relative à l'analyse supérieure, ré-imprimée sur l'édition originale de 1712 avec l'indication des variantes de l'édition de 1722* ..., ed. J.-B. Biot and F. Lefort (Paris, 1856).

Corsi, P., *Science and Religion: Baden Powell and the Anglican Debate, 1800–1860* (Cambridge: Cambridge University Press, 1988).

Crompton, S., 'On the Portraits of Sir Isaac Newton; and Particularly on one of him by Kneller, Painted about the Time of the Publication of the *Principia*, and Representing him as he was in the Prime of Life', *Proceedings of the Literary and Philosophical Society of Manchester* (2 October 1866), pp. 1–7.

Crosland, M., *The Society of Arcueil: A View of French Science at the Time of Napoleon I* (London: Heinemann, 1967).

Crowther, J. G., *Statesmen of Science* (London: Cresset Press, 1965).

Cruikshank, G., 'Sir Isaac Newton's Courtship', in R. Bentley (ed.), *Bentley's Miscellany*, vol. 4 (London, 1838), between pp. 166–7.

Culler, A. D., *The Victorian Mirror of History* (New Haven, CT, and London: Yale University Press, 1985).

Cunningham, A., and P. Williams, 'De-Centring the "Big Picture": *The Origins of Modern Science* and the Modern Origins of Science', *British Journal for the History of Science*, 26 (1993), pp. 407–32.

Daston, L., 'Objectivity and the Escape from Perspective', *Social Studies of Science*, 22 (1992), pp. 597–618.

Daston, L., and P. Galison, 'The Image of Objectivity', *Representations*, 40 (1992), pp. 81–128.

Davie, G. E., *The Democratic Intellect: Scotland and her Universities in the Nineteenth Century* (Edinburgh: Edinburgh University Press, 1961).

Davy, J., *Memoirs of the Life of Sir Humphry Davy*, 2 vols (London: Longman, Rees, Orme, Brown, Green & Longman, 1836).

de la Pryme, A., *The Diary of Abraham de la Pryme, the Yorkshire Antiquary*, ed. C. Jackson, Publications of the Surtees Society, vol. 54 (Durham, London and Edinburgh: Andrews, 1870).

De Morgan, A., 'References for the History of the Mathematical Sciences', *Companion to the Almanac* (1843), pp. 40–65.

—, 'Newton', in C. Knight (ed.), *The Cabinet Portrait Gallery of British Worthies*, 12 vols (London: Charles Knight & Co., 1845–47), vol. 11 (1846), pp. 78–117, reprinted in A. De Morgan, *Essays on the Life and Work of Newton*, ed. P. E. B. Jourdain (Chicago, IL, and London: Open Court Publishing Co., 1914), pp. 3–63, and in R. Iliffe, M. Keynes and R. Higgitt (eds), *Early Biographies of Isaac Newton, 1660–1885*, 2 vols (London: Pickering & Chatto, 2006), vol. 2, pp. 183–211.

—, 'On a Point Connected with the Dispute between Keil and Leibnitz', *Philosophical Transactions*, 136 (1846), pp. 107–9.

—, 'On the Earliest Printed Almanacs', *Companion to the Almanac* (1846), pp. 1–31.

—, *Arithmetical Books from the Invention of Printing to the Present Time, being Brief Notices of a Large Number of Works Drawn up from Actual Inspection* (London: Taylor and Walton, 1847; reprinted London: H. K. Elliott, 1967).

—, 'On the Additions Made to the Second Edition of the *Commercium Epistolicum*', *Philosophical Magazine*, 3rd series, 23 (1848), pp. 446–56.

—, 'A Short Account of Some Recent Discoveries in England and Germany relative to the Controversy on the Invention of the Fluxions', *Companion to the Almanac* (1852), pp. 5–20, reprinted in A. De Morgan, *Essays on the Life and Work of Newton*, ed. P. E. B. Jourdain (Chicago, IL, and London: Open Court Publishing Co., 1914), pp. 67–101.

—, 'On the Authorship of the Account of the Commercium Epistolicum', *Philosophical Magazine*, 4th series, 3 (1852), pp. 440–4.

—, 'On the Early History of Infinitesimals in England', *Philosophical Magazine*, 4th series, 4 (1852), pp. 321–30.

—, 'Lord Halifax and Mrs. Catherine Barton', *Notes and Queries*, 8 (1853), pp. 429–33.

—, 'Fly-Leaves of Books: Reuben Burrow', *Notes and Queries*, 12 (1855), pp. 142–3.

—, 'The Progress of the Doctrine of the Earth's Motion, Between the Times of Copernicus and Galileo; being Notes on the AnteGalilean Copernicans, *Companion to the Almanac* (1855), pp. 5–25.

—, 'Lord Halifax and Mrs. Catherine Barton', *Notes and Queries*, 2nd series, 2 (1856), pp. 161–3.

—, 'Notes on the History of the English Coinage', *Companion to the Almanac* (1856), pp. 5–21.

—, 'Book Dust', *Notes and Queries*, 2nd series, 4 (1857), pp. 301–2.

—, 'Newton's Nephew, the Rev. B. Smith', *Notes and Queries*, 2nd series, 3 (1857), pp. 41–2.

—, 'Publication of Diaries', *Notes and Queries*, 2nd series, 5 (1864), pp. 107–8.

—, *A Budget of Paradoxes ... Reprinted, with the Author's Additions, from the Athenaeum*, ed. S. E. De Morgan (London: Longmans, Green & Co., 1872).

—, *Newton: His Friend: And His Niece*, ed. S. E. De Morgan and A. Cowper Ranyard (London: Elliot Stock, 1885; reprinted London: Dawson, 1968).

—, *Essays on the Life and Work of Newton*, ed. P. E. B. Jourdain (Chicago, IL, and London, Open Court Publishing Co., 1914).

[De Morgan, A.], 'Notices of English Mathematical Writers Between the Norman Conquest and the Year 1600', *Companion to the Almanac* (1837), pp. 21–44.

—, 'Flamsteed, John', in G. Long (ed.), *Penny Cyclopaedia*, 27 vols (London: Charles Knight, 1833–46), vol. 10 (1838), pp. 296–7.

—, 'Review of Whewell's *History of the Inductive Sciences*', *Athenaeum*, 541 (1838), pp. 179–81.

—, 'Review of Halliwell's *A Collection of Letters, Illustrative of the Progress of Science in England*, published by the Historical Society of Science', *Athenaeum*, 719 (1841), pp. 588–9.

—, 'Baily, Francis', in G. Long (ed.), *Penny Cyclopaedia*, 27 vols (London: Charles Knight, 1833–46), Supplement 1 (1845), pp. 166–8.

—, 'Review of Weld's *History of the Royal Society*', *Athenaeum*, 1078 (1848), pp. 621–2, and 1079 (1848), pp. 651–3.

—, 'Sir David Brewster's *Life of Newton*', *North British Review*, 23 (1855), pp. 307–38, reprinted in A. De Morgan, *Essays on the Life and Work of Newton*, ed. P. E. B. Jourdain (Chicago, IL, and London: Open Court Publishing Co., 1914), pp. 119–82, and in R. Iliffe, M. Keynes and R. Higgitt (eds), *Early Biographies of Isaac Newton, 1660–1885*, 2 vols (London, Pickering & Chatto, 2006), vol. 2, pp. 213–44.

—, 'Review of King's *Biographical Sketch of Newton* and Brougham's *Address on Popular Literature*', *Athenaeum*, 1621 (1858), pp. 641–2.

—, 'Review of *The Works of Francis Bacon*', *Athenaeum*, 1611 (1858), pp. 332–4, and 1612 (1858), pp. 367–8.

—, '*Novum Organum Renovatum*. By W. Whewell', *Athenaeum*, 1628 (1859), pp. 42–4.

—, '*The Philosophy of Discovery, Chapters Historical and Critical*. By W. Whewell', *Athenaeum*, 1694 (1860), pp. 501–3.

—, '*History of the Rise and Influence of the Spirit of Rationalism in Europe*. By W. E. H. Lecky', *Athenaeum*, 1960 (1865), pp. 676–7.

—, 'Newton Ousted', *Athenaeum*, 2077 (1867), pp. 209–10, reprinted in J. Rosenblum, *Prince of Forgers* (New Castle, DE: Oak Knoll Press, 1998), pp. 149–53.

—, 'The Pascal Forgeries', *Athenaeum*, 2090 (16 November 1867), pp. 648–9.

—, 'Sir D. Brewster and the *Athenaeum*', *Athenaeum*, 2092 (1867), pp. 724–5.

—, 'Review of Faugère's *Défence de B. Pascal*', *Athenaeum*, 2135 (1868), pp. 398–400.

[De Morgan, A.?], 'Newton', in G. Long (ed.), *Penny Cyclopaedia*, 27 vols (London: Charles Knight, 1833–46), vol. 16 (1840), pp. 197–203.

De Morgan, S. E., *Memoir of Augustus De Morgan ... With Selections from his Letters* (London: Longmans, Green & Co., 1882).

—, *Threescore Years and Ten: Reminiscences of the Late Sophia Elizabeth De Morgan, to which are Added Letters to and from her Husband the Late Augustus De Morgan, and Others*, ed. M. A. De Morgan (London: Richard Bentley & Son, 1895).

Desmond, A. J., 'Lamarkism and Democracy: Corporations, Corruption and Comparative Anatomy in the 1830s', in J. R. Moore (ed.), *History, Humanity and Evolution: Essays for John C. Greene* (Cambridge: Cambridge University Press, 1989), pp. 99–130.

—, *The Politics of Evolution: Morphology, Medicine, and Reform in Radical London* (Chicago, IL, and London: University of Chicago Press, 1989).

—, 'Redefining the X Axis: "Professionals", "Amateurs" and the Making of Mid-Victorian Biology – a Progress Report', *Journal of the History of Biology*, 34 (2001), pp. 3–50.

Dickinson, H. W., 'J. O. Halliwell and the Historical Society of Science', *Isis*, 18 (1932), pp. 127–32.

Dobbs, B. J. T., 'Review of A. Rupert Hall, *Isaac Newton: Adventurer* and R. S. Westfall, *The Life of Isaac Newton*', *Isis*, 85 (1994), pp. 515–17.

[Drinkwater, J. E.], 'Life of Galileo: With Illustrations of the Advancement of Experimental Philosophy', in *Lives of Eminent Persons*, Library of Useful Knowledge (London: Baldwin & Cradock, 1833).

—, 'Life of Kepler', in *Lives of Eminent Persons*, Library of Useful Knowledge (London: Baldwin & Cradock, 1833).

Duff, M. E. G., *A Victorian Vintage: Being a Selection of the Best Stories from the Diaries of the Right Hon. Sir Mountstuart E. Grant Duff*, ed. A. T. Bassett (London: Methuen & Co., 1930).

Edleston, J., *Correspondence of Sir Isaac Newton and Professor Cotes, including Letters of Other Eminent Men, now first Published from the Originals in the Library of Trinity College, Cambridge; together with an Appendix, containing other Unpublished Letters and Papers by Newton; with Notes, Synoptical View of the Philosopher's Life, and a Variety of Details Illustrative of his History* (Cambridge and London: John Deighton and J. W. Parker, 1850; reprinted London: Frank Cass, 1969).

Elkana, Y., 'William Whewell, Historian', *Rivista di storia della scienza*, 1 (1984), pp. 149–97.

Enros, P. C., 'The Analytical Society (1812–1813): Precursor of the Renewal of Cambridge Mathematics', *Historia Mathematica*, 10 (1983), pp. 24–47.

Evans, J., *A History of the Society of Antiquaries* (Oxford: Oxford University Press, 1956).

Fara, P., 'Isaac Newton Lived Here: Sites of Memory and Scientific Heritage', *British Journal for the History of Science*, 33 (2000), pp. 407–26.

—, *Newton: The Making of Genius* (London: Macmillan, 2002).

Farrer, J. A., *Literary Forgeries* (London: Longmans, Green and Co., 1907).

Faugère, M. P., *Défense de B. Pascal, et accessoirement de Newton, Galilée, Montesquieu, &c. Contre les faux documens présentés par M. Chasles à l'Académie des Sciences* (Paris: Hachette, 1868).

Fauvel, J., R. Flood and R. Wilson (eds), *Oxford Figures: 800 Years of the Mathematical Sciences* (Oxford: Oxford University Press, 2000).

Finocchiaro, M. A., *Retrying Galileo, 1633–1992* (Berkeley, CA, and London: University of California Press, 2005).

Fisch, M., and S. Schaffer (eds), *William Whewell: A Composite Portrait* (Oxford: Clarendon Press, 1991).

Fontenelle, B. le B. de, *The Life of Sir I. Newton, with an Account of his Writings* (London: J. Roberts, 1728), reprinted in R. Iliffe, M. Keynes and R. Higgitt (eds), *Early Biographies of Isaac Newton, 1660–1885*, 2 vols (London, Pickering & Chatto, 2006), vol. 1, pp. 109–21.

—, 'Éloge de Neuton', in B. B. de Fontenelle, *Oeuvres* (Paris: Salmon, 1825), trans. A. R. Hall, in A. R. Hall (ed.), *Isaac Newton: Eighteenth-Century Perspectives* (Oxford: Oxford University Press, 1999), pp. 59–74.

Foote, G. A., 'The Place of Science in the British Reform Movement 1830–50', *Isis*, 42 (1951), pp. 192–208.

Forbes, D., *The Liberal Anglican Idea of History* (Cambridge: Cambridge University Press, 1952).

Fox, R., 'The Rise and Fall of Laplacian Physics', *Historical Studies in the Physical Sciences*, 4 (1974), pp. 89–136.

—, 'Laplacian Physics' in R. C. Olby, G. N. Cantor, J. R. R. Christie and M. J. S. Hodge (eds), *Companion to the History of Modern Science* (London: Routledge, 1996), pp. 278–94.

Frankel, E., 'Jean-Baptiste Biot: The Career of a Physicist in Nineteenth Century France' (unpublished PhD thesis, Princeton University, 1972).

—, 'Corpuscular Optics and the Wave Theory of Light: The Science and Politics of a Revolution in Physics', *Social Studies of Science*, 6 (1976), pp. 141–84.

—, 'J. B. Biot and the Mathematization of Experimental Physics in Napoleonic France', *Historical Studies in the Physical Sciences*, 8 (1977), pp. 33–72.

—, 'Career-Making in Post-Revolutionary France: The Case of Jean-Baptiste Biot', *British Journal for the History of Science*, 11 (1978), pp. 36–48.

Fyfe, A., 'Conscientious Workmen or Bookseller's Hacks? The Professional Identities of Science Writers in the Mid-Nineteenth Century', *Isis*, 96 (2005), pp. 192–223.

Galison, P., 'Objectivity is Romantic', in J. Freidman, P. Galison and S. Haack, *The Humanities and the Sciences* (New York: American Council of Learned Societies, 1999), pp. 15–43.

[Galloway, T.], 'Astronomical Society of London – Recent History of Astronomical Science', *Edinburgh Review*, 51 (1830), pp. 81–114.

—, 'French and English Biographies of Newton', *Foreign Quarterly Review*, 12 (1833), pp. 1–27, reprinted in R. Iliffe, M. Keynes and R. Higgitt (eds), *Early Biographies of Isaac Newton, 1660–1885*, 2 vols (London: Pickering & Chatto, 2006), vol. 2, pp. 65–92.

—, 'Life and Observations of Flamsteed – Newton, Halley, and Flamsteed', *Edinburgh Review*, 62 (1836), pp. 359–97, reprinted in R. Iliffe, M. Keynes and R. Higgitt (eds), *Early Biographies of Isaac Newton, 1660–1885*, 2 vols (London: Pickering & Chatto, 2006), vol. 2, pp. 93–132.

Gascoigne, J., 'From Bentley to the Victorians: The Rise and Fall of British Newtonian Natural Theology', *Science in Context*, 2 (1988), pp. 219–56.

—, *Joseph Banks and the English Enlightenment: Useful Knowledge and Polite Culture* (Cambridge: Cambridge University Press, 1994).

—, 'The Scientist as Patron and Patriotic Symbol: The Changing Reputation of Sir Joseph Banks', in M. Shortland and R. Yeo (eds), *Telling Lives in Science: Essays on Scientific Biography* (Cambridge: Cambridge University Press, 1996), pp. 243–65.

Gerard, A., *An Essay on Genius* (London, 1774), reprinted and ed. B. Fabian (Munich and Amsterdam: Wilhelm Fink Verlag, 1966).

Gillispie, C. C., R. Fox and I. Grattan-Guinness, *Pierre-Simon Laplace 1749–1827: A Life in Exact Science* (Princeton, NJ, and Chichester: Princeton University Press, 1997).

Gjertsen, D., *The Newton Handbook* (London and New York: Routledge & Kegan Paul, 1986).

Gleick, J., *Isaac Newton* (London: Fourth Estate, 2003).

Goldstein, D. S., 'History at Oxford and Cambridge: Professionalisation and the Influence of Ranke', in G. G. Iggers and J. M. Powell (eds), *Leopold von Ranke and the Shaping of the Historical Discipline* (Syracuse, NY: Syracuse University Press, 1990), pp. 141–53.

Gordon, M. M., *The Home Life of Sir David Brewster* (Edinburgh: Edmonston & Douglas, 1869).

Gould, S. J., 'Royal Shorthand', *Science*, 251 (1991), p. 142.

Grafton, A., *The Footnote: A Curious History* (London: Faber and Faber, 1997).

Grant, R., *History of Physical Astronomy, from the Earliest Ages to the Middle of the Nineteenth Century: Comprehending a Detailed Account of the Establishment of the Theory of Gravitation by Newton, and its Development by his Successors; with an Exposition of the Progress of Research on all the Other Subjects of Celestial Physics* (London: Baldwin, 1852).

—, 'Letter to the Editor', *The Times*, 20 September 1867, pp. 9c–d, reprinted in J. Rosenblum, *Prince of Forgers* (New Castle, DE: Oak Knoll Press, 1998), pp. 156–62.

Gregory, J. C., 'Notice concerning an Autograph Manuscript by Sir Isaac Newton, Containing Some Notes upon the Third Book of the Principia, and Found Among the Papers of Dr David Gregory ...(Read March 2. 1829)', *Transactions of the Royal Society of Edinburgh*, 12 (1834), pp. 66–76.

Grisenthwaite, W., *On Genius: In which it is Attempted to be Proved, that there is no Mental Distinction among Mankind* (London, 1830).

Grobel, M. C., 'The Society for the Diffusion of Useful Knowledge 1826–1846 and its Relation to Adult Education in the First Half of the XIXth Century' (unpublished MA thesis, University of London, 1933).

Guicciardini, N., *The Development of the Newtonian Calculus in Britain 1700–1800* (Cambridge: Cambridge University Press, 1989).

—, *Reading the Principia: The Debate on Newton's Mathematical Methods for Natural Philosophy from 1687 to 1736* (Cambridge: Cambridge University Press, 1999).

Gully, J. M., *Lectures on the Moral and Physical Attributes of Men of Genius and Talent* (London, 1836).

Hahn, R., 'Laplace's Religious Views', *Archives Internationales d'Histoire des Sciences*, 30 (1955), pp. 38–40.

—, 'Laplace and the Vanishing Role of God in the Physical Universe', in H. Woolf (ed.), *The Analytic Spirit: Essays in the History of Science in Honor of Henry Guerlac* (Ithaca, NY, and London: Cornell University Press, 1981), pp. 85–95.

Hall, A. R. (ed.), *Isaac Newton: Eighteenth-Century Perspectives* (Oxford: Oxford University Press, 1999).

—, 'Cambridge: Newton's Legacy', *Notes and Records of the Royal Society*, 55 (2001), pp. 205–26.

Hall, M. B., *All Scientists Now: The Royal Society in the Nineteenth Century* (Cambridge: Cambridge University Press, 1984).

Halliwell, J. O. (ed.), *A Collection of Letters Illustrative of the Progress of Science in England from the Reign of Queen Elizabeth to that of Charles the Second* (London: Historical Society of Science, 1841).

— (ed.), *The Private Diary of Dr. John Dee, and the Catalogue of his Library of Manuscripts, from the Original Manuscripts in the Ashmolean Museum at Oxford, and Trinity College Library, Cambridge* (London: Camden Society, 1842).

Hankins, T. L., 'In Defence of Biography: The Use of Biography in the History of Science', *History of Science*, 17 (1979), pp. 1–16.

Hays, J. N., 'Science and Brougham's Society', *Annals of Science* 20 (1964), pp. 227–41.

Heilbron, J. L., 'A Mathematicians' Mutiny, with Morals', in P. Horwich (ed.), *World Changes: Thomas Kuhn and the Nature of Science* (Cambridge, MA, and London: MIT Press, 1993), pp. 81–129.

[Hemming, G. W.], 'Newton as a Scientific Discoverer', *Quarterly Review*, 110 (1861), pp. 401–35.

Herschel, C., *Catalogue of Stars, taken from Mr Flamsteed's Observations contained in the Second Volume of the Historia Cœlestis, and not Inserted in the British Catalogue* (London, 1798).

Herschel, J., *A Preliminary Discourse on the Study of Natural Philosophy*, in D. Lardner (ed.), *The Cabinet Cyclopaedia*, 133 vols (London: Longmans, 1830–49), vol. 14 (1831).

—, 'Outlines of Astronomy' (1849), reprinted in I. B. Cohen and H. M. Jones (eds), *Science Before Darwin: An Anthology of British Scientific Writing in the Early Nineteenth Century* (London: Andre Deutsch, 1963), pp. 97–121.

[Herschel, J.], 'Memoir of Francis Baily', *Monthly Notices of the Royal Astronomical Society*, 6 (1844), pp. 89–128.

Higgitt, R., '"*Newton dépossédé!*" The British Response to the Pascal Forgeries of 1867', *British Journal for the History of Science*, 36 (2003), pp. 437–53.

—, (ed.) *Nineteenth-Century Biography of Isaac Newton: Public Debate and Private Controversy*, vol. 2 of R. Iliffe, M. Keynes and R. Higgitt (eds), *Early Biographies of Isaac Newton, 1660–1885*, 2 vols (London: Pickering & Chatto, 2006).

—, 'Discriminating Days? Partiality and Impartiality in Nineteenth-Century Biographies of Newton', in T. Söderqvist (ed.), *The Poetics of Biography in Science, Technology and Medicine* (Aldershot: Ashgate, 2007), pp. 155–72.

Hirst, T. A., 'Letter to the Editor', *The Times*, 1 October 1867, p. 7c.

—, *Natural Knowledge in Social Context: The Journals of Thomas Archer Hirst*, ed. W. H. Brock and R. M. MacLeod (London: Mansell, 1980).

Hobsbawm, E., and T. Ranger (eds), *The Invention of Tradition* (Cambridge: Cambridge University Press, 2000).

Hornberger, T., 'Halliwell-Phillipps and the History of Science' *Huntingdon Library Quarterly*, 12 (1949), pp. 391–9.

[Horsley, Samuel, et al.], *An Authentic Narrative of the Dissensions and Debates in the Royal Society, Containing the Speeches at Large of Dr. Horsley, Dr. Maskelyne, Mr. Maseres, Mr. Poore, Mr. Glennie, Mr. Watson, and Mr. Maty* (London, 1784).

Horton-Smith, L. G. H., *Francis Baily, the Astronomer, 1774–1844* (Newbury: Blacket, Turner & Co., 1938).

—, *The Baily Family of Thatcham and Later of Speen and of Newbury, all in the County of Berkshire* (Leicester: W. Thornley & Sons, 1951).

Howarth, O. J. R., *The British Association for the Advancement of Science: A Retrospect, 1831–1921* (London: BAAS, 1922).

Hyman, A., *Charles Babbage: Pioneer of the Computer* (Oxford: Oxford University Press, 1982).

Iliffe, R., 'A "connected system"? The Snare of a Beautiful Hand and the Unity of Newton's Archive', in M. Hunter (ed.), *Archives of the Scientific Revolution: The Formation and Exchange of Ideas in Seventeenth-Century Europe* (Woodbridge: Boydell Press, 1998), pp. 137–57.

—, 'Isaac Newton: Lucatello Professor of Mathematics', in C. Lawrence and S. Shapin (eds), *Science Incarnate: Historical Embodiments of Natural Knowledge* (Chicago, IL, and London: University of Chicago Press, 1998), pp. 121–55.

— (ed.), *Eighteenth-Century Biography of Isaac Newton: The Unpublished Manuscripts and Early Texts*, vol. 1 of R. Iliffe, M. Keynes and R. Higgitt (eds), *Early Biographies of Isaac Newton, 1660–1885*, 2 vols (London, Pickering & Chatto, 2006).

Iliffe, R., M. Keynes, R. Higgitt (eds), *Early Biographies of Isaac Newton, 1660–1885*, 2 vols (London, Pickering & Chatto, 2006).

Jann, R., *The Art and Science of Victorian History* (Columbus, OH: Ohio State University Press, 1985).

[Jeffrey, F.], 'Memoirs of Sir James Mackintosh', *Edinburgh Review*, 62 (1835), pp. 205–55.

Johns, A., *The Nature of the Book: Print and Knowledge in the Making* (Chicago, IL, and London: University of Chicago Press, 1998).

Kessel, N., 'Genius and Mental Disorder: A History of Ideas concerning their Conjunction', in P. Murray (ed.), *Genius: The History of an Idea* (Oxford and New York: Basil Blackwell, 1989), pp. 196–212.

Keynes, J. M., 'Newton, the Man', in J. M. Keynes, *Essays in Biography ... New Edition with Three Additional Essays*, ed. G. Keynes (London: Rupert Hart-Davis, 1951), pp. 310–23.

King, E. F., *A Biographical Sketch of Sir Isaac Newton. To which are Added Authorized Reports of the Oration of Lord Brougham (with his Lordship's Notes) at the Inauguration of the Statue at Grantham; and of Several of the Speeches Delivered on that Occasion*, 2nd edn (Grantham and London, 1858).

King, P., *The Life of John Locke, with Extracts from his Correspondence, Journals, and Common-Place Books* (London: Colburn, 1829).

Korshin, P. J., 'The Development of Intellectual Biography in the Eighteenth Century', *Journal of English and Germanic Philology*, 73 (1974), pp. 513–23.

Kuhn, T. S., *The Structure of Scientific Revolutions* (Chicago, IL, and London: University of Chicago Press, 1962).

Laudan, R., 'Histories of the Sciences and their Uses: A Review to 1913', *History of Science*, 31 (1993), pp. 1–34.

Levine, G., *Dying to Know: Scientific Epistemology and Narrative in Victorian England* (Chicago, IL, and London: University of Chicago Press, 2002).

Levine, P., *The Amateur and the Professional: Antiquarians, Historians and Archaeologists in Victorian England, 1838–1886* (Cambridge: Cambridge University Press, 1986).

Lewis, G. C., *An Historical Survey of the Astronomy of the Ancients* (London: Parker and Bourn, 1862).

Lindberg, D. C., 'Conceptions of the Scientific Revolution from Bacon to Butterfield: A Preliminary Sketch', in D. C. Lindberg and R. S. Westman (eds), *Reappraisals of the Scientific Revolution* (Cambridge: Cambridge University Press, 1990), pp. 1–26.

Lloyd, C. C., *Mr. Barrow of the Admiralty: A Life of Sir John Barrow 1764–1848* (London: Collins, 1970).

Long, G. (ed.), *Penny Cyclopaedia*, 27 vols (London: Charles Knight, 1833–46).

—, *The Biographical Dictionary of the Society for the Diffusion of Useful Knowledge*, 4 vols (London, 1842–4).

Losee, J., 'Whewell and Mill on the Relation between Philosophy of Science and History of Science', *Studies in the History and Philosophy of Science*, 14 (1983), pp. 113–26.

Lowenthal, D., *The Past is a Foreign Country* (Cambridge: Cambridge University Press, 1985).

[Luard, H. R., G. G. Stokes, J. C. Adams and G. D. Living], *A Catalogue of the Portsmouth Collection of Books and Papers Written by or Belonging to Sir Isaac Newton, the Scientific Portion of which has been Presented by the Earl of Portsmouth to the University of Cambridge, drawn up by the Syndicate Appointed the 6th November 1872* (Cambridge: Cambridge University Press, 1888).

Macaulay, T. B., 'Hallam's *Constitutional History*' (1828), in T. B. Macaulay, *Essays, Critical and Miscellaneous* (New York, D. Appleton and Co., 1861), pp. 67–99.

Maccioni, P. A., 'Guglielmo Libri and the British Museum: A Case of Scandal Averted', *British Library Journal*, 17 (1991), pp. 36–60.

Maccioni Ruju, P. A., and M. Mostert, *The Life and Times of Guglielmo Libri (1802–1869): Scientist, Patriot, Scholar, Journalist and Thief: A Nineteenth-Century Story* (Hilversum: Verloren, 1995).

MacLeod, C., 'Concepts of Invention and the Patent Controversy in Victorian Britain', in R. Fox (ed.), *Technological Change*, Studies in the History of Science, Technology and Medicine (Amsterdam: Harwood Academic Publishers, 1996), pp. 137–53.

Malkin, A. T. (ed.), *Gallery of Portraits: with Memoirs*, 7 vols (London: Charles Knight, 1833–7).

Malkin, B. H., *An Introductory Lecture on History, delivered in the University of London on Thursday, March 11, 1830* (London, 1830).

[Malkin, B. H.], 'Brewster's *Life of Newton*', *Edinburgh Review*, 56 (1832), pp. 1–37.

—, 'Astronomy', in *Natural Philosophy*, Library of Useful Knowledge (London: Baldwin & Cradock, 1829–38), vol. 3 (1834), pp. 1–2.

[Malkin, B. H.?], 'Newton', in A. T. Malkin (ed.), *Gallery of Portraits: with Memoirs*, 7 vols (London: Charles Knight, 1833–7), vol. 1 (1833), pp. 79–88.

Manuel, F. E., *A Portrait of Isaac Newton* (London: Muller, 1980).

Marchand, L. A., *The Athenaeum: A Mirror of Victorian Culture* (Chapel Hill, NC: University of North Carolina Press, 1941).

Mayer, A. K., 'Moralizing Science: The Uses of Science's Past in National Education in the 1920s', *British Journal for the History of Science*, 30 (1997), pp. 51–70.

Merton, R. K., *The Sociology of Science: Theoretical and Practical Investigations*, ed. Norman W. Storer (Chicago, IL, and London: University of Chicago Press, 1973).

Mill, J. S., *The Spirit of the Age*, ed. F. A. von Hayek (Chicago, IL: University of Chicago Press, 1942).

Miller, D. P., 'The Royal Society of London, 1800–1835: A Study of the Cultural Politics of Scientific Organization' (unpublished PhD thesis, University of Pennsylvania, 1981).

—, 'Between Hostile Camps: Sir Humphry Davy's Presidency of the Royal Society of London 1820–1827', *British Journal for the History of Science*, 16 (1983), pp. 1–47.

—, 'Method and the "Micropolitics" of Science: The early years of the Geological and Astronomical Societies of London', in J. A. Schuster and R. R. Yeo (eds), *The Politics and Rhetoric of Scientific Method: Historical Studies* (Dordrecht: Reidel, 1986), pp. 227–57.

—, 'The Revival of the Physical Sciences in Britain, 1815–1840, *Osiris*, 2nd series, 2 (1986), pp. 107–34.

—, '"Into the valley of darkness": Reflections on the Royal Society in the Eighteenth Century', *History of Science*, 27 (1989), pp. 155–66.

—, '"Puffing Jamie": The Commercial and Ideological Importance of Being a "Philosopher" in the Case of the Reputation of James Watt (1736–1819)', *History of Science*, 38 (2000), pp. 1–24.

More, L. T., *Isaac Newton: A Biography* (London and New York: C. Scribner's Sons, 1834).

[Morell, J.], 'Biography', in D. Brewster (ed.), *The Edinburgh Encyclopaedia*, 18 vols (Edinburgh: William Blackwood, 1830), vol. 3, pp. 506–12.

Morrell, J., and A. Thackray, *Gentlemen of Science: Early Years of the British Association of the Advancement of Science* (Oxford: Clarendon Press, 1981).

Morrell, J. B., 'Brewster and the Early British Association for the Advancement of Science', in A. D. Morrison-Low and J. R. R. Christie (eds), *Martyr of Science: Sir David Brewster 1781–1868* (Edinburgh: The Royal Scottish Museum, 1984), pp. 25–9.

—, 'Professionalisation', in R. C. Olby, G. N. Cantor, J. R. R. Christie and M. J. S. Hodge (eds), *Companion to the History of Modern Science* (London: Routledge, 1996), pp. 980–9.

Morrison-Low, A. D., 'Brewster and Scientific Instruments', in A. D. Morrison-Low and J. R. R. Christie (eds), *Martyr of Science: Sir David Brewster 1781–1868* (Edinburgh: The Royal Scottish Museum, 1984), pp. 59–65.

Morrison-Low, A. D., and J. R. R. Christie (eds), *Martyr of Science: Sir David Brewster 1781–1868* (Edinburgh: The Royal Scottish Museum, 1984).

Munby, A. N. L., *The Cult of the Autograph Letter in England* (London: Athlone Press, 1962).

—, *The History and Bibliography of Science in England: The First Phase, 1833–1845* (Berkeley and Los Angeles: School of Librarianship and Graduate School of Library Service, 1968).

Murray, P. (ed.), *Genius: The History of an Idea* (Oxford and New York: Basil Blackwell, 1989).

'Nemo', *Mirror of Literature*, 4 (1824), p. 399.

Nicolas, N. H., *Observations on the State of Historical Literature, and on the Society of Antiquaries, and Other Institutions for its Advancement in England; with Remarks on Record Offices, and on the Proceedings of the Record Commission* (London, 1830).

Osborne, E. A., 'Introduction', in A. De Morgan, *Newton: His Friend: And His Niece*, ed. S. E. De Morgan and A. C. Ranyard (London: Dawson, 1968), pp. v–ix.

Outram, D., 'The Language of Natural Power: The "Eloges" of Georges Cuvier and the Public Language of Nineteenth Century Science', *History of Science*, 16 (1978), pp. 153–78.

Pachter, M. (ed.), *Telling Lives: The Biographer's Art* (Philadelphia, PA: University of Pennsylvania Press, 1981).

Paley, W., *Paley's Natural Theology Illustrated*, ed. H. Brougham and C. Bell, 5 vols (London: Charles Knight, 1835–9).

Paul, C. B., *Science and Immortality: The Éloges of the Paris Academy of Sciences (1699–1791)* (Berkeley, CA, and London: University of California Press, 1980).

Pearce Williams, L., 'The Royal Society and the Founding of the British Association for the Advancement of Science', *Notes and Records of the Royal Society*, 16 (1961), pp. 221–33.

Pepys, S., *Memoirs of Samuel Pepys, Comprising his Diary from 1659 to 1669, deciphered by the Rev. J. Smith, from the Original Short-Hand MS. in the Pepysian Library, and a Selection from his Private Correspondence*, ed. R. Griffin, 2 vols (London: Henry Colburn, 1825).

Phillips, M., 'Macaulay, Scott, and the Literary Challenge to Historiography', *Journal of the History of Ideas*, 50 (1989), pp. 117–33.

Pomata, G., 'Versions of Narrative: Overt and Covert Narrators in Nineteenth Century Historiography', *History Workshop Journal*, 27 (1989), pp. 1–17.

Porter, R., 'Gentlemen and Geology: The Emergence of a Scientific Career, 1660–1920', *Historical Journal*, 21 (1978), pp. 809–36.

Powell, B., *Rational Religion Examined: Or, Remarks on the Pretensions of Unitarianism; Especially as Compared with those Systems which Professedly Discard Reason* (London, 1826).

—, *History of Natural Philosophy, from the Earliest Periods to the Present Time*, in Dionysius Lardner (ed.), *The Cabinet Cyclopaedia*, 133 vols (London: Longmans, 1830–49), vol. 51 (1834).

[Powell, B.], 'An Examination into the Charge of Heterodoxy Brought Against Eminent Men. In a Letter to the Editor of the Christian Remembrancer', *Christian Remembrancer*, 7 (1825), pp. 566–75.

—, 'Sir Isaac Newton and his Contemporaries', *Edinburgh Review*, 78 (1843), pp. 402–37.

—, 'Sir Isaac Newton', *Edinburgh Review*, 103 (1856), pp. 499–534, reprinted in R. Iliffe, M. Keynes and R. Higgitt (eds), *Early Biographies of Isaac Newton, 1660–1885*, 2 vols (London, Pickering & Chatto, 2006), vol. 2, pp. 253–87.

Priestley, J., *The History and Present State of Electricity, with Original Experiments* (London, 1767).

Reed, J. W., *English Biography in the Early Nineteenth Century, 1801–1838* (New Haven, CT, and London: Yale University Press, 1966).

Rice, A., 'Augustus De Morgan: Historian of Science', *History of Science*, 34 (1996), pp. 201–40.

Richards, J. L., 'Augustus De Morgan, the History of Mathematics, and the Foundations of Algebra', *Isis*, 78 (1987), pp. 7–30.

—, '"In a rational world all radicals would be exterminated": Mathematics, Logic and Secular Thinking in Augustus De Morgan's England', *Science in Context*, 15 (2002), pp. 137–64.

Rigaud, J., *Stephen Peter Rigaud: A Memoir* (Oxford, 1883).

Rigaud, S. J., *A Defence of Halley Against the Charge of Religious Infidelity* (Oxford: Ashmolean Society, 1844).

Rigaud, S. P. (ed.), *Miscellaneous Works and Correspondence of The Rev. James Bradley, D.D. F.R.S., Astronomer Royal, Savilian Professor of Astronomy in the University of Oxford &c. &c. &c.* (Oxford: Oxford University Press, 1832).

—, *Supplement to Dr. Bradley's Miscellaneous Works: with an Account of Harriot's Astronomical Papers* (Oxford: Oxford University Press, 1833).

—, 'Biographical Account of John Hadley, Esq. V.P.R.S., the Inventor of the Quadrant ...', *Nautical Magazine*, 4 (1835), pp. 12–22, 137–46.

—, 'Some Particulars respecting the Principal Instruments at Greenwich in the Time of Dr. Halley', *Memoirs of the Royal Astronomical Society*, 9 (1836), pp. 205–27.

—, *Historical Essay on the First Publication of Sir Isaac Newton's Principia* (Oxford, 1838).

— (ed.), *Correspondence of Scientific Men of the Seventeenth Century; Including Letters of Barrow, Flamsteed, Wallis, and Newton, Printed from the Originals in the Collection of the Right Hon. the Earl of Macclesfield*, 2 vols (Oxford, 1841, 1862).

[Rigaud, S. P.], 'Observations on a Note respecting Mr. Whewell, which is Appended to No. CX, of the Quarterly Review', *Philosophical Magazine*, 8 (1836), pp. 218–25.

—, 'Review of *Newton and Flamsteed*. Remarks on an Article in Number CIX of the Quarterly Review by the Rev. William Whewell', *Philosophical Magazine*, 8 (1836), pp. 139–47.

—, *Defence of the Resolution for Omitting Mr. Panizzi's Bibliographical Notes from the Catalogue of the Royal Society* (London: Richard & John E. Taylor, 1838).

[Roebuck, J. A.], 'Life of Mahomet', in *Lives of Eminent Persons*, Library of Useful Knowledge (London: Baldwin & Cradock, 1833).

Rosenblum, J., *Prince of Forgers* (New Castle, DE: Oak Knoll Press, 1998).

Ross, S., '"Scientist": The Story of a Word', *Annals of Science*, 18 (1962), pp. 65–85.

Rouse Ball, W. W., *An Essay on Newton's "Principia"* (London: Macmillan & Co., 1893).

Rouse Ball, W. W., and J. A. Venn (eds), *Admissions to Trinity College, Cambridge, 1801–1850*, 5 vols (London: Macmillan & Co., 1911–16).

Royle, E., 'Mechanics' Institutes and the Working Classes, 1840–1860', *Historical Journal*, 14 (1971), pp. 305–21.

Rumker, C., 'Astronomical Observations made at the Observatory at Paramatta in New South Wales', *Philosophical Transactions*, 119 (1829), pp. 1–152.

Ruse, M., 'William Whewell: Omniscientist', in M. Fisch and S. Schaffer (eds), *William Whewell: A Composite Portrait* (Oxford: Clarendon Press, 1991), pp. 87–116.

S., C., 'On Whiston, Halley, and the Quarterly Reviewer of the "Account of Flamsteed"', *Philosophical Magazine*, 8 (1836), pp. 225–6.

Schaffer, S., 'Scientific Discoveries and the End of Natural Philosophy', *Social Studies of Science*, 16 (1986), pp. 387–420.

—, 'Priestley and the Politics of Spirit', in R. G. W. Anderson and C. Lawrence (eds), *Science, Medicine and Dissent: Joseph Priestley (1733–1804)* (London: Wellcome Trust, 1987), pp. 39–53.

—, 'The Consuming Flame: Electrical Showmen and Tory Mystics in the World of Goods', in J. Brewer and R. Porter (eds), *Consumption and the World of Goods* (London and New York: Routledge, 1993), pp. 489–526.

Secord, J. A., 'Progress in Print', in M. Frasca-Spada and N. Jardine (eds), *Books and the Sciences in History* (Cambridge: Cambridge University Press, 2000), pp. 369–89.

—, *Victorian Sensation: The Extraordinary Publication, Reception, and Secret Authorship of Vestiges of the Natural History of Creation* (Chicago, IL, and London: University of Chicago Press, 2000).

Shapin, S., 'Brewster and the Edinburgh Career in Science', in A. D. Morrison-Low and J. R. R. Christie (eds), *Martyr of Science: Sir David Brewster 1781–1868* (Edinburgh: The Royal Scottish Museum, 1984), pp. 17–23.

—, 'The Philosopher and the Chicken: On the Dietetics of Disembodied Knowledge', in C. Lawrence and S. Shapin (eds), *Science Incarnate: Historical Embodiments of Natural Knowledge* (Chicago, IL, and London: University of Chicago Press, 1998), pp. 21–50.

—, *Science as a Vocation: Scientific Authority and Personal Virtue in Late Modernity* (forthcoming; unpublished draft, 2004).

Shapin, S., and B. Barnes, 'Head and Hand: Rhetor ical Resources in British Pedagogical Writing, 1770–1850', *Oxford Review of Education*, 2 (1976), pp. 231–54.

—, 'Science, Nature and Control: Interpreting Mechanics' Institutes', *Social Studies of Science*, 7 (1977), pp. 31–74.

Shapiro, A. E., *Fits, Passions and Paroxysms: Physics, Method, and Chemistry and Newton's Theories of Colored Bodies and Fits of Easy Reflection* (Cambridge: Cambridge University Press, 1993).

Sheets-Pyenson, S., 'New Directions for Scientific Biography: The Case of Sir William Dawson', *History of Science*, 28 (1990), pp. 399–410.

Shortland, M., and R. Yeo, 'Introduction', in M. Shortland and R. Yeo (eds), *Telling Lives in Science: Essays on Scientific Biography* (Cambridge: Cambridge University Press, 1996), pp. 1–44.

Smeaton, W. A., 'History of Science at University College London', *British Journal for the History of Science*, 30 (1997), pp. 25–8.

Smith, H., *Society for the Diffusion of Useful Knowledge 1826–1846: A Social and Bibliographical Evaluation* (London: The Vine Press, 1974).

Söderqvist, T., 'Existential Projects and Existential Choice in Science: Science Biography as an Edifying Genre', in M. Shortland and R. Yeo (eds), *Telling Lives in Science: Essays on Scientific Biography* (Cambridge: Cambridge University Press, 1996), pp. 45–84.

Somerville, M., *Queen of Science: Personal Recollections of Mary Somerville*, ed. D. McMillan (Edinburgh: Canongate, 2001).

[South, J.], *Refutation of the Numerous Mistatements and Fallacies contained in a Paper Presented to the Admiralty by Dr. Thomas Young (Superintendent of the Nautical Almanac) and Printed by Order of the House of Commons* (London, 1829).

[Southey, R.], 'Hayley's *Life and Posthumous Writings of William Cowper*', *Annual Review*, 2 (1804), pp. 457–62.

Spevack, M., *James Orchard Halliwell-Phillipps: The Life and Works of the Shakespearian Scholar and Bookman* (New Castle, DE, and London: Oak Knoll Press, 2001).

Stauffer, D. A., *The Art of Biography in Eighteenth-Century England* (Princeton, NJ: Princeton University Press, 1941).

Stewart, D., 'Dissertation on the Progress of Metaphysical, Ethical, and Political Philosophy', in M. Napier (ed.), *Supplement to the fourth, fifth, and sixth editions of the Encyclopaedia Britannica*, 6 vols (Edinburgh: A. Constable and Co., 1824).

Stewart, L., 'Seeing through the Scholium: Religion and Reading Newton in the Eighteenth Century', *History of Science*, 34 (1996), pp. 123–65.

Storella, E. A., '"O, What a World of Profit and Delight": The Society for the Diffusion of Useful Knowledge' (unpublished PhD thesis, Brandeis University, 1969).

Stukeley, W., *Memoirs of Sir Isaac Newton's Life: Being Some Account of his Family and Chiefly the Junior Part of his Life*, ed. A. G. H. White (London: Taylor & Francis, 1936).

Theerman, P., 'Unaccustomed Role: The Scientist as Historical Biographer – Two Nineteenth-Century Portrayals of Newton', *Biography*, 8 (1985), pp. 145–62.

Thomas, W. K., and W. U. Ober, *A Mind For Ever Voyaging: Wordsworth at Work Portraying Newton and Science* (Edmonton: University of Alberta Press, 1989).

Todhunter, I., *William Whewell, D.D., Master of Trinity College, Cambridge. An Account of his Writings with Selections from his Literary and Scientific Correspondence*, 2 vols (London: Macmillan, 1876).

Toole, B. A., *Ada, the Enchantress of Numbers: Prophet of the Computer Age* (Mill Valley, CA: Strawberry Press, 1998).

Topham, J. R., 'Science and Popular Education in the 1830s: The Role of the *Bridgewater Treatises*', *British Journal for the History of Science*, 25 (1992), pp. 397–430.

—, 'Scientific Publishing and the Reading of Science in Nineteenth-Century Britain: A Historiographical Survey and Guide to Sources', *Studies in the History and Philosophy of Science*, 31 (2000), pp. 559–612.

Turnbull, H. W., J. F. Scott, A. R. Hall and L. Tilling (eds), *The Correspondence of Isaac Newton*, 7 vols (Cambridge: Cambridge University Press, 1959–77).

Turner, D., *Thirteen Letters from Sir Isaac Newton to John Covel D.D. – from Original Manuscripts in the Library of Dawson Turner, Esq., Yarmouth* (Norwich, 1848).

Turner, F. M., 'John Tyndall and Victorian Scientific Naturalism', in W. H. Brock, N. D. McMillan and R. C. Mollan (eds), *John Tyndall: Essays on a Natural Philosopher* (Dublin: Royal Dublin Society, 1981), pp. 169–80.

—, *Contesting Cultural Authority: Essays in Victorian Intellectual Life* (Cambridge: Cambridge University Press, 1993).

Turnor, E., *Collections for the History of the Town and Soke of Grantham* (London: W. Miller, 1806).

Tyndall, J., *Faraday as a Discoverer* (London: Longmans, Green and Co., 1868).

—, 'On the Scientific Use of the Imagination', in J. Tyndall, *Fragments of Science: A Series of Detached Essays, Lectures, and Reviews*, 3rd edn (London: Longmans, Green and Co., 1871), pp. 125–62.

—, *Address Delivered Before the British Association Assembled at Belfast, With Additions* (London: Longmans, Green & Co., 1874).

Venn, J. A., *Alumni Cantabrigienses*, Part 2 (Cambridge: Cambridge University Press, 1951).

Verrier, U. Le, 'Examen de la discussion soulevée au sein de l'Académie des Sciences, au sujet de la découverte de l'attraction universelle', *Comptes rendus*, 68 (1869), pp. 1425–33.

Wallis, P., and R. Wallis, *Newton and Newtoniana, 1672–1975: A Bibliography* (Folkestone: Dawson, 1977).

Warner, B., and N. Warner, *Maclear and Herschel: Letters and Diaries at the Cape of Good Hope, 1834–1838* (Cape Town: A. A. Balkema, 1984).

Warwick, A., *The Masters of Theory: Cambridge and the Rise of Mathematical Physics* (Chicago, IL, and London: University of Chicago Press, 2003).

Weld, C. R., *History of the Royal Society, with Memoirs of the Presidents. Compiled from Authentic Documents*, 2 vols (London, 1848).

[Weld, C. R.], 'Review of Edleston's *Correspondence of Sir Isaac Newton and Professor Cotes*', *Athenaeum*, 1217 (1851), pp. 211–12.

—, 'Review of Brewster's *Memoirs of Sir Isaac Newton*', *Athenaeum*, 1442 (1855), pp. 697–9.

Westfall, R. S., 'Short-Writing and the State of Newton's Conscience, 1662 (I)', *Notes and Records of the Royal Society of London*, 18 (1963), pp. 10–16.

—, 'Introduction', in D. Brewster, *Memoirs of the Life, Writings and Discoveries of Sir Isaac Newton*, The Sources of Science, vol. 14 (New York and London: Johnson Reprint Corporation, 1965).

—, *Never at Rest: A Biography of Isaac Newton* (Cambridge: Cambridge University Press, 1980).

—, 'Newton and his Biographer', in S. H. Baron and C. Pletsch (eds), *Introspection in Biography: The Biographer's Quest for Self-Awareness* (Hillsdale, NJ: Analytic Press, 1985), pp. 175–89.

Whewell, W., *Newton and Flamsteed: Remarks on an Article in Number CIX of the Quarterly Review* (Cambridge and London: J. & J. J. Deighton, 1836), reprinted in R. Iliffe, M. Keynes and R. Higgitt (eds), *Early Biographies of Isaac Newton, 1660–1885*, 2 vols (London, Pickering & Chatto, 2006), vol. 2, pp. 133–43.

—, 'Remarks on a Note on a Pamphlet Entitled "Newton and Flamsteed"', *Philosophical Magazine*, 8 (1836), pp. 211–18.

—, *History of the Inductive Sciences, from the Earliest to the Present Time*, 3rd edn, 3 vols (London, 1857).

Whiston, W., *The Longitude Discovered by the Eclipses, Occultations, and Conjunctions of Jupiter's Planets* (London: J. Whiston, 1738).

Wilkinson, T. T., 'Publication of Diaries', *Notes and Queries*, 2nd series, 5 (1864), pp. 215–16.

Wilson, D. B., 'Herschel and Whewell's Version of Newtonianism', *Journal of the History of Ideas*, 35 (1974), pp. 79–97.

Wormhoudt, A., 'Newton's Natural Philosophy in the Behmenistic Works of William Law', *Journal of the History of Ideas*, 10 (1949), pp. 411–29.

Worrall, J., 'Thomas Young and the "Refutation" of Newtonian Optics: A Case-Study in the Interaction of Philosophy of Science and History of Science', in C. Howson (ed.), *Method and Appraisal in the Physical Sciences: The Critical Background to Modern Science, 1800–1905* (Cambridge: Cambridge University Press, 1976), pp. 107–79.

Wright, T., *Popular Treatises on Science Written During the Middle Ages in Anglo-Saxon, Anglo-Norman and English* (London: Historical Society of Science, 1841).

Yeo, R., 'An Idol of the Marketplace: Baconianism in Nineteenth-Century Britain', *History of Science*, 23 (1985), pp. 251–98.

—, 'Genius, Method and Morality: Images of Newton in Britain 1760–1860', *Science in Context*, 2 (1988), pp. 257–84.

—, *Defining Science: William Whewell, Natural Knowledge and Public Debate in Early Victorian Britain* (Cambridge: Cambridge University Press, 1993).

—, 'Introduction', in W. Whewell, *Collected Works of William Whewell*, ed. R. Yeo, 16 vols (Bristol: Thoemmes, 2001).

Young, E., *Conjectures on Original Composition* (1759; Leeds: Scholar Press, 1966).

Young, R. M., 'Biography: The Basic Discipline for Human Sciences', *Free Associations*, 11 (1988), pp. 108–30.

INDEX

Académie des Sciences, Paris, 8, 20, 22, 23, 171–82 *passim*
   proceedings of (*Comtes rendus*), 172, 175
Account of Flamsteed see under Baily, Francis; controversies in Newtonian biography
Adams, John Couch, 171, 238n109
Airy, George Biddell, 36, 84, 86, 87, 206n73, 217n19
analytical mathematics, 2, 25, 36
   Analytical Society, 36, 206n80
   calculus dispute *see under* controversies in Newtonian biography
Arago, François, 22, 24, 40, 178
Ashmolean Society, 102
Ashworth, William J., 69–70, 72, 74, 76, 90–1, 93, 97
Aston, Francis, 27
Athenaeum, 114, 175, 177, 178, 180, 181, 183, 191
   and Pascal forgeries, 174, 175
Aubrey, John, 103
Austin, Sarah, 32, 205n59
   'Life of Carsten Neibuhr', 32–3

Babbage, Charles, 50, 51–2, 53, 55, 84, 106, 206n80
Reflections on the Decline of Science, 50–1, 53, 56, 76, 84
Bacon, Francis, 118, 119
   Baconian method, 64–6, 67, 109, 138, 187, 209n4, 210n19, 215n108
   and Brewster *see under* Brewster, David
Bailly, Jean-Sylvain, 36, 126
Baily, Francis, 13–15, 72–3, 96–7, 100, 101, 103, 106, 108, 116, 117, 119, 123, 139, 166, 176, 184, 187, 190, 192, 216n13, 220n75, 221n90

Account of Flamsteed, 13, 30, 66, 68, 69–74, 76–80, 97–8, 99, 101, 129, 138, 157, 165, 168
   and Admiralty, 76, 80–1, 89
   British Catalogue, 71, 72, 76, 80, 85, 89, 92
   distribution of, 80–3, 89, 221n94
   errors in, 217–18n33
   reception of, 84–8, 89, 97, 106, 191
   reviews of, 88–94
   as a sensation, 88–9
   see also controversies in Newtonian biography
   historical method/style, 14–15, 69, 75–8, 80, 91, 96, 98, 104, 106, 125, 127, 140–1, 171, 188–9, 218n34, 226n20
   and reform of science, 50, 84–5
   scientific method, 14–15, 72–3, 75, 76, 88, 98
   and star catalogues, 72, 74–5, 216n13, 217n26
   Supplement to the Account of Flamsteed, 94–6, 97, 131
   see also under Royal Astronomical Society
Ball, Walter William Rouse, 193
Banks, Joseph, 50, 84, 90, 97, 221n101
Barrow, Isaac, 240n16
Barrow, John, 80, 221n101, 221–2n106
   see also under controversies in Newtonian biography
Barton, Catherine, 8, 163–4, 236n81
   biographical writing on Newton, 9
   relationship with Montagu, 88, 93, 95, 104, 120, 122, 163–171, 183
Barton, Ruth, 2

Beaufort, Francis, 74, 84, 85, 92
Beaumont, Elie de, 174
Beg, Ulugh, 72
Bensaud-Vincent, Bernadette, 157
Bentley, Richard, 11
  Newton's letters to, 11, 57, 61
biography, 4–6, 9, 10, 11, 17, 32–4, 134, 167, 193, 194, 197n18, 207n91
  as hagiography, 4, 14, 17, 20, 134, 189
  'impartiality' in, 62, 67, 134, 148, 188–9, 192
  use of manuscript sources in, 4, 11, 13, 17, 75–6, 78, 96, 131, 188–9, 196n15, 200n58, 218n44
  reputational studies, 1, 6, 193
  scientific biography, 4–5, 18, 20, 33, 33–5, 38–9, 99, 193
  suppression in, 4, 15, 129, 142, 147–9
  see also historiography, nineteenth-century; history of science, nineteenth-century
Biot, Jean-Baptiste, 13, 20–1, 100, 103
  and Brewster see Brewster, David
  Catholicism of, 30, 61, 203n40
  as editor of *Commercium Epistolicum*, 30, 60
  and Laplace see Laplace, Pierre-Simon
  'Life of Newton' (translation), 12, 20, 30–1, 32, 35–40, 44, 45, 66, 98
  on Newton and religion, 28–30, 39, 42, 61
  'Newton', *Biographie universelle*, 12–13, 19, 23–30, 42, 63
  Newton's genius see genius, scientific
  theory of light, 13, 16, 20–4, 40–1, 42, 136, 187, 201n22
  review of Brewster's *Life of Newton*, 27, 30, 44, 59–61, 66
  review of Brewster's *Memoirs of Newton*, 135, 147
  Romanticism of, 12, 26–8, 42, 56, 59–60
  scientific methodology, 21, 23–4
  Traité de physique mathématique et expérimentale, 21
  see also under controversies in Newtonian biography
Birch, Thomas, 9, 11, 19, 36
Blake, William, 198n30

Bliss, Philip, 101, 104, 106
Board of Longitude, 50, 84, 90, 92
Bodleian Library, Oxford, 78
Boehme, Jacob, 146, 237n96
Bonaparte, Napoleon, 29
Bond, Edmund, 182
Bordier, Henri, 175
Bostock, John, 86
Boswell, James, 4
Boyle, Robert, 38, 39, 48, 172, 179
  and alchemy, 146
  and Baconianism, 65, 67
Bradley, James, 95, 181, 217n19
  Rigaud on *see under* Rigaud, Stephen Peter
Brahe, Tycho, 48, 56, 72, 146
Braybrooke, Lord *see* Griffin, Richard
Brewster, David, 13, 14, 43–5, 51–5, 67, 103, 111, 130, 160, 191, 212nn61, 63
  and BAAS, 13, 44, 52–3, 56, 136–7, 211n42
  and Baconianism, 13, 44, 47–9, 53, 64–5, 67, 68, 137–8, 187
  on Catherine Barton, 167–8, 169, 170, 183
  correspondence with Babbage, 50, 51–2, 55
  correspondence with Brougham, 41, 51, 54, 132, 133, 139, 144, 147, 150, 151, 235n49
  and De Morgan *see* De Morgan, Augustus
  historical method/style, 15, 44, 58, 59, 62–4, 65, 66–7, 97, 126, 129, 131–5, 138, 147–9, 188
  Life of Newton, 13, 35, 39, 43–7, 48–9, 50, 52, 55–9, 78, 79, 99, 115, 129–30, 133, 157, 187, 188, 192, 209nn5, 7, 11
  reviews of, 44, 56, 59–66, 66–8, 94, 98, 116, 124, 129
  Martyrs of Science, 56, 130, 146
  Memoirs of Newton, 15, 44, 99, 129–57, 163, 164, 169, 187, 192, 235n41
  reviews of, 122, 127, 129, 132, 134–5, 145, 147, 149–53, 157, 159, 162, 167, 184

'Newton', *Encyclopaedia Britannica*, 130, 131, 233n11
  on Newton and religion, 39, 47, 58–9, 61, 63, 129, 132, 133, 138, 143–6, 236nn73, 84
  and Newton's alchemy, 129, 132, 133, 138, 145–7
  on Newton's genius *see under* genius, scientific
  and Pascal forgeries *see under* Pascal forgeries
  publishers of *see* Constable, Thomas; Murray, John
  and reform of science, 13, 43–4, 47, 50–9, 97
  religious beliefs, 47, 55, 130, 144–6, 190
  response to Biot, 13, 43, 45, 47, 49, 56–9, 133, 135, 138, 176
  and St Andrews, 53, 54, 55, 130, 132
  and theory of light, 16, 41, 129–30, 135–7, 187, 235n41, 235n49
  as a Whig, 45, 189
  see also under controversies in Newtonian biography
Brisbane, Thomas, 74, 88
British Association for the Advancement of Science (BAAS), 2, 52, 160, 185
  and Brewster *see under* Brewster, David
  and Pascal forgeries, 175, 177, 181
British Museum, 78, 105, 106, 108, 177, 182
Britton, John, 76, 87, 106, 191
Brodie, Benjamin, 160, 162
Brougham, Henry Peter, 13, 32, 35, 36–8, 41–2, 43, 62, 108, 143, 160–3, 168, 178, 190, 191
  Analytical View of the Principia, 41, 147, 160
  and Brewster *see under* Brewster, David
  Discourse on Natural Theology, 39–40
  Great Reform Act 1832, 31
  Lives of Eminent Men of Letters, 37–8
  review of King's *Life of Locke*, 40, 63
  on scientific education, 31–2
  and SDUK, 31–2, 35, 39–40, 67
  and theory of light, 20, 40–2, 135–6, 208n105, 235n49
  as translator of Biot's 'Newton', 35, 205n68

Burgess, Thomas, 143, 236n73
Burrow, Ruben, 120
Butterfield, Herbert, 199n56

calculus dispute *see under* controversies in Newtonian biography
Cambridge, University of, 15, 26, 36, 52, 55, 69, 85, 86, 87, 88, 92, 95, 98, 106, 107, 108, 110, 111, 115, 119, 143, 147, 160, 162, 189, 190, 228n70
  'Cambridge Network', 81, 84
  manuscript collections of, 11–12, 18, 29, 37, 65, 78, 107, 108, 110, 130, 192–3
  Newton and, 9, 10, 30, 43, 48, 59–60, 111, 147, 177
  see also Trinity College, Cambridge
Camden, William, 87
Camden Society, 106, 108
Cannon, Susan, 3
Cantor, Geoffrey, 21–2, 207n91
Carlyle, Thomas, 17, 45, 106–7
Cassini, Giovanni, 181
Chasles, Michel, 100, 107–8, 120
  see also Pascal forgeries
Christie, J. R. R., 133, 138, 144, 235n41
Church of England, 3, 87, 101
  and Newton, 16, 18, 42, 187, 190, 193
  and universities, 15, 98, 191
Cobbett, William, 31, 87
Cockburn, Henry, Lord Cockburn, 55
Cohen, H. Floris, 99–100
Colenso, John William, 191, 245n5
Comte, Auguste, 5, 100, 126
Conduitt, Catherine *see* Barton, Catherine
Conduitt, John, 163
  biographical writing on Newton, 8–11, 131, 151, 199n44
Constable, Thomas, 130
controversies in Newtonian biography
  calculus dispute, 19, 25, 41, 64, 162
  Biot on, 19, 24–6, 30, 38, 42, 60, 64
  Brewster on, 60, 62–3, 64, 67, 129, 132–5 *passim*, 138, 141–2, 148, 149, 176, 235nn57, 62
  Commercium Epistolicum, 25, 60, 103, 141, 144
  De Morgan on, 114, 116, 121, 122, 123, 124, 132, 138, 141–2, 148

Edleston on, 114
Rigaud on, 103
Flamsteed/Newton dispute, 13–14,
  70–1, 73–4, 85–8, 149, 150, 189–90,
  191
  Baily on, 13–14, 66, 69, 71–2, 74,
    76–80, 94–6, 115, 139, 140, 149
  Barrow on, 90–1, 92, 93, 97
  Biot on, 94
  Brewster on, 76, 78, 79, 97, 129,
    138–41, 147, 148–9, 176
  De Morgan on, 71, 73, 80–1, 85, 89,
    114, 116, 119, 122, 123, 124, 138,
    140–1, 164
  Edleston on, 111, 114–16, 229nn83,
    88
  Galloway on, 88, 93–4
  and political views, 88, 97, 101, 191
  and reform of science, 84–5, 87–8,
    90, 98
  and religious belief, 87, 88, 93, 97–8,
    101, 191
  and RAS, 75–6, 81, 84–5, 98, 189
  and Royal Society, 70, 74, 84, 87, 90,
    91, 97, 140
  Rigaud on, 85, 91–3, 95, 101, 104, 168
  as theory versus practice, 69–70, 73–4,
    85, 88, 91, 93–4, 97, 111, 139–40
  Whewell on, 69, 79, 85, 87, 91–3, 94,
    95, 119, 168, 222n107
  see also Baily, Francis; Flamsteed, John
Newton's breakdown or madness, 19, 30,
  31, 37–9, 57, 150, 207n86
  Biot on, 12–13, 19, 26–9, 45, 56, 61,
    223n140
  Brewster on, 43–4, 45, 47–8, 49, 56–9,
    68, 115, 132, 135, 148, 210n13
  Brougham on, 35
  Edleston on, 115–16
  Rigaud on, 103
Cooper, Charles Purton, 108
Copernicus, Nicolas, 48, 105, 118–19
Craig, John, 199n44
Crompton, Samuel, 153, 229n80
Crosland, Maurice, 29
Cumming, James, 110
Curll, Edmund, 169

Darwin, Charles, 8, 191
Daston, Lorraine, 124, 192
Daunou, Pierre-Claude-François, 29,
  202n34, 208n100
Davies, Thomas Stephens, 108, 109
Davy, John, *Memoirs of the Life of Sir Humphry Davy*, 67
De Morgan, Augustus, 14–15, 41, 60, 67,
  72–3, 85, 87–8, 98, 99, 103, 108, 111,
  114, 116–27, 129, 133, 134, 138, 139,
  147, 149, 176, 189, 191, 192
  on Catherine Barton, 88, 120, 122, 148,
    165–8, 184
    Newton: His Friend: and His Niece, 15,
    163–5, 166, 169–71, 240n26
  as a bibliophile/bibliographer, 107, 116,
    117, 118, 141, 179
  and Brewster, 15–16, 66, 116, 122, 123,
    127, 133, 144–5, 157, 159–60, 164,
    167–71, 189
    review of Brewster's *Memoirs of Newton*, 132, 134, 135, 138–43 *passim*,
    147–52 *passim*, 59, 167
  'Budget of Paradoxes', 175, 178
  historical method/style, 14–15, 66, 100,
    102, 105, 109–10, 117–23, 124–7,
    188, 241n48
  history of mathematics, 116, 118
  history of science, importance of,
    116–20, 178–9, 182–3
  and 'impartial' history, 14, 123, 125, 134,
    167, 192
  on 'myths' and hero-worship, 15–16, 66,
    118–19, 120–3, 162–3, 170–1, 187,
    189, 192, 238n127
  'Newton', *Cabinet Portrait Gallery*,
    122–4, 143–4
  'Newton', *Penny Cyclopaedia*, 66, 122
  on Newton's genius *see under* genius,
    scientific
  on Newton's Grantham statue, 15, 159,
    162–3
  on Newton's religious beliefs, 143–4, 145
  on Pascal forgeries *see under* Pascal
    forgeries
  political views, 121
  on pre-Newtonian science, 116, 117–19

and religion, 14, 15, 84–5, 143, 145, 170, 190, 191, 236n79, 241–2n59
and Royal Society, 121–2
on scientific theories, development of, 117, 119, 123, 179, 240n16
and SDUK, 66, 190
publications, 116, 120
and UCL, 15, 98, 145, 170–1, 190, 191, 241–2n59
on Whewell's *History of the Inductive Sciences*, 117–18, 119, 230nn100, 102
see also under controversies in Newtonian biography
De Morgan, Sophia, 123, 145, 164–5, 170
Derham, William, 199n44
Descartes, René, 9, 10, 25, 180
Desmaizeaux, Pierre *see under* Pascal forgeries
Desmond, Adrian, 39, 207–8n97
Dobbs, B. J. T., 12
Drinkwater, John Elliot, 33–5, 36, 49, 119, 206n73
'Life of Galileo', 33–5
'Life of Kepler', 33–5
Dullier, Fatio de, 25

Edinburgh Encyclopaedia, 4, 54
Edinburgh Journal of Science, 51, 54
Edinburgh Review, 62, 130
Brougham and, 40, 62
Galloway's review of *Account of Flamsteed*, 93–4
Malkin's review of *Life of Newton*, 62–6, 67
Powell, 'Sir Isaac Newton' *see under* Powell, Baden
see also Napier, Macvey
Edinburgh, University of, 36, 54–5
Edleston, Joseph, 14–15, 106, 110–11, 116, 131, 133, 139, 142, 153, 160, 189, 190, 228n74
Correspondence of Newton and Cotes, 99, 110–16, 129, 131, 134, 147, 153, 162
historical method/style, 14, 100, 110–11, 114–16, 117, 125, 127, 188, 189
see also under controversies in Newtonian biography
Ekins, Jeffrey, 144, 167, 236n81

Ellis, Henry, 106, 219n59
Elphinstone, Howard (senior), 36
Elphinstone, Howard, 30, 35–6, 42, 206n73
as translator of Biot's 'Newton', 30, 35, 36, 205n68
Essays and Reviews (1860), 191, 245n5

Fara, Patricia, 5, 7, 12
Faugère, Prosper, 174, 175, 180
Fellowes, Henry, 96, 131, 140, 148–9, 218n45, 219n59, 234n19
Fellowes, Newton, 4th Earl of Portsmouth, 96–7, 131, 219n59
Fermat, Pierre de, 179
Flamsteed, John, 11, 58, 103, 122, 147, 149, 150, 165, 168, 182, 187, 220n70
assessments of work of, 13–14, 71–3, 74–5, 78–9, 86, 91, 93–5, 139, 182
correspondence with Newton, 58, 71, 79, 140
correspondence with Sharp, 71–2, 75, 76
Flamsteed/Newton dispute *see under* controversies in Newtonian biography
and Halley, 71, 79–80, 90, 93, 94, 115, 140
Historia Coelestis, 71, 75, 80, 216n7
on Newton, 71, 73, 79, 148
observations of, 70–2, 74, 76, 86, 91, 93–5, 115, 139, 181
and Royal Observatory *see* Royal Observatory, Greenwich
see also under controversies in Newtonian biography
Fontenelle, Bernard le Bovier de, 8
'Éloge de Newton', 8, 9, 10, 19, 25
Forbes, Duncan, 126
Forbes, James David, 48–9, 54–5
Foster, Henry, 75, 217n31
Fourier, Joseph, 23
Fox, Robert, 20
France, 3, 10, 30, 45, 47, 50, 51, 108, 120, 177
historiography in, 29, 59
mathematics in, 36
physics in, 16, 20–5, 40
see also Laplace, Pierre-Simon

see also Académie des Sciences, Paris;
   Pascal forgeries
Franklin, Benjamin, 153
Frend, William, 84, 121
Fresnel, Augustin Jean, 20, 22, 23, 24, 40, 51

Galilei, Galileo, 33, 48, 56, 100, 105,
   118–19, 172, 176, 203n40, 231n107
   see also under Drinkwater, John Elliot
Galison, Peter, 124, 192
Galloway, Thomas, 44, 66, 78, 88, 93, 189,
   192, 220n75
   'French and English biographies of Newton', 66–7, 68, 116
   review of *Account of Flamsteed*, 93–4, 96
   review of RAS *Memoirs*, 93
Gascoigne, William, 105
Gay-Lussac, Joseph Louis, 40
George, Prince of Denmark, 70
genius, scientific, 5–6, 6–8, 9, 12, 14, 16, 18,
   19, 68, 72–3, 162, 205n60, 217n20
   Biot on, 26–8, 42, 56, 59–60
   Brewster on, 13, 44, 47–9, 52–3, 56,
      67–8, 129, 137–8, 210n21, 211nn37,
      40
   De Morgan on, 123–4
   Rigaud on, 102–3
   and SDUK, 20, 33–4, 38
Geological Survey, 2
Gerard, Alexander, 7, 49
Gerhardt, C. J., 114
Gilbert, Davies, 74
Giles, Edmund, 72
Gordon, Maria (née Brewster), 43, 54, 55,
   130, 133, 138, 144, 145, 176
Grant, Robert, 176
   History of Physical Astronomy, 78–9, 99,
      127, 244n109
   see also under Pascal forgeries
Grantham, 11
   Newton's statue in, 15, 41, 111, 159,
      160–3, 184, 190, 239n4
Gregory, David, 58, 111
Gregory, James C., 58
Griffin, Richard, 3rd Lord Braybrooke, 12,
   56, 57, 212–13n69
Gully, J. M., 8

Hadley, John, 101, 104
Halifax, Lord *see* Montagu, Charles
Hall, A. Rupert, 8
Hall, Basil, 89
Hallam, Henry, 125
Halley, Edmund, 69, 70, 72, 78, 92, 93, 94,
   95, 101, 103–4, 222–3n126
   see also under Flamsteed, John
Halliwell, James Orchard (later Halliwell-Phillipps), 105, 106, 107–10, 116, 130,
   133, 227n52
   see also Historical Society of Science
Harcourt, Vernon, 52
Harriot, Thomas, 101
Haynes, Hopton, 143
Helvelius, Johannes, 72
Hemming, George, 184
Herschel, Caroline, 85, 216n11, 220n73
Herschel, John, 50, 61, 76, 85, 86, 109,
   124–5, 141, 168, 178, 181, 183,
   206n80, 207n85, 209n4
   on Baily, 72, 75, 76–7, 81
Hill, John, 118
Hirst, Thomas Archer, 176, 185
   see also under Pascal forgeries
Historical Society of Science (HSS), 14, 99,
   108–10, 114, 116, 228n61
historiography, nineteenth-century, 3–4
   antiquarianism, 10–11, 59, 106–8, 117,
      174
   use of footnotes, 11, 60, 103–4, 108,
      226n31
   'impartiality' in, 13, 33, 37–8, 42, 44,
      66–7, 78, 98, 101, 104, 119–20,
      125–7, 188–9, 192
   use of manuscript sources, 3–4, 11–12,
      17, 106–8, 119, 188
   forgery and theft of, 107–8, 174–5,
      227n55, 244n127
   see also Pascal forgeries
   professionalization and history *see under*
      professionalization
   'scientific history', 3–4, 118, 189, 245n2
   see also biography
history of science, nineteenth-century, 2,
   4–5, 99–100, 119, 127, 192

development of expertise in, 6, 14–15, 16, 17–18, 100, 105, 108–10, 111, 119, 127, 187–8, 192–4
'inductive' approach to, 14–15, 75, 98, 102, 104, 107, 109–10, 98, 125, 127, 188–9
as means of countering specialization, 109, 120
research on British scientific tradition, 72, 101, 103, 105, 108
see also historiography, nineteenth-century
Hooke, Robert, 23, 25, 34–5, 88, 234n21
and Newton, 24, 103, 132, 147, 149, 150, 76
and optics, 23, 135, 137
and scientific method, 23, 24, 34, 103
Horrox, Jeremiah, 87, 101, 105
Horsley, Samuel, 36, 143
Howitt, William, 87
Hume, David, 144
Hunter, Joseph, 76, 96, 106, 108, 166, 219n59, 220n75, 221n90, 227n40
Hussey, Thomas John, 50, 106
Hutchinson, John, 88
Hutton, Charles, 202n29n31, 220–1n87
Huygens, Christiaan, 26, 172, 176, 179, 181
and Newton's breakdown, 26, 37, 40, 57, 61, 63, 115

Innes, George, 87

Jeffrey, Francis, 196n15, 207n90
Johns, Adrian, 70
Johnson, Samuel, 4
Jones, Owen, 182
Jones, William, 12, 103

Keats, John, 198n30
Kepler, Johann, 33, 48, 49, 56, 105, 179, 212n67
see also under Drinkwater, John Elliot
Keynes, John Maynard, 147, 193
King, Edmund, 162
Biographical Sketch of Newton, 162–3
King, Peter, 7th Baron King, 37
collections of, 12, 37
Life of John Locke, 37, 58, 213n74

review of *Life of John Locke see under* Brougham, Henry Peter
Knight, Charles, 116
Cabinet Portrait Gallery of British Worthies, 122
Companion to the Almanac, 116, 164, 168, 240n26
Knight's Cyclopaedia of Biography, 105

Laertius, Diogenes, 67
Lamé, Gabriel, 181
Laplace, Pierre-Simon, 13, 20–1, 22, 23, 24, 25, 31–2, 36, 41, 42, 61, 63
Laplacian circle, 20, 22, 36
Laplacian Programme, 20–1, 32
Mécanique céleste, 21, 29, 36
Newton's breakdown, effects of, 29–30, 42, 45, 47, 58, 61
on Newton's religion, 29–30, 202–3n38, 207n85
Lavoisier, Antoine-Laurent, 157
Law, William, 146, 237n96
Le Verrier, Urbain, 171, 174, 180, 182
Leibniz, Gottfried Wilhelm, 19, 25, 38, 182
see also controversies in Newtonian biography
Lemon, Robert, 76, 106, 219n59
Levine, George, 125–6
Levine, Phillippa, 17
liberal Anglican history, 126
Libri, Guglielmo, 100, 107–8, 120, 168, 178–9
Life of Newton see under Brewster, David
Locke, John, 12, 116, 146, 149
correspondence with Newton, 12, 37, 40, 58, 61, 63–4, 146, 150, 190
London, University of *see* University College London
Lovelace, Ada, 148–9
Lubbock, Ellen Frances, 182
Lubbock, John, 109
Lucas, Vrain-Denis *see under* Pascal forgeries

Mabille, Emile, 175
Macaulay, Thomas Babbington, 17, 59, 62, 106–7, 125, 168
Macclesfield, Earl of, 177
Macclesfield Collection, 9, 12, 96, 103, 131

MacLaurin, Colin, 36, 43
Maclear, Thomas, 87, 216n13
Madden, Frederick, 108, 177
madness
    and genius, 8, 38, 47
    Newton's see under controversies in Newtonian biography
Malkin, Arthur, 62, 214n92
Malkin, Benjamin Heath (senior), 62, 214n92, 215n111
Malkin, Benjamin Heath, 44, 62, 66, 67, 78, 214n92
    'Newton' in *Gallery of Portraits*, 62, 64, 124, 189, 214n106
    review of *Life of Newton*, 44, 62–6, 68, 116, 215n107
Malkin, Frederick, 62
Manley, Delarivier, 165
Manuel, Frank E., 164, 184
Mason, Thomas, 199n44
'Mathematicians' Mutiny', 84
'Mathematical Practitioners', 81, 84, 88, 97
Mayer, Tobias, 72
Memoirs of Newton see under Brewster, David
Merton, Robert K., 191, 192
methodology, historical *see under individuals' entries*
methodology, scientific, 7, 8, 16, 17, 33–5, 69–70, 73–4, 75–6, 123–4, 188
    Baconian methodology *see under* Bacon, Francis
    Newtonian methodology, 23–4, 37, 42
Milbourne, William, 105
Mill, John Stuart, 3, 125
Miller, David P., 84, 88, 90
Millington, John, correspondence regarding Newton's breakdown, 57–8, 61, 212–13n69n72
Milne, Joshua, 119, 231n111
Moigno, Abbe François, 183
Montagu, Charles, Lord Halifax, 56, 163–4, 168
    and Catherine Barton *see* Barton, Catherine
    as Newton's patron, 56
Munby, A. N. L., 100
Murchison, Roderick, 52

Murray, John, 130, 206n81, 233n13
    Family Library, 45, 209nn8, 10

Napier, Macvey, 63, 64–5, 124, 131, 215n107
Napier, Mark, *Memoirs of John Napier*, 99
natural theology, 3, 17, 32, 39, 42, 47, 67, 185, 190
Neptune, discovery of, 2, 171
Newton, Humphrey, 153, 199n44, 237n96
Newton, Isaac, 9
    and alchemy and chemistry, 10, 132, 133, 146–7, 187, 188, 193, 237n101, 245n1
    apple anecdote, 7, 9, 26, 48, 59, 64, 102–3, 185, 198n26, 202n30, 210n18
    as an autodidact, 9, 38–9
    breakdown or madness of *see under* controversies in Newtonian biography
    abilities after 1692/3, 10, 11, 27–9, 40, 41, 43, 58, 63–4, 116, 138, 150
    at Cambridge *see under* Cambridge, University of
    character, assessments of, 10, 26, 27, 47, 66, 68, 79, 86, 88, 89, 91–2, 95, 97, 116, 123–4, 129, 132, 148–53, 162, 166, 188, 202n31
    interest in, 1, 3, 8, 16, 18, 69–70, 93, 94, 96, 100, 104, 111, 140, 163, 164–5, 168, 170, 187, 189, 191–2
    publishing, dislike of, 9, 27, 68, 143, 162, 163, 166, 198n37
    childhood, 9, 10, 11, 38, 48, 59–60, 131, 177
    chronology, writings on, 10, 28, 29, 39, 68
    dog (Diamond), 26, 40, 202n29, 208n103
    and Flamsteed *see under* controversies in Newtonian biography
    fluxions, 9, 25, 60, 103, 114, 133–4, 141–2, 147, 172
    see also controversies in Newtonian biography
    genius of *see* genius, scientific
    and Hooke *see* Hooke, Robert
    'love letter' to Lady Norris, 131, 133, 151
    lunar theory, 70, 79, 86, 94–5, 111, 150

manuscripts of, 8–12, 131–2
  see also entries for individual collections
mother, Hannah Smith, 177, 178
and the Mint, 9, 28, 50, 59, 79, 122, 163, 165, 169, 170, 229n83
  neglect of, 56–8
nephew, Benjamin Smith, 120–1, 168
niece *see* Barton, Catherine
Opticks, 24, 29, 30, 40, 41, 58, 136, 142, 203n39
portraits and statues, 153–6
  see also under Grantham
Principia, 23, 25, 27, 37, 58, 59, 70, 71, 88, 102–4, 111, 123, 133, 142, 148, 153, 160, 168, 181
  removal of acknowledgments from, 25, 64, 94, 116, 149
and Catherine Storer, 151
theological writings, 10–11, 28–30, 39, 40, 43, 58, 63, 132, 133, 143–5, 149, 236n72
  see also Bentley, Richard
theory of light, 9, 12, 23–4, 58, 60, 135–6
theory of fits, 21, 22, 24, 41, 201n24, 208n
theory of universal gravitation, 9, 21, 28, 102, 111, 141, 147, 163
  Hooke on, 25, 103
  Pascal and *see* Pascal forgeries
  and Whiston *see under* Whiston, William
Nicolas, Nicholas Harris, 106, 219n59, 227nn40, 42
Niebuhr, Barthold, 4, 33, 104, 127
Niebuhr, Carsten, 32–3, 38
Notes and Queries, 164
  De Morgan and, 120, 164, 166, 167
North British Review, 130, 132
Nonconformism, 14, 15, 16, 31, 39, 44, 88, 143, 145, 190–1, 221n90

optics, 13, 20, 21–2, 23–4, 41, 42, 45, 58, 60, 208n105
  corpuscular (particle) theory of light, 20, 23–4, 40–1, 130, 136–7, 187
  history of, 21–2, 129
  reinists, 23, 136, 201n22

undulatory (wave) theory of light, 20, 55, 135–7
  see also under individuals' entries
Osborne, E. A., 164
Oughtred, William, 105
Owen, Richard, 160
Oxford, University of, 15, 98, 120, 130
  collections of, 78, 130
  Rigaud and, 15, 34, 69, 93, 101, 192

Paine, Thomas, 87, 220n85
Palgrave, Francis, 108
Panizzi, Anthony, 105, 118
Pappus of Alexandria, 101
Paramatta Observatory, 74, 81
Pascal, Blaise, 27–8, 61
Pascal forgeries, 16, 159, 171–83, 185, 191, 242n71, 243n93
  Brewster on, 159, 175–8, 183
  De Morgan on, 159, 175–6, 178–80, 181, 182–3
  Desmaizeaux as author of, 177–8, 181, 183
  Grant on, 176, 180–1, 182, 185
  Hirst on, 176, 181–2, 185
  Lucas as author of, 174, 175, 176, 179
Peacock, George, 36, 95, 107, 108, 147, 208n105
  history of mathematics, 118, 119, 120, 171, 230n105
Pemberton, Henry, 9, 36
Pepys, Samuel, 12
  correspondence regarding Newton's breakdown, 57, 61, 150, 212–13nn69, 72
Pettigrew, Thomas, 108, 109
Phillips, Richard, 87–8, 191, 220–1nn85, 86, 87
Philosophical Magazine, 91–2
Philosophical Transactions see under Royal Society
Picard, Jean, 102
Plutarch, 59
Pond, John, 87, 220n82
Pope, Alexander, 149

Portsmouth, Earls of, 96, 130, 177, 192
  Hurtsbourne Park, 11, 97, 130, 131, 140, 148
  Portsmouth Papers, 10, 11, 12, 15, 18, 37, 65, 78, 96–7, 129, 131–2, 134, 135, 140, 143, 144, 146, 148–9, 163, 192–3, 198–9n42, 200n57, 218n45, 234n19
  see also Fellowes, Newton
Pound, James, 181
Powell, Baden, 88–9, 108, 109, 189, 190, 236n72
  biography of Hooke, 34
  History of Natural Philosophy, 68, 99, 127
  review of Rigaud's works, 99, 103–4, 105, 226n30
  'Sir Isaac Newton', 134–5, 147, 150, 151, 234n37
Priestley, Joseph, 7–8, 75, 87, 124, 190, 221n90
professionalization
  and history, 4, 17, 106–7
  and history of science, 4, 193, 195–6n5, 211nn34, 35
  and science, 1, 2, 52, 84, 184, 187, 191–2
Pryme, Abraham de la, 37, 229n90
  account of Newton's madness, 37, 40, 57, 61, 63, 115, 148
Pryme, Charles de la, 115
Pryme, George, 115
Ptolemy, 72
publication, cheap, 31, 45, 88, 208n8, 219n53

Quarterly Review, 51, 90–3 *passim*, 97, 184

Ranke, Leopold von, 4, 104, 226n31
revolution in science, 118
  in analytical mathematics, 2
  in optics, 2, 20
  second scientific revolution, 2
Rice, Adrian, 116, 164
Richards, Joan, 116, 119, 164
Rickman, John, 86
Rigaud, Stephen Peter, 14–15, 96, 101–5, 108, 111, 129, 131–5 *passim*, 189, 190
  on Catherine Barton, 88, 93, 95, 104, 166, 168–9
  Correspondence of Scientific Men, 101, 103, 131
  Historical Essay on the Principia, 95, 99, 101, 102–5, 111, 114, 226n22
  on Halley, 92, 93, 95, 101, 102, 104–5, 222–3n126
  historical method/style, 14, 101–2, 103–6, 117, 119, 125, 127, 168–9, 188–9, 226nn20, 29
  and Oxford *see* Oxford, University of
  on Newton's genius *see* genius, scientific
  sons (Stephen Jordan, John, Gibbes), 108, 226n29
  Works and Correspondence of Bradley, 99, 101–2, 104, 226n36
  see also under controversies in Newtonian biography
Roebuck, John Arthur, 33, 35, 37
  'Life of Mahomet', 33, 37
Routh, Edward, 160, 239n7
Royal Astronomical Society (RAS), 13, 14, 15, 66, 69, 72, 78, 84, 85, 90, 92, 93, 98, 101, 190, 221n101
  and Baily, 13, 14, 69, 70, 81
  see also under controversies in Newtonian biography
Royal Observatory, Greenwich, 70, 71, 80, 84, 93
  collections of, 11, 72, 81, 85
Royal Society, 14, 23, 57, 74, 80, 81, 85, 90, 92, 95, 97, 101, 105, 106, 121–2, 131, 148, 165, 171, 181, 232n126, 243n93
  and Banks *see* Banks, Joseph
  and calculus dispute, 25, 30, 121, 141
  collections of, 9, 10, 97, 115
  and Flamsteed *see under* controversies in Newtonian biography
  motto, 121, 231–2n122
  Newton as President of, 9, 28, 59, 70
  and Newton's statue, 160, 162
  and Pascal forgeries, 175, 177, 181
  Philosophical Transactions, 40, 121
  reform of, 50–1, 84, 189
Royal Society of Edinburgh, 144
Rumker, Charles, 74–5, 217n26

Sabine, Edward, 177
Schaffer, Simon, 53

'Scientific Servicemen', 81, 88, 97, 190
Scientists' Declaration, 191, 237n94
Scott, Walter, 59
secularization, 2, 3, 184
    and BAAS, 51
    of Newton's legacy, 18, 192
    and progress of science, 185, 192
    and UCL, 119, 191
Sedgwick, Adam, 86, 87
Shakespeare, William, 87
Shapin, Steven, 191, 192
Shapiro, Alan, 40
Sharp, Abraham, 87, 216n11
    see also under Flamsteed, John
Skirrow, Walker, 96
Sloane, Hans, 148
Smiles, Samuel, 33, 192, 207n91
Smyth, William Henry, 89, 121, 189–90
Society for the Diffusion of Useful Knowledge (SDUK), 12–13, 20, 31, 35–45 *passim*, 62, 64, 66–7, 70, 78, 97–8, 190, 204nn45, 53, 205n63
    and biography, 30–5, 205n58
    and Brougham *see under* Brougham, Henry Peter
    and De Morgan *see under* De Morgan, Augustus
    Gallery of Portraits, 32, 62, 64, 116
    Library of Entertaining Knowledge, 32, 62
    Library of Useful Knowledge, 32, 62
    'Life of Newton' *see under* Biot, Jean-Baptiste
    Lives of Eminent Persons, 32–5
    Penny Cyclopaedia, 32, 66, 72, 116, 122
    and religion, 39–40, 207–8n97
    and translations, 35, 206n71
    see also under genius, scientific
Society of Antiquaries, 106, 108
Somerville, Mary, 36, 206n81
Somerville, William, 36
South, James, 84, 87
Southey, Robert, 196n15, 218n44
Stewart, Dugald, 58, 213n74
Stratford, William Samuel, 96
Stukeley, William, 9, 10, 11, 87, 199nn44, 45
Surtees Society, 106, 115

Swift, Jonathan, 166, 168
Swinden, Jan Hendrik van, 26

Theed, William, 160
Theerman, Paul, 164, 191
Thierry, Augustin, 59
Thiers, Louis Adolph, 172
Times, The, 35, 97, 108, 131
    and Pascal forgeries, 174, 175, 177, 180, 181, 183
    review of *Account of Flamsteed*, 89
    review of *Memoirs of Newton*, 151–3
Trinity College, Cambridge, 15, 35, 36, 62, 106, 107, 110–11, 115, 153, 160, 162, 190
    manuscript collections of, 11, 37, 78, 108, 110–11, 115, 131
Turner, Dawson, 107, 108, 111, 166, 219n59, 228–9n76, 239n133
    Thirteen Letters from Newton to John Covel, 107, 131
Turnor, Charles, 199n49
Turnor, Edmund, 10–11
    Collections for the History of Grantham, 10–11, 19, 234n18
Tyndall, John, 185

University College London (UCL), 16, 39, 62, 176, 191
    see also under De Morgan, Augustus

Vestiges of the Natural History of Creation (Robert Chambers), 145
Vinci, Leonardo da, 48
Voltaire, François-Marie Arouet de, 165

Wallace, William, 87, 88
Wallis, John, 105
Weld, Charles Richard, 106, 114, 150, 151, 208n103
    History of the Royal Society, 99, 127, 165
Westfall, Richard S., 133, 157
Whewell, William, 15, 37, 56, 87, 104, 108–9, 110, 111, 116, 131, 136, 143, 160, 162, 184, 189, 190, 191, 195n4, 208n105, 228n75
    on Catherine Barton, 168
    and history of science, 5, 99–100, 109, 119, 126, 127, 210n21

History of the Inductive Sciences, 95–6, 99–100, 117–18, 126, 224n151, 225n9, 230n101, 235n41
   see also under controversies in Newtonian biography
Whiston, William, 87, 118, 143, 145
   and Newton, 79, 88, 91, 116, 122, 123, 143
White, Walter, 160
Willis, Robert, 108

Winter, Thomas, 160, 162, 229n79
Wood, J., 88
Woodward, John, 148
Woolsthorpe, Manor of, 9, 10, 11, 45
Wright, Thomas, 107, 108, 109

X-Club, 185, 191

Yeo, Richard, 5, 7, 49, 65, 69, 93, 97, 191–2
Young, Thomas, 40–1, 84, 135–6, 208n105